# BIOMEDICAL APPLICATIONS OF INTRODUCTORY PHYSICS

**JACK A. TUSZYNSKI**

*Department of Physics*
*University of Alberta*
*Edmonton, Alberta, Canada*

**JOHN M. DIXON**

*Department of Physics*
*University of Warwick*
*Coventry, United Kingdom*

**JOHN WILEY & SONS, INC.**

*This book is dedicated to
our wives, our children,
and our students*

Acquisitions Editor    *Stuart Johnson*
Senior Production Editor    *Elizabeth Swain*
Senior Marketing Manager    *Robert Smith*
Cover Designer    *Madelyn Lesure*
Illustration Editor    *Anna Melhorn*
Photo Editor    *Sara Wight*

This book was set in 10/12 Times Roman by Matrix Publishing Services and printed and bound by Hamilton. The cover was printed by Phoenix.

This book is printed on acid-free paper. ♾

To order books or for customer service, call 1(800)-CALL-WILEY (225-5945).

*Library of Congress Cataloging in Publication Data:*
Tuszynski, J. A.
    Biomedical applications of introductory physics / Jack A. Tuszynski, John M. Dixon.
      p.  cm.
    Includes bibliographical references and index.
    ISBN 0-471-41295-3 (pbk. : alk. paper)
      1. Biophysics.  2. Medical physics.  3. Physics.  I. Dixon, J. M., Dr.  II. Title.
  QH505.T875  2001
  571.4—dc21                               2001046683

Printed in the United States of America

10 9 8 7 6 5 4 3

# PREFACE

For a long time we have felt that almost all introductory physics texts are inadequate in terms of their presentation of bio-medical phenomena and we believe that many other instructors share our concern. We have written this book to help address this problem. Our goal is to provide instructors and students with a rich selection of topics, worked examples, and problems. The book can be used in conjunction with any standard two-semester introductory physics text. The topic sequence follows the standard syllabus, but the instructor can pick and choose as appropriate. The mathematical level is that of a typical algebra-based course though a handful of examples require the use of integration and differentiation.

Each chapter begins with a brief overview of the key physical concepts. The purpose of this overview is not to present complete exposition of concepts, but rather to help the student make the connection to the basic text. However, in some chapters a need arose to expand on some physical background beyond the typical presentation at the first-year physics level. For instance, the reader will find an introduction to osmotic pressure and chemical bonds.

Following the physical background section, each chapter covers several bio-medical topics that are either examples or applications of the physics principles. These topics are largely descriptive but with sufficient biological, anatomical or physiological detail to make both the application of physical quantities and the understanding of the biology possible. Our intention here is not to use biology as an excuse for practicing physical concepts on animate matter systems. Rather, the objective is to understand biology better as a result of applying a physical insight. This, clearly, is the power of an exact science like physics. Finally, each chapter ends with a number of problems that the instructor can use in assignments or exams.

Many of the introductory level applications of physics to biology can be grouped according to biological systems and processes. Accordingly, the book presents numerous applications of physics to the strength of bones, the elasticity of muscles, the flow of blood, the functioning of the lungs, the electrical signal propagation in the nerve cells, the optical properties of the eye, and the sound generation of the vocal cords. In addition, there are numerous other topics that may pique the interest of the student. These include the capillary rise in trees, generation of DNA defects due to radiation, and an assortment of topics in the area of the physics of sport and health.

It is hoped that this book will be a small step towards a wider appreciation of physics in the burgeoning field of the bio-medical sciences. The authors strongly believe that physics has an important role to play both as a scientific methodology and as a conceptual framework to be utilized by biology and medicine in their quest to become predictive and quantitative sciences. In order to succeed in this quest, new generations of students should be properly exposed to the value of physics as an applied science.

*Jack Tuszynski*
*John Dixon*

# ACKNOWLEDGMENTS

The authors wish to thank Stuart Johnson, Elizabeth Swain, Anna Melhorn, and Sara Wight of John Wiley & Sons for their continued encouragement, guidance, and support during the process of planning and production of this text. We wish to thank Michelle McCurdy for her dedicated technical assistance in the preparation of the manuscript. Andrew Kobos is gratefully acknowledged for his editorial help. Finally, our wives, Eileen Dixon and Ela Tuszynska, deserve a big thank you for their patience and support during the process of work on the manuscript, which took us away either physically or mentally from our families.

*Edmonton, Coventry, and Brussels*
*March/April 2001*

# CONTENTS

# CHAPTER 1

# INTRODUCTION

The connection between the physical and biomedical sciences has been developing rapidly over the past few decades, especially since the groundbreaking discoveries in molecular genetics. There is clearly a need for a continuing dialogue and a cross-fertilization between these two groups of sciences.

Nowadays, it is generally accepted that the laws of physics do apply to living organisms as much as to inanimate matter. Attempts at applying physical laws to living systems can be traced to the early creators of modern science. Galileo analyzed the structure of animal bones using physical principles, Newton applied his optics to color perception, Volta and Cavendish studied animal electricity, and Lavoisier demonstrated that the process of respiration is just another example of an oxidative chemical reaction. Robert Mayer was inspired by physiological studies to formulate the first law of thermodynamics. A particularly fruitful area of application to physiology is the physics of fluid flows, in which, for example, blood flow was analyzed by Poiseuille using the principles of hydrodynamics, and air flow in the lungs has been described consistently with the laws of aerodynamics. An important figure in the history of biophysics is that of the German physicist and physiologist Hermann von Helmholtz, who laid the foundation for the fundamental theories of vision and hearing. The list of physicists who made a large impact on biology and physiology is very long, so we name only a few of the most well-known figures who have crossed this now-disappearing boundary. Delbrück, Kendrew, von Bekesy, Crick, Meselson, Hartline, Gamow, Schrödinger, Hodgkin, Huxley, Fröhlich, Davydov, Cooper, and Szent György have undoubtedly pushed the frontier of the life sciences in the direction of exact quantitative analysis. It is both our expectation and hope that the process will accelerate in the twenty-first century.

While in many areas of physiology, biology, and medical research physics is helpful in providing deeper insight into the phenomena studied by these sciences, in some fields of investigation physics has actually provided the primary stimulus for development. One such area is electrophysiology, in which membranes of nerve cells are characterized by a voltage gradient called the action potential. The propagation of action potentials along the axons of nerve cells is the key observation made in the investigation of brain physiology. A physical theory of action potential propagation was developed by Huxley and Hodgkin, who earned a Nobel Prize for their discovery. Likewise, the structure of deoxyribonucleic acid (DNA), discovered by Crick and Watson, which ushered in a new area of molecular biology, would not have been possible without both experimental and theoretical tools developed by physicists. In this case it was X-ray crystallography that revealed the double-helix structure of DNA. More recently, investigations of DNA sequences have been pursued in the hope of revealing a molecular basis of genetically inherited diseases. Gel electrophoresis and fluorescent labeling techniques are the crucial methods perfected by biophysicists and biochemists for the studies of DNA sequences. Many other techniques that originated in physics laboratories have made their way to become standard equipment used by molecular biologists and chemists. Here, we can list such techniques as mass spectroscopy, laser spectroscopy, atomic force microscopy, nuclear magnetic resonance (NMR), ultrasonography, and, of course, X-ray and electron crystallography.

From their original inventions as probes of physical phenomena, these experimental techniques are then frequently adapted to molecular level biology, and eventually some of them are further transformed into diagnostic and therapeutic tools in modern medicine. X-ray machines are used for the detection of abnormalities, and NMR, now called MRI (magnetic resonance imaging) by medical practitioners, helps in the detection of tumor growth, which in turn can be treated by radiation from radioactive sources. Electrical manifestations of the activity of the heart are monitored by the cardiologist, who uses electrocardiography (ECG), and likewise brain activity is studied through the use of electroencephalography (EEG). Ultrasound has found applications in both diagnostics (e.g., fetal development) and therapeutics (gall and kidney stone shattering). Finally, optical fibers are being used in the examination of internal organs without the need for surgery.

The purpose of this publication is twofold. First and foremost, we wish to convince the often reluctant student of physics that physical methods and analyses can be made very relevant in the realm of living systems and processes. We hope that some of the students may be inspired to continue in more advanced studies of biophysics. Second, this book is a compilation that physics instructors can find useful as a resource material, especially when teaching physics to biology majors and premedical students. In what follows we have gathered a large collection of examples of biological phenomena into which better insight can be gained through the application of elementary physics. Each of the 31 chapters begins with a very brief overview of the key physical concepts used. We follow the structure of one particular first-year textbook, namely *Physics* by Cutnell and Johnson (Wiley, 5th edition, 2000), but any comparable text could be used as a companion to this manual. Following a summary of physical concepts we discuss a number of topics which are descriptive in nature but require physical reasoning. The number of these topics greatly varies from chapter to chapter, and this fact is naturally linked to the applicability of particular areas of physics to biology. Within each chapter we then continue with a selection of solved problems aimed at a quantitative level of analysis. Again the number of these given examples varies from chapter to chapter. Finally, we close each chapter with a set of unsolved exercises that the instructor can use as assignment or examination problems. The book also contains a bibliography of sources which were used in compiling the material here. We begin with two topics which show how simple mathematical concepts such as scaling and vector algebra help in elucidating some biological properties.

## TOPIC 1.1   Scaling

The size of a freestanding mammal cannot exceed certain limits using the material available in nature. There are two restricting limits: static and dynamic. First consider the static limit. The resistance of the supporting parts of the body, the limbs, to a crushing stress is proportional to their cross-sectional area. Thus if a small animal with leg radius $r_1$ were scaled up in all dimensions by a scaling factor $L$ such that the new leg radius were $Lr_1$, then the area would have increased by a factor $L^2$ and the strength would have increased by $L^2$. It can be shown that leg strength is proportional to $L^2$. The legs, however, have to hold up a body the volume and weight of which will increase by $L^3$. If an animal were scaled up by $L = 10$, then legs 100 times as strong as before would have to hold up 1000 times the weight and collapse would be inevitable. As animal bodies get larger, the legs must get relatively thicker with the diameter increasing as $L^{3/2}$. Thus, the small antelope can have very slim and spindly legs whereas short and stocky legs occupy much of the space under the hippopotamus. However, if a counterforce balances the body weight, for example, the buoyant force of water, mammals can attain a greater size than the strength of body materials would normally permit. Thus the enormous size of some whales can be easily explained.

However, the purely static argument does not provide much of a limit to size. The space under the animal could be completely filled to create a treelike animal the ultimate height of which would be limited to about 10 km by the ultimate compressive strength of the material. Wind loading would tend to topple such an animal unless it had the shape of an inverted cone of diameter 10 km at the base.

The dynamic effects are even more important than static considerations in limiting the size of an animal. An animal must move if it is to survive. To move a limb, an animal must do work. The work required to move a limb of mass $m$ a distance $s$ is proportional to $ms$. Since $m$ is proportional to $L^3$, and because a limb can only be moved a distance comparable to its own size, then $s$ is of the order $L$ and, consequently the work required scales as $W$ or as $ms$ or equivalently as $L^4$. To provide this work, a muscle must exert a force $F$ through a distance $s$. But the force exerted by a muscle will be proportional to $L^2$, and $s$ again is of the order of $L$. Therefore the work that can be done will increase as $W$ or as $Fs$ or consequently as $L^3$. Here, too, as the animal's size is scaled up, problems soon arise, since the work required to move the limb increases faster than the ability of the animal to provide the work. For physical background on work and force, the reader is referred to Chapters 6 and 4, respectively.

## TOPIC 1.2   Vector Representation of DNA

Deoxyribonucleic acid is a biomolecule that contains genetic information and is found in the nucleus of cells. It consists of two strands of polynucleotides running in opposite directions and twisted about a central axis to form a double helix, as shown in Figure 1.1. If one assumes that the two helices are smooth and circular, it is possible to approximate DNA geometry as a circular cylinder whose coordinates $(\rho, \theta, z)$ are related to Cartesian or rectangular coordinates $(x, y, z)$ by

$$x = \rho \cos \theta \tag{1.1}$$

$$y = \rho \sin \theta \tag{1.2}$$

$$z = z \tag{1.3}$$

As the variables related to DNA geometry, $\rho$ is the radius of the double helix ($=1.0$ nm), $\theta$ is the rotation or screw angle of the DNA backbone as it winds about the central axis given in radians, and $z$ is the distance along the central axis. The periodicity along the $z$-axis is 3.4 nm. Since there are two strands of nucleotides in the double helix, two vectors are required to represent the structure of the DNA molecule.

For the strand that is moving in the positive $z$ direction (bottom to top) denoted by $\mathbf{S}_1$, the vector can be expressed as

$$\mathbf{S}_1 = S_{1x}\hat{i} + S_{1y}\hat{j} + S_{1z}\hat{k} = [(1.0 \text{ nm}) \cos \theta]\hat{i} + [(1.0 \text{ nm}) \sin \theta]\hat{j} + (3.40 \text{ nm})\hat{k} \tag{1.4}$$

The other strand is similar to the previous one except that (1) it is moving in the negative $z$ direction and (2) it is out of phase or offset by an angle of two radians. Keeping this in mind, we see that the vector for the other strand, denoted by $\mathbf{S}_2$, can be expressed as

$$\mathbf{S}_2 = S_{2x}\hat{i} + S_{2y}\hat{j} + S_{2z}\hat{k} = [(1.0 \text{ nm}) \cos (\theta - 2)]\hat{i} + [(1.0 \text{ nm}) \sin (\theta - 2)]\hat{j} + [(-3.4 \text{ nm})(\theta - 2)]\hat{k} \tag{1.5}$$

It is possible to further refine these vectors by calculating the value for the rotation or screw angle $\theta$ and substituting into the above expressions for $\mathbf{S}_1$ and $\mathbf{S}_2$.

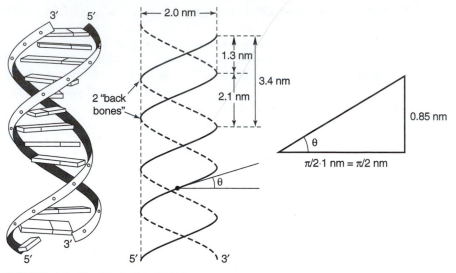

**FIGURE 1.1**   Double helix of DNA.

From the geometric considerations presented in the figure, the magnitude of the opposite side of the triangle must be calculated, followed by calculation of $\theta$. Setting up ratios of distance, we have

$$\frac{360°}{3.4 \text{ nm}} = \frac{90°}{y \text{ nm}} \tag{1.6}$$

Solving for $y$ yields

$$y = 0.85 \text{ nm} \tag{1.7}$$

Solving for the rotation angle $\theta$ gives

$$\tan \theta = \frac{y}{x} = 0.54 \tag{1.8}$$

or

$$\theta = \tan^{-1} \quad 0.54 = 28° \tag{1.9}$$

## SOLVED PROBLEMS

**1.1**   Verify the formula

$$V(H) = aH^3 \tag{1.10}$$

for an isometric series of animals with the body plan of (i) a cube and (ii) a sphere. Determine the constant $a$ for these two cases.

**Solution**   For an isometric cubic series of animals, the natural choice for size is the length $L$ of the cube. The volume of a cubic animal is $V(L) = L^3$, so the constant $a$ is equal to 1. For a spherical animal, the radius $R$ is a good choice for size. The volume of a spherical animal is $V(R) = \frac{4}{3}\pi R^3$, so the constant $a$ is $\frac{4}{3}\pi$ in this case.

**1.2** Derive a relationship between the surface area $S$ and the weight $W$ of an isometric series of animals.

**Solution** The dimension of surface area is

$$[S] = [L]^2 \tag{1.11}$$

If $L$ is the linear size of this series, then

$$S(L) = bL^2 \tag{1.12}$$

where $b$ is a constant that depends on the body plan. To express $S$ in terms of the weight, we use the fact that for an isometric series, $W(L) = a\rho g L^3$. Then we invert this relation to find $L$ as a function of $W$:

$$L(W) = \left(\frac{W}{a\rho g}\right)^{1/3} \tag{1.13}$$

Now we insert this expression into the formula for $S(L)$ and obtain

$$S(W) = cW^{2/3} \quad \text{where} \quad c = b\left(\frac{1}{a\rho g}\right)^{2/3} \tag{1.14}$$

The above formula predicts that the animal's surface area increases with the body weight $W$ as the two-thirds power of $W$. Therefore, large animals have less surface area per unit weight than do small animals.

**1.3** If the body dimensions of a dog were increased by a factor of 5, by what factor should the leg bone diameter be increased to maintain the same relative strength?

**Solution** The dog in the problem would have a volume and weight increased by $L^3 = 5^3 = 125$ times. The cross-sectional area would therefore have to increase by 125 times for the same relative strength. Since diameter is proportional to the square root of the area, the diameter would increase by $\sqrt{125} = 11.2$ times.

**1.4** A dog owner uses 100 g of material to make a little coat for a 1-kg lap dog. How much of the same material would be needed for a 50-kg Labrador?

**Solution** The mass of the large dog is 50 times that of the small. Therefore the volume of the large dog will be 50 times that of the small. Assuming they are scaled similarly, $L^3 = 50$ and $L = 3.684$. The surface area to be covered on the large dog increases by $L^2$ relative to the small dog. Therefore the surface of the large dog is greater by a factor of $L^2 = 3.684^2 = 13.6$. Since the thickness of the material does not change, the mass of material will be proportional to the area. Therefore the large dog will require $13.6 \times 100$ g $=$ 1.36 kg of material.

# EXERCISES

**1.1** It has been found that the leg length $L(W)$ of various African mammals scales with $W^{0.33}$, where $W$ is the weight of the animal. Explain this result using concepts of geometric similarity.

**1.2** Typically, large bacteria have the shape of cylinders of fixed radius $a$ and variable length $L$.

(a) Demonstrate that the surface-to-volume ratio of these bacteria does not depend on the length of the bacterium.

(b) Explain why a large cylindrical bacterium can feed itself more efficiently than a large spherical bacterium. Assume that the bacteria feed by absorbing nutrients through their surface. Compare the surface areas of cylindrical and spherical bacteria with the same volume for the same food requirement.

**1.3** Consider a cell culture $A$ with a population of 1000 and a culture $B$ with a population of 1500. Assume that culture $A$ has a growth constant of 0.05 day$^{-1}$ while culture $B$ doubles in size every 3 days. How long will it be until culture $B$ has 3 times as many cells as culture $A$?

**1.4** Suppose that you work with two cell cultures, $X$ and $Y$. Culture $X$ has a growth constant of 0.0125 day$^{-1}$ and a present population of 1200. Culture $Y$ has a decay constant of magnitude 0.180 day$^{-1}$ and a population of 1600.

(a) Plot a graph of the populations of both cultures over the next 10 days.

(b) From your graph, estimate when the populations will be equal.

**1.5** A cell culture, called $Z$, has a growth constant of 0.2 day$^{-1}$ when supplied with proper nutrients. Without nutrients, the population decays by a factor of 3 every 6 days. Exactly 1 day ago, culture $Z$ was provided with enough nutrients for 4 days. Its present population is 800.

(a) Plot a graph of the population, starting with time $t = 0$ at 1 day ago and ending 10.0 days from now.

(b) From your graph, estimate when the population will have declined by 50% from its present value.

# CHAPTER 2

# KINEMATICS IN ONE DIMENSION

## PHYSICAL BACKGROUND

Displacement is a vector that points from an object's initial position toward its final position. The magnitude of the displacement is the shortest distance between the two positions. The average speed of an object is the distance traveled by the object divided by the time required to cover the distance. The average velocity of an object is defined as the object's displacement vector $\Delta \mathbf{x}$ divided by the elapsed time $\Delta t$:

$$\mathbf{v} = \frac{\Delta \mathbf{x}}{\Delta t} \tag{2.1}$$

Average velocity is a vector that has the same direction as the displacement. Instantaneous velocity is a measure of velocity at a particular time and in the limit as $\Delta t$ approaches zero is given by

$$\mathbf{v} = \lim_{\Delta t \to 0} \frac{\Delta \mathbf{x}}{\Delta t} = \frac{d\mathbf{x}}{dt} \tag{2.2}$$

Average acceleration $\mathbf{a}$ is a vector that is equal to the change in the velocity $\Delta \mathbf{v}$ divided by the elapsed time $\Delta t$:

$$\mathbf{a} = \frac{\Delta \mathbf{v}}{\Delta t} \tag{2.3}$$

When $\Delta t$ is infinitesimally small, the average acceleration becomes equal to the instantaneous acceleration $\mathbf{a}$. Acceleration is the rate at which the velocity is changing:

$$\mathbf{a} = \lim_{\Delta t \to 0} \frac{\Delta \mathbf{v}}{\Delta t} = \frac{d\mathbf{v}}{dt} = \frac{d^2\mathbf{x}}{dt^2} \tag{2.4}$$

When an object moves with a constant acceleration along a straight line, its displacement $\Delta x = x - x_0$, final velocity $v$, initial velocity $v_0$, acceleration $a$, and elapsed time $t$ are related by the equations,

$$x = \bar{v}t = \tfrac{1}{2}(v_0 + v)t \quad \text{and} \quad v = v_0 + at \tag{2.5}$$

$$v^2 - v_0^2 = 2a(x - x_0) \tag{2.6}$$

and

$$x = x_0 + v_0 t + \tfrac{1}{2}at^2 \tag{2.7}$$

assuming that $x = x_0$ at $t = 0$.

In free-fall motion, an object experiences negligible air resistance and a constant acceleration due to gravity $g$ which is directed toward the center of the earth and has a magnitude of approximately 9.81 m/s$^2$.

## TOPIC 2.1   Mass Variations over Time

Linear kinematics provides the basic mathematical vocabulary for discussing motion along all types of curve. Let us look at some simple examples. Suppose we plot a curve of the average mass of an animal as a function of time and find that not only does the average mass itself increase with time but the increase in mass or growth rate also changes with time (see Figure 2.1). That is, the tangent to the curve of mass versus time changes as time increases. A second curve of growth rate versus time could be nonlinear or linear depending on the time period (see Figure 2.2), in the latter case (the middle time interval) we may conclude that the rate of change of the growth rate is constant. These curves could, in principle, describe a biological experiment, and we can use the corresponding equations of motion given in this chapter but we must change the physical interpretation of the symbols. Thus $x$ must be interpreted as the mass of the animal and not the position of a particle. The symbol $v$ becomes the growth rate instead of the velocity of the particle. Similarly, $a$ becomes the rate of change of the growth rate instead of acceleration. The basic mathematical structure, however, is identical. We might say that the average mass decreases with brood size at a constant rate and the use of the word *rate* is a simple example of this interpretation of the slope of graphs in general.

We can discuss position–time curves and derive from them the corresponding velocity–time curves. It is also possible to start with a velocity–time curve and obtain from it the change in displacement or position over any interval. We can do this by measuring the area underneath the velocity–time curve over the same interval. In the life sciences there are many examples where we are similarly interested in the area under various curves.

## TOPIC 2.2   The Volume of Respiratory Dead Space

Let us consider a specific example from pulmonary physiology. The respiratory tract is made up of two main parts. The first is the conducting airway (i.e., the nose, pharynx, larynx, trachea, bronchi, and bronchioles) and the second are the alveoli. The lungs have the major function of exchanging oxygen and carbon dioxide between the air intake and the pulmonary capillary blood flow (see Chapter 11 for a more detailed exposition). Only the air in the alveoli is involved in this exchange. The air in the conducting airway does not take part in the gas exchange, and so the volume of this airway is referred to as the respiratory dead space. This particular volume is of very great interest to pulmonary physiologists. To measure the dead space, the individual is asked to take one breath of pure

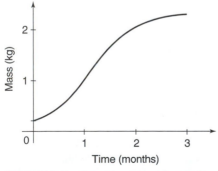

**FIGURE 2.1**   Average mass of an animal as a function of time.

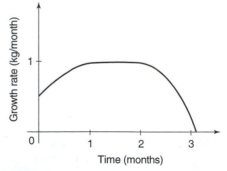

**FIGURE 2.2**   Growth rate or mass per unit time of an animal as a function of time.

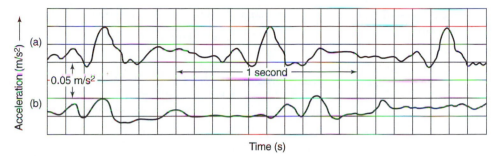

**FIGURE 2.3** Ballistocardiograms for (*a*) a healthy young individual and (*b*) one who is recovering from a heart attack.

oxygen and then breathe out through the mouth. Not only the volume rate of flow but also the concentration of nitrogen, or amount of nitrogen per unit volume, are continuously measured over the whole period of expiration. From these latter two quantities one can determine the volume of the dead space via the area under the volume rate curve and the average rate at which oxygen is utilized by the subject. The analogy drawn here is between the displacement and the volume.

## TOPIC 2.3   The Ballistocardiogram

Another example of a displacement–time curve is the ballistocardiogram. The corresponding instrument which generates these curves is the ballistocardiograph and is designed to make a quantitative assessment of the mechanical pumping action of the heart. The graph which records the horizontal position of the slider as ordinate against time horizontally is called a ballistocardiogram.

The idea of the ballistocardiogram is to obtain, from the position–time curve of the sliding table, information about cardiac malfunction. This means that only those features concerned with heart movement are to be recognized. Thus unwanted movement has to be suppressed by recording the acceleration of the table or slide rather than its displacement. This will exaggerate any rapid vibrational movement such as heart beat at the expense of any much slower movement such as that due to breathing and slow vibrations due to the stabilizing springs. A very sensitive accelerometer measures accelerations of the table as small as $10^{-5}$ m/s². In Figure 2.3 we show a typical ballistocardiogram for a young normal adult (top) compared with that of an individual who is recovering from a heart attack (bottom). For the healthy individual the acceleration has a range of 0.06 m/s², whereas in the recovering individual it is only 0.03 m/s². Interestingly, it has been found that the acceleration range is a good indicator of the chance of the patient's recovery, and in fact, the bigger this range, the better the prognosis.

## SOLVED PROBLEMS

**2.1**   A jaguar can reach speeds of 100 km/h. The fastest man is capable of reaching a velocity of 35 km/h. Suppose a man and a jaguar are 0.5 km apart. Assuming that both man and jaguar are constantly running at their top speed, how long does it take the jaguar to catch up to the man?

**Solution**   The first step is to sketch a diagram showing the initial and final conditions.

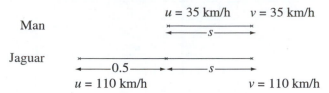

The man covers the distance $s$ in $t$ hours, the time it takes for the jaguar to move a distance $s + 0.5$. This is a special case where $a = 0$. Therefore $v = u$, and we obtain

$$s = ut = vt \tag{2.8}$$

For the man $s = 35t$, and for the jaguar $s + 0.5 = 100t$. Solving for $t$, we get

$$t = \frac{s + 0.5}{100} \quad \text{or} \quad t = \frac{35t + 0.5}{100} \tag{2.9}$$

Hence $100t - 35t = 0.5$. Consequently, we obtain

$$t = \frac{0.5}{65}h = 28 \text{ s} \tag{2.10}$$

**2.2**   A 2-kg falcon flies horizontally with a velocity of 10 m/s at a height of 50 m. It suddenly drops down toward its prey. When does the falcon hit the ground? How far from the point at which it folds its wings does the bird hit the ground?

**Solution**   We choose the zero for the $x$ axis to be the point where the falcon folds its wings, while the zero for the $y$ axis corresponds to the ground level. If $t = 0$ is the time that the falcon folds its wings, then $x(0) = 0$, and the motion along the $x$ axis is given by

$$x(t) = (10 \text{ m/s})t \tag{2.11}$$

The motion along the $y$ axis is free fall from a height of 50 m. The initial position $y(0)$ is thus 50 m, and the initial vertical velocity $v_y(0)$ is zero. The time $t^*$ at which the falcon hits the ground is given by

$$t^* = \sqrt{\frac{2y(0)}{g}} = 3.2 \text{ s} \tag{2.12}$$

We now determine how far the falcon travels horizontally during its fall: $x(t^*) = 10 \text{ ms}^{-1} \times 3.2 \text{ s} = 32 \text{ m}$.

# EXERCISES

**2.1**   Suppose blood in the aorta is accelerated by the action of the heart and it increases its velocity from zero to 0.35 m/s over a distance of 0.02 m. Calculate the value of the corresponding acceleration. How long does the event of acceleration take?

**2.2**   Assume that while pecking on a tree the woodpecker's head comes to a stop from an initial velocity of 0.7 m/s over a distance of 0.003 m. Evaluate the magnitude of the corresponding deceleration in units of the acceleration due to gravity $g$. What is the stopping time?

# CHAPTER 3

# KINEMATICS IN TWO DIMENSIONS

## PHYSICAL BACKGROUND

Motion that occurs in two dimensions can be described in terms of the time $t$ and the $x$ and $y$ components of three vectors: the displacement, the velocity, and the acceleration. The motion can be analyzed by treating the $x$ and $y$ components of the vectors separately. The directions of these components are conveyed by assigning a plus ($+$) or a minus ($-$) sign to each one. When the acceleration vector is constant, the $x$ and $y$ components of these vectors are related by the equations of kinematics.

Projectile motion is a kind of two-dimensional motion that occurs when the moving object (the projectile) experiences only the acceleration due to gravity, which acts in the vertical direction. The acceleration of the projectile has no horizontal component ($a_x = 0$), the effects of air resistance being negligible. The vertical component of the acceleration is $a_y$, and it equals the acceleration due to gravity, $g$ ($a_y = -g$). The kinematic equations of motion in two dimensions are given in general as

$$x = x_0 + v_{0x}t + \tfrac{1}{2}a_x t^2 \qquad y = y_0 + v_{0y}t + \tfrac{1}{2}a_y t^2 \tag{3.1}$$

$$v_x = v_{0x} + a_x t \qquad v_y = v_{0y} + a_y t \tag{3.2}$$

$$a_x = \text{const} \qquad a_y = \text{const} \tag{3.3}$$

supplemented by

$$v_x^2 = v_{0x}^2 + 2a_x x \qquad v_y^2 = v_{0y}^2 + 2a_y y \tag{3.4}$$

where, at $t = 0$, $x = x_0$, $y = y_0$ and $\mathbf{v}_0 = (v_{0x}, v_{0y})$.

## TOPIC 3.1   Maximum Range of Broadjumping

To estimate the range of broadjumping $R$, we can simply assume that the jumper is a projectile particle in the presence of gravity $g$ (see Figure 3.1). From equation (3.1) we have, for the $x$ component,

$$R = v_{0x}t \tag{3.5}$$

and for the $y$ component

$$0 = v_{0y}t - \tfrac{1}{2}gt^2 \tag{3.6}$$

Since $v_{0x} = v_0 \cos\theta$ and $v_{0y} = v_0 \sin\theta$, we solve the first equation, (3.5), for $t$ and substitute to the second, (3.6), which then yields the formula for $R$. Thus the range $R$ of a point particle is given by

$$R = \frac{v_0^2}{g}\sin 2\theta \tag{3.7}$$

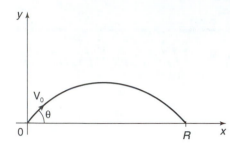

**FIGURE 3.1**  Motion of a projectile in the presence of gravity.

The optimal angle $\theta = 45°$ for the maximum range $R$, and taking the takeoff velocity $v_0$ to be approximately 10 m/s, this gives

$$R \cong \frac{(10 \text{ m/s})^2}{9.81 \text{ m/s}^2} \cong 10 \text{ m} \tag{3.8}$$

while the world record set at the 1968 Mexico City Summer Olympics by Bob Beamon of the United States was 8.9 m.

## SOLVED PROBLEMS

**3.1**  A shot putter can putt a shot of mass $m = 7.2$ kg to a maximum distance $x_{max,s} = 25.0$ m if he or she launches it at height $h_s = 2.5$ m from the ground. What is the initial velocity required to reach the above maximum distance?

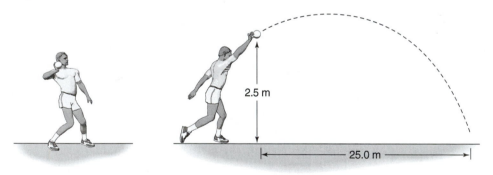

**Solution**  We first determine the optimal angle $\alpha$ of the putt. If the shot is putted at this angle, it will cover the maximum distance in the horizontal plane which requires an initial height $h$ and a velocity $v$. After the launch with a velocity $v$ at an angle $\alpha$, the shot moves with constant speed $v \cos \alpha$ horizontally and an initial speed of $v \sin \alpha$ vertically. As a function of time $t$ after the start, the $x$ and $y$ coordinates of the shot are

$$x = vt \cos \alpha \tag{3.9}$$

and

$$y = h + vt \sin \alpha - \tfrac{1}{2}gt^2 \tag{3.10}$$

The shot falls to the ground at time $t = t_0$, at which point $y(t = t_0) = 0$. Hence

$$0 = h + vt_0 \sin \alpha - \tfrac{1}{2}gt_0^2 \tag{3.11}$$

Introducing the range $r$ as

$$r = \sqrt{\frac{2gh}{v^2}} \qquad (3.12)$$

and reexpressing equation (3.11) with the help of equation (3.9) to eliminate $t$ yields

$$x = \frac{v^2 \cos \alpha}{g}(\sin \alpha + \sqrt{r^2 + \sin^2 \alpha}) \qquad (3.13)$$

This provides a relationship between the distance and the initial angle of the putt. The angle at which the distance of the putt is maximum is given by the condition $dx/d\alpha = 0$. Applying this to equation (3.13) gives

$$\tan \alpha_{max} = \frac{1}{\sqrt{1 + r^2}} \qquad (3.14)$$

The maximum distance of the putt is therefore

$$x_{max} = \frac{v}{g}\sqrt{v^2 + 2gh} \qquad (3.15)$$

The initial velocity $v_0$ required to reach the maximum distance can be found from equation (3.15) after rearrangement as

$$v_0 = \sqrt{gh\left(\sqrt{1 + \frac{x_{max}^2}{h^2}} - 1\right)} \qquad (3.16)$$

Substituting the numerical values given above, we obtain

$$v_0 = \sqrt{9.81 \text{ m/s}^2 \times 2.5 \text{ m}\left(\sqrt{1 + \frac{25^2}{2.5^2}} - 1\right)} = 14.90 \text{ m/s} \qquad (3.17)$$

**3.2** A honey bee is visiting a clover patch. Its average speed between flowers is $v = 1$ m/s. Starting at one flower, it travels north for 0.4 s to another flower, then east for 0.6 s to another flower, and finally southeast for 0.2 s to another flower. What is the bee's final displacement and direction from the first flower?

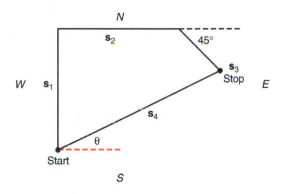

**Solution** We first find the displacements

$$s_1 = vt_1 = 0.40 \text{ m} \qquad (3.18)$$
$$s_2 = vt_2 = 0.60 \text{ m} \qquad (3.19)$$
$$s_3 = vt_3 = 0.20 \text{ m} \qquad (3.20)$$

The east–west component is

$$s_4 = s_2 + s_3 \cos 45° = 0.74 \text{ m} \tag{3.21}$$

The north–south component is

$$s_4 = s_1 - s_3 \sin 45° = 0.26 \text{ m} \tag{3.22}$$

The bee's displacement from the start is given by

$$s_4 = \sqrt{0.74^2 + 0.26^2} = 0.78 \text{ m} \tag{3.23}$$

The bee's direction from the start is given by

$$\theta = \tan^{-1}\frac{0.26}{0.74} = 19° \text{ north of east} \tag{3.24}$$

## EXERCISES

**3.1** A basketball player is approaching the basket at a velocity of 6 m/s when he jumps into the air to perform a slam dunk. Calculate the vertical velocity component required to lift the player 1 m above the ground. Determine the distance from the basket needed for the start of the jump in order for the player to reach the maximum height at the position of the basket.

**3.2** A seagull flies at a speed of 10 m/s directly into the wind. Knowing that it takes the bird 25 min to travel 8 km relative to the earth, find the velocity of the wind. How long will it take the seagull to travel a distance of 12 km with the wind?

**3.3** An owl accidentally drops a mouse it was carrying while flying horizontally at a speed of 5 m/s. The mouse drops to the ground 10 m below. What is the mouse's impact velocity and direction?

**3.4** A bird is flying due north with a speed of 20 m/s. A train is traveling due east with a speed of 25 m/s. What are the magnitude and direction of the velocity of the bird from the point of view of a passenger in the train?

**3.5** A bird is migrating nonstop a distance of 2500 km. Its destination is due south of its present position. In still air, the bird can fly with a speed of 40 km/h. However, there is a wind of 20 km/h from the west. How long will it take the bird to reach its destination?

**3.6** A butterfly is flying in a garden. First it travels due south for 5 m and then 30° west of north for 15 m. Assuming that the positive $x$ direction is east and the positive $y$ direction is north, (a) what are the coordinates of the butterfly's final position relative to its initial position and (b) what are the magnitude and direction of the displacement of the butterfly (relative to its initial position)?

# FORCES AND NEWTON'S LAWS OF MOTION

## PHYSICAL BACKGROUND

Newton's first law of motion or law of inertia states that an object continues in a state of rest or in a state of motion at a constant speed along a straight line unless compelled to change that state by a net force. Inertia is the natural tendency of an object to remain at rest or in motion at a constant speed along a straight line. The mass of a body is a quantitative measure of inertia and is measured in an SI (International System of Units) unit called the kilogram (kg). Force is a mechanical "push" or "pull" exerted on an object. Two or more external forces can act on an object without directly affecting its physical state as long as they cancel each other. Newton's second law of motion states that when several forces act on a body, their vector sum, $\sum \mathbf{F}$, produces an acceleration $\mathbf{a}$ of the object of mass $m$ and is given by

$$\sum \mathbf{F} = m\mathbf{a} \tag{4.1}$$

The SI unit of force is the newton (N = kgm/s²). A free-body diagram is a diagram that represents the object and the forces acting on it.

Newton's third law of motion, often called the action–reaction law, states that whenever one object exerts a force on a second object, the second object exerts an oppositely directed force of equal magnitude on the first object. For two forces $\mathbf{F}_A$ and $\mathbf{F}_B$ acting between two objects A and B, $\mathbf{F}_A = -\mathbf{F}_B$.

Newton's law of universal gravitation states that every particle in the universe exerts an attractive force on every other particle. For two particles that are separated by a distance $r$ and which have masses $m_1$ and $m_2$, the law states that the magnitude of this attractive force is

$$F = \frac{Gm_1m_2}{r^2} \tag{4.2}$$

while its direction lies along the line between the particles. The constant $G$ has a value of $6.673 \times 10^{-11}$ Nm²/kg² and is called the universal gravitational constant. The weight of an object on the earth is the gravitational force that the earth exerts on the object, namely $mg$.

A surface exerts a force on an object with which it is in contact. The component of the force perpendicular to the surface is called the normal force.

Friction is a force generated by the properties of the interface between the object and the surface and acts to resist the object's motion. A frictional force is defined as

$$F_f = \mu N \tag{4.3}$$

where $\mu$ is a coefficient of friction and $N$ is the normal force exerted by the surface on the object. Friction is independent of the sliding speed of the object and the area of contact between the object and the surface while it is directly proportional to the normal force exerted by the object on the surface.

There are two types of frictional forces corresponding to the physical state of motion of an object. If the object is stationary, then a frictional force (static friction) is acting on the object to prevent motion, described by

$$F_{f,s} \leq \mu_s N \qquad (4.4)$$

where $\mu_s$ is the coefficient of static friction. Force $F_s$ can increase to a maximum value of $\mu_s N$ to ensure static equilibrium of the object. Once this force is overcome and the object is set in motion, the object is subject to kinetic (sliding) friction, which is a force that acts to retard or slow down motion of the object and is defined as

$$F_{f,k} = \mu_k N \qquad (4.5)$$

where $\mu_k$ is the coefficient of kinetic friction.

The term *tension* means the tendency of a rope to be pulled apart due to forces that are applied at each end. Because of tension, a rope transmits a force from one end to the other.

An object is in equilibrium when the object moves at a constant velocity, that is, when it is not accelerating. The sum of the forces that act on an object in equilibrium is zero. Under equilibrium conditions in two dimensions, the separate sums of the force components in the $x$ direction and in the $y$ direction must each be zero: $\sum F_x = 0$ and $\sum F_y = 0$. If an object is not in equilibrium, Newton's second law accounts for the acceleration: $\sum F_x = ma_x$ and $\sum F_y = ma_y$.

## TOPIC 4.1   Applications of Newton's Laws to Muscles and Joints

Physical principles such as Newton's second and third laws can be of great use in studying the forces on muscles and bones. In general a muscle is attached, via tendons, to at least two different bones (Figure 4.1 shows muscles in the human arm). At a joint, two or more bones are flexibly connected (e.g., the elbow, knee, and hip). A pull is exerted by a muscle when its fibers contract under stimulation by a nerve. Muscles which bring two limbs closer together, like the biceps muscle in the human arm, are called flexor muscles. On the other hand, those that extend a limb outward, such as the triceps, are called extensor muscles. Flexor muscles are used, for example, when the upper arm is used to lift an object in the hand, whereas the extensor muscle is used when throwing a ball. More generally, the human skeleton is a very sophisticated device which transmits forces to and from various parts of the body. It is the muscles which move the parts and generate forces, use up chemical energy, and hence perform work. Thus they provide the power for movement in most many-cell organisms.

Muscles generate forces by contracting after they have been stimulated electrically. The tendons, or ropelike attachments, experience a net tension, after a series of these contractions, which increases with the number of electrical stimuli per second. The important function of tendons is to connect muscles to the limbs. Muscles try to shorten the distance between the attachment points of the tendons, but they cannot push these points apart. Thus pairs of muscles are necessary to operate a limb; for example, when the knee is bent, the hamstring muscle, at the base of the thigh, shortens. To straighten the knee, the quadriceps muscle at the front of the thigh shortens. Figure 4.2 gives examples of counteracting muscle pairs in the human arm and leg.

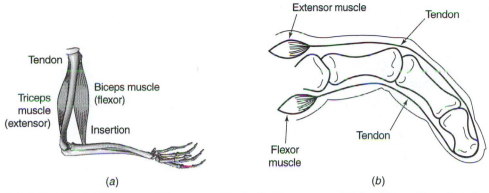

**FIGURE 4.1** (a) Triceps and biceps muscles in the human arm. (b) Forefinger. Note that the tendons carrying the forces exerted by the muscles go over joints that change the direction of the force.

Other types of muscles may join back on themselves and cause the constriction of an opening when they contract. Such muscles, called sphincters, serve several functions. The sphincter at the lower end of the esophagus prevents the backflow of stomach fluids. Another sphincter muscle in the eye changes the curvature of the lens of the eye to allow clear vision of near and distant objects (see Chapter 26).

Several muscles act simultaneously in concert in the shoulder to produce the total force exerted on the arm. The graphical method of adding forces yields the expected result: The upward and downward components of the forces cancel, leaving a large horizontal force (see Figure 4.2b).

Single muscles may be extracted from any organism like a frog and put into salt solution. These may be electrically stimulated by applying a voltage pulse to the solution. The length of the muscle may be kept fixed by clamping the ends of the muscle. When we measure the tension of a muscle of fixed length, we speak of an isometric force measurement. On the other hand, an isotonic force measurement is performed for a fixed load. We find that the maximum isometric force $F_{max}$ which a muscle can generate is proportional to its cross-sectional area $A$. Thus

$$F_{max} = \sigma A \qquad (4.6)$$

If $F_{max}$ is in newtons, then the constant of proportionality has units of newtons per square meter (or pascals, Pa) if $A$ is in square meters. A remarkable result is that $\sigma$ is approximately the same for all vertebrate muscles and its value is $\sigma \cong 0.3 \times 10^6$ Pa. Muscles consist of fiber bundles where each fiber can contract with a given force. From this we see why the total contractile force of a muscle must be proportional to its cross-sectional area.

As an example, consider the cross-sectional area of the hamstring muscle of a young human male. This area is about $1.5 \times 10^{-2}$ m$^2$ and produces a maximum force of $\cong 4500$ N. The hamstring muscle must be capable of supporting a man's weight. If we take a man whose mass is 70 kg, then his weight will be 690 N, and hence we have found that the maximum force of the hamstring muscle is about 6 times larger than a man's weight, which is large enough to allow a man to run or jump.

Standing obviously requires a direct interaction with the ground, and the person concerned exerts a force equal to his or her weight on the ground. By Newton's third law, the

*Arm*

*Leg*

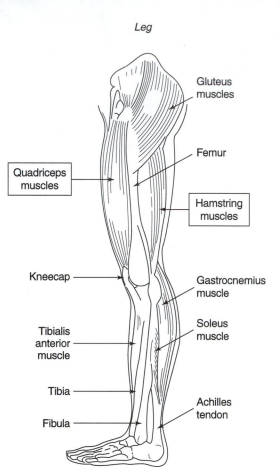

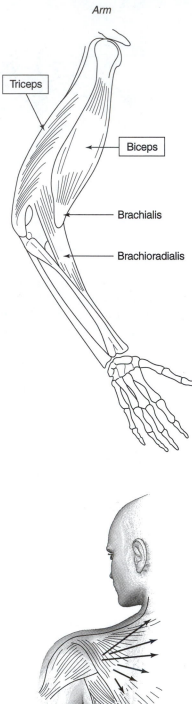

(*a*)

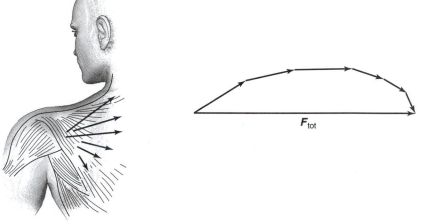

(*b*)

**FIGURE 4.2** (*a*) Counteracting muscle pairs. (*b*) Principle of simultaneous action of several muscles and their net force.

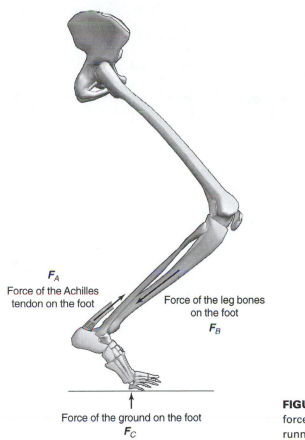

$F_A$
Force of the Achilles
tendon on the foot

Force of the leg bones
on the foot

$F_B$

Force of the ground on the foot
$F_C$

**FIGURE 4.3**   Illustration of the three
forces acting on the foot during
running.

ground exerts an upward force equal in magnitude to the person's weight but in the opposite direction.

The foot is subject to three forces during running activities: (1) the force of the ground acting upward to counteract the downward force on the foot, which is equal in magnitude to the weight of the runner's body; (2) the force of the Achilles tendon on the foot; and (3) the force of the leg bones acting downward on the foot. Each of these forces can be represented by a vector, as shown in Figure 4.3.

## TOPIC 4.2   Mechanics of Raising the Arm

The arm is raised principally by the forces exerted by the deltoid muscle. This muscle is connected to the upper side of the shoulder and extends over the humerus of upper arm and is again attached near the elbow. In raising the arm, three forces are involved. The first is the force exerted by the shoulder, $F_s$, and acts on the humerus in the positive $x$ direction. Second, the weight of the arm, $F_a$, acts vertically downward; and third, the force of the deltoid muscle, $F_d$, acts at approximately $\theta = 15°$ along the negative $x$ axis, as shown in Figure 4.4. If we assume that the weight of the arm, $F_a$, is 15 N, we can obtain the magnitudes $F_d$ and $F_s$. Resolving forces in the $x$ direction, we must have, in static equilibrium,

$$F_d \cos 15° = F_s \tag{4.7}$$

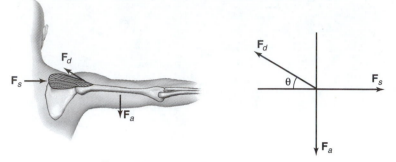

$F_a$ = weight of arm acting downward
$F_d$ = force of deltoid muscle acting at angle $\theta$
$F_s$ = force exerted by shoulder on humerus

**FIGURE 4.4**   Forces acting on the arm.

In the $y$ direction, we have

$$F_d \sin 15° = F_a \tag{4.8}$$

Hence from equation (4.8) we have

$$F_d = \frac{15}{0.259} = 57.92 \text{ N} \tag{4.9}$$

and hence, in equation (4.7), $F_S = F_d \cos 15° = (57.92 \text{ N})(0.966) = 55.95 \text{ N}$.

## TOPIC 4.3   The Skeletal Mechanics of the Hip

A group of three independent muscles control the motion of the leg. Collectively, they are called the hip abductor muscles and are attached to the pelvis. Their principal action is to swing the leg sideways relative to the hip (see Figure 4.5). Defining the horizontal to be the $x$ axis, the forces exerted by the abductor muscles may be approximated by the following vectors, the angles being all relative to the positive $x$ axis:

$\mathbf{F}_1$: 75 N at an angle $\alpha_1 = 90° - \theta_1 = 86°$.
$\mathbf{F}_2$: 220 N at an angle $\alpha_2 = 90° - \theta_2 = 78°$.
$\mathbf{F}_3$: 100 N at an angle $\alpha_3 = 90° - \theta_3 = 48°$.

Here, we have selected the numerical values for illustration purposes only. The resultant force in the positive $x$ direction, $F_x$, and the resultant force in the positive $y$ direction, $F_y$, may be written as

$$F_x = (75 \text{ N}) \cos 86° + (220 \text{ N}) \cos 78° + (100 \text{ N}) \cos 48° = 117.88 \text{ N} \tag{4.10}$$
$$F_y = (75 \text{ N}) \sin 86° + (220 \text{ N}) \sin 78° + (100 \text{ N}) \sin 48° = 364.31 \text{ N} \tag{4.11}$$

The magnitude of the resultant force $R$ is given by

$$R = \sqrt{F_x^2 + F_y^2} = 382.91 \text{ N} \tag{4.12}$$

and the angle $\theta$ that this makes with the $x$ axis is

$$\theta = \tan^{-1}\left(\frac{F_y}{F_x}\right) = 72.1° \tag{4.13}$$

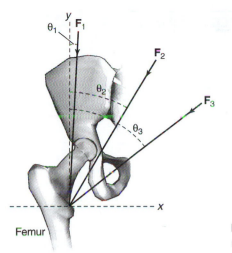

**FIGURE 4.5** Forces exerted by abductor muscles.

## TOPIC 4.4   Forces in Chewing

A group of muscles which control the position and motion of the upper (maxilla) and lower (mandible) jaws are operative in the process of chewing or mastication. The forces generated arise from two groups of muscles, the largest of which is the masseter muscle, which lowers the mandible or opens the mouth. The other muscle involved is the temporal muscle, which assists the masseter muscle in raising the mandible or closing the mouth. The forces generated by these two muscles can be represented by vectors. The resultant vectors are illustrated in Figure 4.6.

## TOPIC 4.5   Traction Systems and Treatment of Broken Bones

Sometimes it is necessary, when treating broken bones, to prevent the affected parts from moving. In Figure 4.7, we present a schematic diagram of three pulleys and a rope arrangement to keep a patient's head stationary. The pulley labeled 2 is fixed above the patient and pulley 1 to the patient's head. Pulleys 3 and 4 are fastened to the ceiling. The rope is

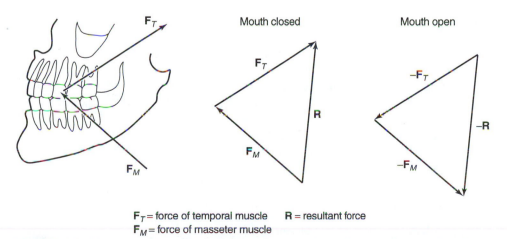

$F_T$ = force of temporal muscle     $R$ = resultant force
$F_M$ = force of masseter muscle

**FIGURE 4.6** Vectors representing masseter and temporal muscles and their resultant forces.

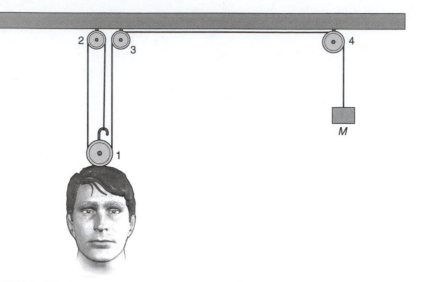

**FIGURE 4.7** Pulley and rope arrangement for head injuries.

fastened to the top of pulley 1 via a hook, threads through pulleys 2, 1, 3, and 4, and is then attached to a mass $M$. The force exerted at any point in the rope is called the tension at that point, and the rope will break if this tension exceeds a certain critical value.

Let $T$ be the constant tension in the rope. An upward vertical force of magnitude $T$ is exerted on the head by each of the two segments of rope which connect pulleys 1 and 3. However, that section of the rope between pulleys 1 and 3 also exerts an upward force of magnitude $T$. Thus the total force exerted by the rope on the head will be $F$, where

$$F = 3T \tag{4.14}$$

Figure 4.8 shows a patient in traction in a Stryker frame for an injury to the spinal vertebrae. It is necessary to stretch the spinal column along its length for the vertebrae to heal. The force applied is equal to the weight hanging on the right. The wire transmits the tension to the patient; the frictionless pulley simply changes the direction of the wire. Because a wire is used, the force is in the same direction as the wire at the point where it is attached to the patient. In the Stryker frame, the wire exerts a force parallel to the spine, as desired. Friction between the patient and the bed prevents the system from sliding to

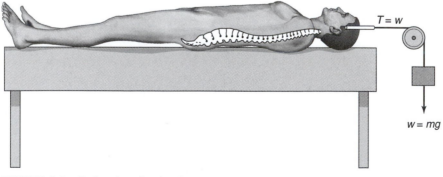

**FIGURE 4.8** Patient in a Stryker frame.

the right. Any traction system can be analyzed by keeping two things in mind. First, the force applied is in the direction of the wire at the point where it is attached to the patient; second, the force is equal to the weight hung on the wire. If several forces act at one point, the graphical method of vector addition can be used to analyze the system.

When a thigh bone or femur becomes fractured, a potential problem arises in the healing process. There is a tendency for the major leg muscles to pull together and mis-align the two segments of the broken bone at the fracture point, thus sometimes causing shortening and ultimately a limp. To prevent this, traction is used. Traction is a procedure in which part of the body is placed under tension to correct the alignment of the two ad-joining structures or to hold them in position. In Figure 4.9 the Russell system of traction

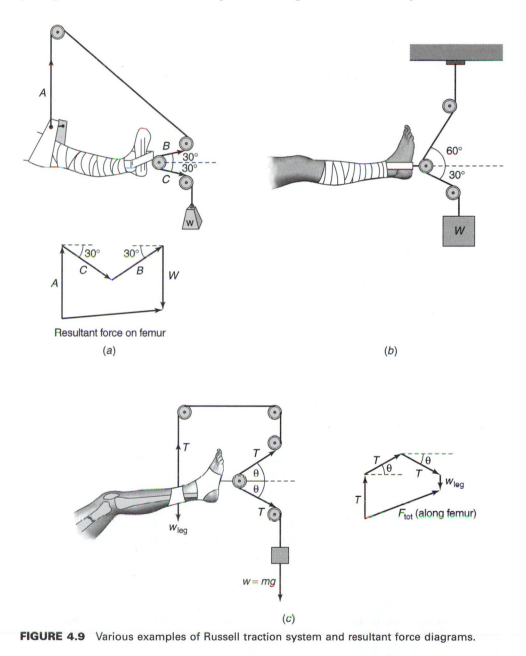

**FIGURE 4.9**  Various examples of Russell traction system and resultant force diagrams.

is shown, the objective being to generate a force along the axis of the femur. The lower leg bone, or tibia, is thus stabilized by a traction cord which is attached to weights in a pulley system. Assuming that there is negligible friction on the corresponding pulleys, the force exerted on the leg by the traction cord is solely due to the weight acting downward.

## TOPIC 4.6   Osteoarthritis and Friction at Skeletal Joints

When two bones are connected at a joint, the bones do not touch each other at this point because they are covered by cartilage, which permits low-friction movement, and are surrounded by a space filled with the synovial fluid. This fluid has the consistency of water, and in human joints the coefficient of friction $\mu$ ranges from 0.005 to 0.02.

Let us consider, as a simple example, the frictional force exerted on the hip. The frictional force $F_f$, which will clearly depend on the motion and the corresponding speed, exerted on the hip is approximately

$$F_f = 2.5 \; \mu W = 2.5 \; \mu mg \tag{4.15}$$

where $W$ is the weight of the person. For a normal hip and a person with mass 70 kg, the frictional force is

$$F_f \geq (2.5)(0.005)(70)(9.8 \text{ m/s}^2) = 8.58 \text{ N} \tag{4.16}$$

The aging process and osteoarthritis modify the immunological conditions of a joint and adversely affect the composition of the synovial fluid. The result is to diminish its lubricating capacity and permit direct contact between the ends of the connecting bones. The force generated by normal movements, without the lubricant, is translated into heat energy, which further acts to destroy the joint. In addition, inflammation, swelling, and pain are caused. Thus, if we assume an increased coefficient of friction of 0.5, the frictional force exerted on the arthritic hip becomes much larger and we find that

$$F_f = (2.5)(0.5)(70)(9.8 \text{ m/s}^2) = 858 \text{ N} \tag{4.17}$$

which explains the cause of pain the patient experiences.

## TOPIC 4.7   Application of Newton's Laws to Molecules

Newton's third law applies to molecules, and an important illustration of this is the protein–protein interaction, as shown in Figure 4.10. When two proteins having outside surfaces which are complementary in shape (i.e., surfaces which fit together like a "lock and key"; see Figure 4.10) are moving toward each other, they interact via a series of weak forces between the atoms near the contact surface. These forces may include hydrogen bonding, the van der Waals force, the electrostatic interaction, and the hydrophobic interaction (see Chapter 19 for a detailed exposition). As the surfaces of the molecules are complementary, these many weak forces may add up to provide a strong attractive force. On the other hand, if the surfaces are not complementary, the two proteins may not be attracted strongly enough to form a bond. This mechanism of selective bonding is termed lock-and-key recognition. Powerful computers using highly specialized molecular dynamics (MD) software can now calculate the contact force between two proteins. It is found that the total force exerted by protein A on protein B and the total force exerted by protein B on protein A form an action–reaction pair. In other words, these two forces satisfy Newton's third law and are equal in magnitude and opposite in direction.

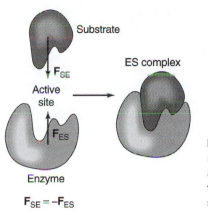

Substrate

ES complex

Active
site

Enzyme

$\mathbf{F}_{SE} = -\mathbf{F}_{ES}$

**FIGURE 4.10** Diagram illustrating the lock-and-key mechanism for two molecules: a substrate protein (S) and an enzyme protein (E). The force acting on S attracting it to E, $\mathbf{F}_{SE}$, is equal in magnitude and opposite to the force on E attracting it to S, $\mathbf{F}_{ES}$.

## TOPIC 4.8   Newton's Third Law and Locomotion

The octopus, a highly developed mollusk (cephalopod), spends the majority of the day in seclusion behind rock crevices along the ocean floor. It moves along, when hunting its prey, in a gliding motion, propelled by a jet of water from its gill chamber. This jet, according to Newton's third law, exerts a force equal in magnitude and opposite in direction to its motion, thus allowing the octopus to move forward. When alarmed, the octopus is well adapted to escape at high speeds by increasing the magnitude and direction of its water jet.

## SOLVED PROBLEMS

**4.1**   Patients suffering from a neck injury have their cervical vertebrae kept under tension by means of a traction device, as Figure 4.8 illustrates. The device creates tension in the vertebrae by pulling to the right on the head with a force $T$, which, in effect, is applied to the first vertebra at the top of the spine. This vertebra remains in equilibrium because it is simultaneously pulled to the left by a force $F$ that is supplied by the next vertebra in line. The force $F$ comes about in reaction to the pulling effect of the force $T$, in accord with Newton's third law. If it is desired that $F$ have a magnitude of 40 N, how much mass $m$ should be suspended from the rope?

**Solution**   The condition for equilibrium is that the net force in the $x$ direction is zero:

$$\sum F_x = F - T = 0 \tag{4.18}$$

so that $F = T$. Since $T = mg$, it follows that the necessary mass is

$$m = \frac{F}{g} = \frac{40 \text{ N}}{9.80 \text{ m/s}^2} = 4.1 \text{ kg} \tag{4.19}$$

**4.2**   The neck of the thigh bone is loaded by a weight $G = 300$ N. What frictional forces arise in the joints of a healthy and an arthritic patient if the coefficients of friction are $\mu = 0.003$ and $\mu = 0.03$ in normal and arthritic joints, respectively?

**Solution**   The frictional force $F$ opposes the movement of one surface over another and is proportional to the normal force $F_n$ at right angles to the two surfaces: $F = \mu F_n$. As the weight is perpendicular to the surface of the head of the femur, the normal force $F_n = G = 300$ N, and thus frictional forces of 0.9 and 9 N are exerted in the joints of healthy and arthritic patients, respectively.

**4.3**   The quadriceps tendon passes over the kneecap and presses the femur. The tension in the tendon is $F = 1500$ N. What are the magnitude and direction of the resultant force exerted on the femur in the configuration shown in the diagram?

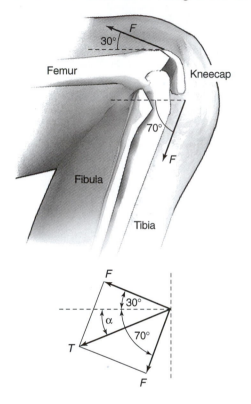

**Solution**   The femur is loaded by a resultant tension $T$ enclosing an angle $\alpha$ with the axis of the femur. In equilibrium, the vectorial sum of the forces targeting the femur should be zero. The balance of the components for the horizontal direction is

$$F \cos 30° + F \cos 70° = T \cos \alpha \tag{4.20}$$

and that for the vertical direction is

$$F \sin 30° - F \sin 70° = T \sin \alpha \tag{4.21}$$

Adding the squares of the two equations gives

$$2F^2(1 + \cos 30° \cos 70° - \sin 30° \sin 70°) = T^2 \tag{4.22}$$

where the identity $\sin^2 \alpha + \cos^2 \alpha = 1$ was used. On substitution of the numerical values, $T = 1928$ N. After insertion of this value into the first equation, we obtain $\cos \alpha = 0.94$, or $\alpha = 20°$.

# EXERCISES

**4.1**   A person of mass 75 kg is standing on the tiptoes of both feet. The heel is lifted just above the ground, so the foot can be treated as horizontal. There are three forces acting on each foot; the upward vertical force **N** exerted by the floor on the toes, the downward vertical force $\mathbf{F}_2$ exerted by the leg bone (tibia) on the ankle (talus), and the upward vertical force $\mathbf{F}_3$ exerted by the Achilles tendon on the heel. The distance from the toes to the ankle is 12.5 cm, and the distance from the ankle to the heel is 6.5 cm.

**(a)** Draw a free-body force diagram for the foot, showing the relevant forces and distances.

**(b)** Applying the conditions of force equilibrium, find the magnitudes of $\mathbf{F}_2$ and $\mathbf{F}_3$.

**4.2** An animal biting with an upward vertical force $\mathbf{F}_B$ on food experiences a reaction force $\mathbf{F}_N$ on the jaw, as shown in the diagram. A muscular force $\mathbf{F}_M$ directed 40° above the horizontal is applied on the jaw at point 0. The jaw hinges at the joint $J$, which feels a reaction force $\mathbf{F}_R$. The magnitude of $\mathbf{F}_B$ is 1200 N. The distances are $x_1 = 0.35$ m and $x_2 = 0.25$ m.

**(a)** Draw a free-body force diagram showing all the forces acting on the jaw.

**(b)** Set up equations describing the balance of forces.

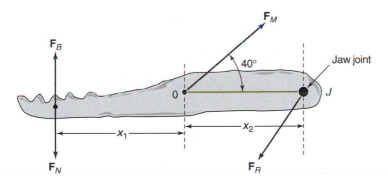

**4.3** In the accompanying figure, you see a record of the force exerted by a runner's foot on the ground during a stride of speed 4.0 m/s. Try to explain how the force on the runner's foot can exceed his body weight $Mg$.

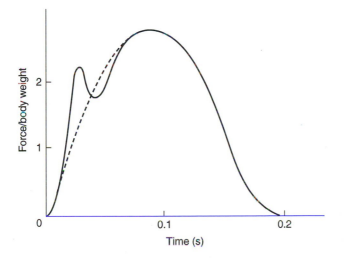

**4.4** A volleyball player jumps into the air by first lowering his body by 0.4 m and then accelerating by straightening his legs rapidly. As a result, his body is lifted 0.8 m above the floor level. What is the velocity of the volleyball player at the moment of leaving the ground? What is the corresponding acceleration in units of the constant due to gravity $g$? If his mass is 75 kg, what is the force exerted by his feet on the floor?

**4.5**   What force is exerted on the tooth by the braces in the figure if the tension in the wire is 30 N?

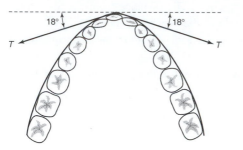

**4.6**   Two muscles in the back of the leg, called the medial and lateral heads of the gastrocnemius muscle, pull upward on the Achilles tendon, as shown in the figure. Find the magnitude and direction of the total force on the Achilles tendon.

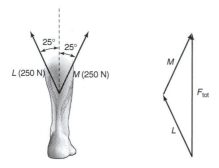

**4.7**   When resting, the head is normally held as shown in the figure. The forces acting are the weight of the head, $W$, the force of the neck muscles, $\mathbf{F}_M$, and the force supplied by the upper vertebrae, $\mathbf{F}_V$. These forces add to zero when the head is stationary. Calculate the magnitude and direction of $\mathbf{F}_V$ needed to accomplish this.

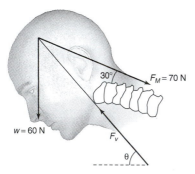

**4.8**   What force is applied by the traction setup shown in the figure?

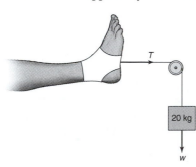

# CHAPTER 5

# DYNAMICS AND UNIFORM CIRCULAR MOTION

## PHYSICAL BACKGROUND

In uniform circular motion, an object of mass $m$ travels at a constant speed $v$ on a circular path of radius $r$. The period $T$ of the motion is the time required to make one revolution. The speed, the period, and the radius are related according to

$$v = \frac{2\pi r}{T} \tag{5.1}$$

The velocity vector is tangential to the circular pattern in such motion and hence is always changing direction. Therefore, an acceleration exists. This acceleration is called centripetal acceleration, and its magnitude, $a_c$, can be shown to satisfy the formula

$$a_c = \frac{v^2}{r} \tag{5.2}$$

while its direction is toward the center of the circle. To create this acceleration, a net force $F_c$ pointing toward the center of the circle is needed. This net force is called the centripetal force, and its magnitude is

$$F_c = \frac{mv^2}{r} \tag{5.3}$$

An equal and opposite force acting on the mass $m$ is called the centrifugal force, which is a reaction force which tends to throw the mass outward when released.

## TOPIC 5.1  Walking and Running

When a person is walking, it is found that the hips move along a circular arc whose radius $R$ is approximately equal to the length of the leg, the center of motion being the foot of the person in contact with the ground. The vertical position of the center of mass of the person is close to the hip and moves along a circular path whose radius is comparable to $R$. Thus, we model the dynamics of walking as the motion of a mass $M$, the person's mass, along a circle of radius $R$. If the walking velocity is $v$, then a radially inward force of $F = Mv^2/R$ must be exerted on it to make the motion possible. The foot exerts a force equal to the weight of the person, that is, $Mg$, on the ground. By Newton's third law, the reaction force exerted by the ground on the foot is also equal in magnitude to $Mg$. This latter force balances the force $F$ so we have

$$Mg = \frac{Mv^2}{R} \tag{5.4}$$

**29**

The maximum walking velocity $v_m$ is therefore given by

$$v_m = \sqrt{gR} \tag{5.5}$$

where $v_m$ is about 3 m/s for an adult with a leg length of order 0.9 m. Such a person switches from walking to running a little below this threshold when the velocity is about 2.5 m/s. We define the dimensionless Froude number, Fr, of an animal with speed $v$ and leg length $L$ by

$$\text{Fr} = \frac{v^2}{gL} \tag{5.6}$$

When the Froude number is of order 1, an upright animal must change from walking to running. Interestingly, it turns out that animals with the same Froude number tend to walk in a dynamically similar way. This is the principle of dynamic similarity. From equation (5.6), as Fr remains close to 1 at the walking–running threshold, then for two quadruped animals whose leg lengths $L$ differ by a factor of 9, the larger animal will change from walking to running at three times the velocity of the small animal; for example, cats change from walking to running at about 1 m/s, but camels (whose leg length is about nine times longer) change at approximately 2.9 m/s.

# EXERCISE

**5.1**  Assume a man's arm is 0.6 m long and that he can rotate it about his shoulder in a vertical circle once per second. Compare the centripetal force developed with the gravitational force on the blood in his hand.

# CHAPTER 6

# WORK AND ENERGY

## PHYSICAL BACKGROUND

The work $W$ done by a constant force acting on an object is

$$W = Fs \cos \theta \qquad (6.1)$$

where $F$ is the magnitude of the force, $s$ is the magnitude of the displacement, and $\theta$ is the angle between the force and displacement vectors. Work is a scalar quantity and can be positive or negative, depending on whether the angle $\theta$ is less than or greater than $90°$, respectively. The work is zero if the force is perpendicular ($\theta = 90°$) to the displacement.

The kinetic energy KE of an object of mass $m$ and speed $v$ is defined by

$$\text{KE} = \tfrac{1}{2}mv^2 \qquad (6.2)$$

The work–energy theorem states that the work $W$ done by the net external force acting on an object equals the difference between the object's final kinetic energy $\text{KE}_f$ and initial energy $\text{KE}_i$:

$$W = \text{KE}_f - \text{KE}_i \qquad (6.3)$$

Since the force of gravity is perpendicular to the earth's surface, the work done by the force of gravity on an object of mass $m$ is

$$W_{\text{gravity}} = mg(h_i - h_f) \qquad (6.4)$$

where $h_i$ and $h_f$ are the initial and final heights of the object, respectively. Gravitational potential energy, PE, is the energy that an object has by virtue of its position above the surface of the earth and is given by

$$\text{PE} = mgh \qquad (6.5)$$

where $h$ is the height of the object relative to an arbitrary zero level.

All forces can be categorized as conservative or nonconservative. A conservative force is one that, in moving an object between two points, does the same work, independent of the path taken between the points. Alternatively, a force is conservative if the work it does moving an object around any closed path is zero. A force is a nonconservative force if the work it does on an object moving between two points depends on the path of the motion between the points. Examples of conservative forces are the gravitational force near the earth's surface, the elastic force of a spring, and the electrostatic force of an electric charge. An example of a nonconservative force is the force of friction.

The total mechanical energy $E$ is the sum of the kinetic energy and the potential energy:

$$E = \text{KE} + \text{PE} \qquad (6.6)$$

The principle of conservation of energy states that energy of an isolated system can be neither created nor destroyed but can be transformed to other forms of energy. In other words, the following must hold:

$$E_{\text{before}} = E_{\text{after}} \qquad (6.7)$$

Average power $\bar{P}$ is the work done per unit time:

$$\bar{P} = \frac{\text{work}}{\text{time}} \tag{6.8}$$

or the rate at which work is done.

The efficiency of a device is defined by

$$e = \frac{W_{\text{out}}}{E_{\text{in}}} \tag{6.9}$$

where $W_{\text{out}}$ is the useful work produced by the device and $E_{\text{in}}$ is the energy input. Since work done on a system results in energy being transferred into that system, it is equally acceptable to define efficiency in terms of energy alone:

$$e = \frac{E_{\text{out}}}{E_{\text{in}}} \tag{6.10}$$

where $E_{\text{out}}$ is the useful energy output of the device. Efficiency also can be defined in terms of power input and output. If both energies in the last expression are divided by time, then

$$e = \frac{E_{\text{out}}/t}{E_{\text{in}}/t} = \frac{P_{\text{out}}}{P_{\text{in}}} \tag{6.11}$$

where $P_{\text{out}}$ is the useful power output and $P_{\text{in}}$ is the power input for the device.

## TOPIC 6.1  Work and Power of Muscles

As discussed in Chapter 4, animal and human posture and motion are controlled by forces generated by muscles. A muscle is composed of a large number of fibers which can be contracted under direct stimulation by nerves. A muscle is connected to two bones across a joint and is attached by tendons at each end. The contraction of a muscle produces a pair of action–reaction forces between each bone and the muscle. The work $W$ performed by a contracting muscle is the product of the force of contraction, the cross-sectional area $A$, and the shortening distance $\Delta\ell$:

$$W = \sigma A \, \Delta\ell \tag{6.12}$$

where $\sigma$ is the tensile stress produced by the muscle, that is, the force of contraction per unit area. The power developed by the muscle is then

$$P = \sigma A \frac{\Delta\ell}{\Delta t} = \sigma A v \tag{6.13}$$

and the speed of shortening, $v = \Delta\ell/\Delta t$, appears to be a constant across species, since as the length of a muscle fiber increases, the time of contraction increases by the same factor. Thus, it has been found that the power generated is regulated by the value of the cross-sectional area $A$.

## TOPIC 6.2  Energy and Power When Running

The major work done in running is the acceleration and deceleration of the legs for each stride. Lifting a leg off the ground, it is brought up to a speed $v$ which is approximately equal to the speed of the body. During this process the muscles in the leg do work equal

**TABLE 6.1   Efficiency of Body and Mechanical Devices**

| Source of work | Efficiency $e$ (%) |
| --- | --- |
| Body | |
|   Cycling | 20 |
|   Swimming, surface | 2 |
|   Swimming, submerged | 4 |
|   Shoveling | 3 |
| Steam engine | 17 |
| Gasoline engine | 38 |
| Nuclear power plant | 35 |

to the change in kinetic energy, that is, $\frac{1}{2}mv^2$, where $m$ is the mass of the leg. As the leg is brought back to rest, the muscles also do the same amount of work, so for each stride the leg muscles do work approximately equal to $mv^2$.

As an example, consider a 65-kg man running at 2.5 m/s. Each of his legs has a mass of about 9 kg, so the work done on each stride is

$$W = mv^2 = (9 \text{ kg})(2.5 \text{ m/s})^2 = 56 \text{ J} \qquad (6.14)$$

If the man takes 1.5 strides per second with each leg, the total power input to his two legs is

$$P = 2(56 \text{ J/stride})(1.5 \text{ strides/s}) = 168 \text{ W} \qquad (6.15)$$

The efficiency $e$ of a muscle is only about 0.25, so the rate of energy expenditure is

$$\frac{P}{e} = \frac{168 \text{ W}}{0.25} = 672 \text{ W} \qquad (6.16)$$

Our estimate should have come out smaller than the measured value because we have not taken into account the energy consumed in all the other bodily functions, such as breathing and pumping blood. The speed of running is proportional to the stride length and the frequency of strides. The world record for male 100-m dash sprinters stands at 9.8 s, which translates into 10.1 m/s, close to 11 m/s, a calculated maximum speed for humans. (Thus the limit for this record is 9.1 s.) What our calculation does show is that the energy involved in motor activity can be approximately calculated using simple physical principles. Table 6.1 compares some aspects of human efficiency to mechanical devices.

## TOPIC 6.3   Maximum Height of a Vertical Jump

To calculate the height of a jump, $H$, we use conservation of energy whereby

Work done on the human body = change in gravitational potential energy

The work done on the body is done by the ground exerting a force on the jumper through a certain crouch length which is approximately equal to the length of one's legs. The legs of most people can support their own weight and an additional "dead load" equal to their weight. Conservatively estimating the leg strength, we take their maximum additional force used for jumping as $F = \alpha W$, where $\alpha \cong 0.5$ and $W = mg$ is the person's weight. Therefore, the work done on the body is $\alpha mg \times$ crouch length, and the energy equation becomes

$$\alpha mg \times \text{crouch length} = mgH \tag{6.17}$$

giving

$$H \cong \alpha \times \text{crouch length} \tag{6.18}$$

Crouch length is approximately 1 m, so $H \cong 0.5$ m. In fact, for a vertical jump from rest the values from studies range from 0.3 to 0.6 m. Compared to its size, the greatest jumper in the animal world is a 2-mm-long flea, which is able to elevate its 0.5-mg mass by 0.2 m with a takeoff speed of 2 m/s, achieving an acceleration of approximately 245 times the acceleration due to gravity versus the human's 1.5 times gravity in a vertical jump. Amazingly, if not for a significant air resistance, the flea would be able to jump twice as high in a vacuum.

## TOPIC 6.4 Mechanical Efficiency of the Heart

The power of the heart is the rate at which it does work (see also Topic 11.4). The work done by each ventricle of the resting heart is approximately $W_{LV} = 1.11$ J for the left ventricle and $W_{RV} = 0.22$ J for the right ventricle. Thus the total average work done by the heart, $W_T$, during a single contraction is

$$W_T = W_{LV} + W_{RV} = (1.11 + 0.22)\ \text{J} = 1.33\ \text{J} \tag{6.19}$$

The power output is then calculated as the rate of change with time of $W_T$. Over the time elapsed during each heart beat the time interval $t_c$ has the approximate value of 1.0 s. Hence

$$\text{Power output} = \frac{W_{LV} + W_{RV}}{t_c} = \frac{(1.11 + 0.22)\ \text{J}}{1.0\ \text{s}} = 1.33\ \text{W} \tag{6.20}$$

The mechanical efficiency of the heart is probably less than 10%, and in its resting state, the rate at which it uses energy is about 10 times its power output, or approximately 12 to 15 W. The total metabolic rate for a resting person is about 100 W, so the power output of the heart constitutes a small fraction of the total power output of the body.

## TOPIC 6.5 Treadmill Exercise and Cardiac Stress

To assess a patient who is suspected of having heart disease, the physician must examine the cardiac function when (a) the patient is at rest with a heart beating at a normal pace and (b) when under stress, for example, after exercise, when the heart is beating rapidly. To simulate the conditions of stress, the patient exercises by walking a treadmill to increase heart beat and sustain high levels of cardiac stress. To reach this level of stress, the average times required are typically 20 to 30 minutes. The work done in achieving the above level of stress can be determined by analyzing the forces acting on the patient. We use Figure 6.1 to do this.

Denoting the external forces exerted by the patient by $F_{ext}$ whose direction is up the slope of the treadmill, the weight of the patient by $W = mg$, the normal force exerted upward by the treadmill as $N$, the angle of inclination of the treadmill by $\theta$, and the coeffi-

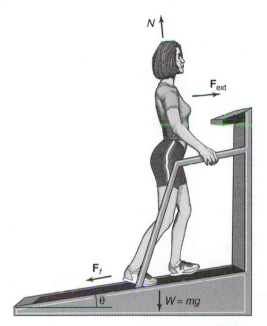

**FIGURE 6.1**  Treadmill and forces acting on the patient.

cient of static friction by $\mu_s$, we have the following balance of forces. For the forces up the slope of the treadmill,

$$F_{\text{ext}} - F_f - mg \sin \theta = F_{\text{ext}} - mg \sin \theta - \mu_s N \tag{6.21}$$

with the normal reaction $N$ being perpendicular to the slope of the treadmill. For the forces perpendicular to the slope of the treadmill,

$$N - mg \cos \theta = 0 \tag{6.22}$$

since the net force perpendicular to the slope is zero. From equations (6.21) and (6.22), the net force up the slope is

$$F_{\text{net}} = F_{\text{ext}} - (mg \sin \theta + \mu_s mg \cos \theta) \tag{6.23}$$

and thus, if $d$ is the distance the patient moves parallel to the treadmill, the work done is given as

$$W = F_{\text{net}}d = [F_{\text{ext}} - (mg \sin \theta + \mu_s mg \cos \theta)]d \tag{6.24}$$

Taking as representative numbers $F_{\text{ext}} = 300$ N, $m = 60$ kg, $\theta = 20°$, $\mu_s = 0.1$, and $d = 2000$ m, we get

$$W = 8.68 \times 10^4 \text{ J} \tag{6.25}$$

## TOPIC 6.6   Dynamics of Bumblebee Flight

When a bumblebee begins feeding at a suitable flower for nectar, it hovers about the flower, as shown in the Figure 6.2. The average force exerted by the two wings during a single downward stroke is

$$F = F_{\text{wing1}} + F_{\text{wing2}} = W + W = 2W \tag{6.26}$$

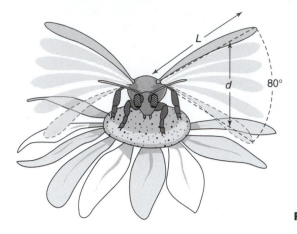

**FIGURE 6.2** A hovering bumblebee.

We will assume that the mass of the bumblebee is m $= 2.5 \times 10^{-4}$ kg and its corresponding weight to be $W = mg$. Each downward stroke involves movement of the wing through a vertical distance $d = r\theta$, where $r$ is the length of the wing and $\theta$ is the angle swept out by the wing in radians. We shall assume the length of each wing is $L = 0.01$ m and sweeps out an angle of $80° = 1.4$ rad during each stroke. So,

$$d = (1.0 \text{ cm})(1.4 \text{ rad}) = 1.4 \text{ cm} = 0.014 \text{ m} \tag{6.27}$$

The work done by the bumblebee in hovering about the flower is

$$\text{Work} = 2Wd = 2mgd = 2(0.25 \times 10^{-3} \text{ kg})(9.8 \text{ m/s}^2)(0.014 \text{ m})$$
$$= 0.0686 \times 10^{-3} \text{ J} \tag{6.28}$$

per downward stroke. If we also assume that the bumblebee averages 100 downward strokes per second, the power output of the bumblebee is

$$\text{Power} = \frac{\text{work}}{\text{time}} = (686 \times 10^{-7} \text{ J/stroke})(100 \text{ strokes/s}) = 686 \times 10^{-5} \text{ W} \tag{6.29}$$

## TOPIC 6.7 Force Generation in Cells:
### Hill's Law, Muscles, and Motor Proteins

Muscles in the body are not constantly generating the maximum force $F_{max}$ they can exert. It is a matter of common experience that the smaller a weight $W$ you hold, the faster you can lift it. Defining the contraction velocity $v$ of a muscle to be the shortening length of a muscle divided by the shortening time, one can fit the results of an isotonic measurement (tension fixed) to the relation

$$v = b \frac{F_{max} - W}{W + a} \tag{6.30}$$

where $a$ and $b$ are constants, the former with the dimensions of force and the latter of velocity. This relation is known as Hill's law. According to equation (6.30), $v = 0$ when the load $W$ is equal to the maximum force $F_{max}$. In Figure 6.3, we plot the contraction velocity $v$ against the load $W$, and it can be seen that as the load is reduced, the muscle con-

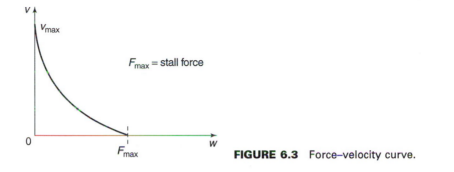

**FIGURE 6.3** Force–velocity curve.

tracts more rapidly. In terms of the parameters in equation (6.30) the maximum contraction velocity for zero load ($W = 0$) is given by

$$v_{max} = \frac{b}{a} F_{max} \tag{6.31}$$

The plot in Figure 6.3 is called a force–velocity curve. When force–velocity curves are measured for car engines, automotive engineers call $F_{max}$ the stall force.

In terms of the molecular structure of muscles, we have seen earlier in Chapter 4 that muscles are made up of fiber-like bundles called fascicles. Bundles such as these are made up of very long cells with a diameter of approximately 0.4 mm and lengths of about 40 mm. The forces the fibers can exert are additive so that the total force generated is proportional to the number of fibers and hence the cross-sectional area, as was pointed out in earlier sections. Fiber cells come in two types: thin filaments called actin protein fibers and thick filaments known as myosin protein fibers (see Figure 6.4).

Fibers are arranged in an alternating pattern of thick and thin filaments which may slide over each other. When we contract our muscles, motor proteins (see Figure 6.5) or molecular cross-bridges pull the thin filaments over the thick ones, and when the muscles are extended, filaments do not overlap. Motor proteins are fixed to myosin fibers and move

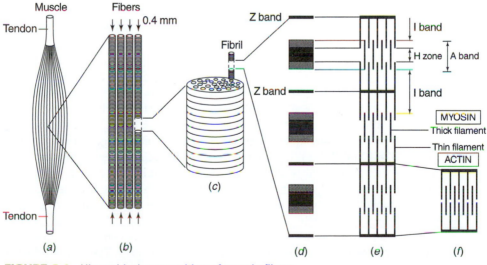

**FIGURE 6.4** Hierarchical composition of muscle fibers.

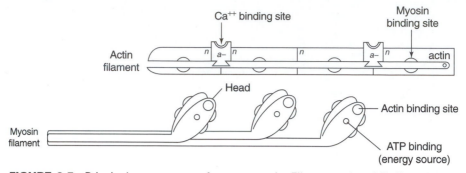

**FIGURE 6.5**   Principal components of motor protein. Fibers consist of fibrils which have light and dark bands. Myosin and actin filaments are designated.

over the actin fiber, very like a collapsing telescope. The muscle force generated is larger the more activated motor proteins there are. Force–velocity curves have been measured for individual motor proteins and are quite similar to Hill's law in equation (6.30), so there is a molecular basis for this law.

## TOPIC 6.8   Metabolic Energy

Almost all metabolic energy of animals and humans comes from the conversion of adenosine triphosphate (ATP) into ADP (adenosine diphosphate) and AMP (adenosine monophosphate). The energy to convert ADP and AMP back to ATP is supplied by the oxidation of carbohydrates, fats, and proteins. Metabolism requires a constant supply of oxygen, which determines the metabolic rate. Oxygen is supplied by blood pumped by the heart at a variable rate depending on the type of activity pursued. For example, a person completely at rest consumes 15 L of oxygen per hour. The energy production is linked to the oxygen supply via the relationship

$$E = 2 \times 10^4 \text{ J/L of oxygen supplied} \tag{6.32}$$

Thus, at rest the metabolic power generated is approximately

$$P = \frac{15 \text{L}}{\text{h}} \frac{2 \times 10^4 \text{ J}}{\text{L}} \frac{1 \text{ h}}{3600 \text{ s}} = 83 \text{ W} \tag{6.33}$$

Food energy is the energy stored by chemical bonds and has traditionally been measured in food calories, but in science the joule is used as a measure of energy. One food calorie is equivalent to 1000 physics calories or 4186 J. To measure the energy content of food, the food is burnt in oxygen and the measured heat that is produced is the energy content. The human body combines food with oxygen, that is, it burns food, to produce $CO_2$.

As shown above, the basic metabolic rate of an average person, when resting, is about 80 to 100 W. For resting each of us would require the same energy as a 100-W bulb. Walking at 2.5 km/h takes another 100 J/s and walking faster at 5 km/h takes another 100 J/s. Surprisingly, at either speed, it takes approximately the same energy to walk 1 km (although walking at 5 km/h takes more energy per second, it takes less time to walk 1 km at this faster speed).

About three-fourths of all the energy generated is converted to heat, and the remainder can be used for activities like walking and running. Taking 100 W as an average metabolic rate for an adult translates into a dietary requirement of 2600 cal/day (recall that 1 cal $\cong$ 4200 J). During vigorous physical exercise such as cycling the metabolic rate may increase 5-fold to over 400 W or even 10-fold under extreme conditions. It is interesting to compare these numbers to the power rating of common household appliances. For example, a stereo consumes 100 W, a TV 350 W, a self-defrosting fridge 600 W, a dishwasher 1200 W, a clothes dryer 5000 W, and a cooking range 12,000 W.

## TOPIC 6.9    Scaling Relationships Involving Metabolic Rates

A person's metabolic rate depends on his or her size and the level of physical activity, as mentioned in Topic 6.8. The normal metabolic rate under standardized conditions is proportional to the total surface area of the person's body. The proportionality of basic metabolic rate to body area is an empirical scaling law. It is found that the maximum metabolic rate of similar animals is also proportional to their total surface area, so that metabolic rate scales as $L^2$. However, it was also found that the work done by a muscle scales as $L^3$.

Although the work done by a muscle scales as $L^3$, the speed $v$ with which a limb moves is independent of $L$. The time it takes the limb to move a distance $d$ is $d/v$, and hence time scales as $L$. The power achieved scales as work divided by time, or as $L^2$. Since the oxygen required for metabolism is supplied by the blood, the metabolic rate is proportional to the volume of blood pumped per second by the heart. This is proportional to the volume $V$ of the heart times the heart rate $r$ (the number of heart beats per second). We have then that

$$P \propto Vr \tag{6.34}$$

so that $r$ scales as $P/V$. Since $P$ scales as $L^2$ and $V$ scales as $L^3$, $r$ scales as $L^2/L^3 = 1/L$. Thus the heart of a large animal beats slower than the heart of a small animal.

## TOPIC 6.10    Energy Management in the Human Body

Chemical reactions within the body are responsible for storing, releasing, absorbing, and transferring the energy humans need to, for example, move, breathe, and pump blood. Those reactions which require energy are called endothermic reactions, while those which release energy are termed exothermic reactions. The energy source for endothermic reactions is the food we eat. The energy required for muscle contraction is brought about through a network of chemical reactions within which there are two types: input and output ingredients. Input ingredients come from the air we breathe and the food we eat.

The two outputs are the carbon dioxide we exhale and the water we produce as a by-product. To be more specific, in the lungs the body removes oxygen from inhaled air and transports it to muscle cells via hemoglobin in the bloodstream. The body digests food in the mouth, stomach, and intestines, processing some of it into glucose ($C_6H_{12}O_6$), part of which is transported to muscle cells. Here oxygen and glucose combine to form water, carbon dioxide, and an exothermic reaction to also produce energy, $E_{out}$. Thus, in terms of a chemical reaction

$$C_6H_{12}O_6 + 6O_2 \longrightarrow 6CO_2 + 6H_2O + E_{out} \tag{6.35}$$

One can measure the net oxygen inhaled and/or the net carbon dioxide exhaled and the energy released can then be deduced from the glucose oxidation reaction.

Necessarily there is a time taken to breathe, transport oxygen, and deliver glucose to muscle cells, but the muscle fibers need not wait until glucose oxidation provides the energy required. This is because the body has the ability to store energy. We have already mentioned the role of ATP, ADP, and AMP in this process in Topic 6.8, and as long as ATP is available, immediate muscle contraction can take place. When the power demand of muscles increases abruptly, the energy required cannot be supplied by oxidation, since there is a time required to move to a new equilibrium state. This energy gap is provided by energy stored in the muscle cells in the form of existing ATP which does not require oxidation. After the transition time, respiration will gradually increase ATP until a new steady state is attained. During exercise, for approximately the first 2 min energy is supplied anaerobically (without oxygen) and aerobically thereafter via respiration. When the body uses up energy faster than respiration can support ATP production, available energy declines, a condition called oxygen debt. Thus, at the end of vigorous exercise, when the demand for energy is returning to normal, the "debt" is repaid when, for example, the sprinter gasps for air.

Experiments have demonstrated that, to run a given distance, the power required $P$ is nearly proportional to the running speed $v$. Thus

$$P \propto v \tag{6.36}$$

On the other hand, the power required to walk a given distance is related to velocity $v$ via

$$P \propto v^x \quad \text{with} \quad x > 1 \tag{6.37}$$

Since power is the rate of change of energy with time, we see that the energy per unit distance for running is approximately independent of speed, but to walk, this energy per unit time increases with walking speed. Very roughly, a rule of thumb gives

$$\frac{\text{Running energy}}{\text{Distance}} \cong 0.93 \text{ kcal/km} \tag{6.38}$$

Using the relation in equation (6.38), we can compare the energy content of food with the energy expended by running. A person whose mass is 80 kg burns approximately 120 kcal running 1.6 km at any speed. The energy equivalent of 0.5 kg of fat is about 4200 kcal. If a person's goal was to lose 1 kg of fat in 30 days, he or she could either (a) run 3 km a day or (b) eat four fewer slices of bread a day; that is, exercise is a poor substitute for dieting as a method of losing weight! The best strategy is to do both as regular exercise increases metabolic rate. The main idea behind aerobic exercise is to attain a good cardiovascular condition. This is a constructive process provided the aerobic exercise is regular and not harmfully intensive, for example, running can put a harmful stress on the skeleton, such as the joint problems long-term distance runners can experience.

The athlete, to increase glycogen stored beyond normal concentration, consumes no carbohydrates for a week or so and then eats food with a high concentration of carbohydrates for a few days before a race. The idea is to trick the body into overstocking muscle cells with glycogen, which will provide extra fuel late in the race. This strategy has been very effective in helping marathon runners over "the wall" after running about 30 km. Training for such events as swimming the English Channel, competing in the triathlon, or long-distance cycling normally involves special high-energy diets. In extreme cases body fat may be metabolized to support the body.

Depending on metabolic rate, a 70-kg person requires about 2400 kcal/day for normal bodily functions. Obviously, the power output rises during exercise and falls during rest. The energy produced in the body is used to generate heat and perform work. When a muscle contracts, the force generated moves a part of the body through a distance and work is performed. For example, using only the upper body muscles, a skilled rope climber can climb 5 m in 2 s. Thus for an 80-kg climber, about 2 kW of power is developed. Extra power may be generated, up to 5 to 6 kW, for short bursts of activity such as climbing or jumping. However, the longer a person works, the lower is the rate at which chemical energy is converted into mechanical work.

## SOLVED PROBLEMS

**6.1** A person who normally requires an average of 3000 kcal of food energy per day but instead consumes 4000 kcal/day will steadily gain weight. How much bicycling per day is required to work off this extra 1000 kcal?

**Solution** Note that 5.7 kcal/min is used when cycling at a moderate speed. The time required to work off 1000 kcal at this rate is then

$$\text{Time} = \frac{\text{energy}}{\text{energy/time}} = \frac{1000 \text{ kcal}}{5.7 \text{ kcal/min}} = 175 \text{ min}$$

**6.2** Thermal energy is equal to the excess amount between food energy consumed (assume this to be 1.4 kcal = 5880 J) and work done.

**(a)** How much thermal energy is generated by a 60-kg person climbing stairs such that $h = 2$ m?

**(b)** What is the rate of thermal energy production in watts when the person walks up the stairs taking 7 s in the process?

**(c)** What is the rate when the person runs up the stairs in 2 s?

**Solution**

**(a)** Identifying the various forms of energy involved and using the conservation-of-energy principle yields

$$KE_i + PE_i + \Delta E_i = KE_f + PE_f + \Delta E_f$$
$$0 + 0 + \text{food energy} = 0 + mhg + \text{thermal energy}$$

so that

$$\text{Thermal energy} = \text{food energy} - mgh$$

The food energy is 5880 J (1.40 kcal), and $mgh$ can be easily calculated to be 1176 J. Hence

$$\text{Thermal energy} = 5880 \text{ J} - 1176 \text{ J} = 4704 \text{ J}$$

**(b)** The rate of thermal energy production is

$$P = \frac{\text{thermal energy}}{\text{time}}$$

If it took the person 7.0 s to walk up the stairs,

$$P = \frac{4704 \text{ J}}{7.0 \text{ s}} = 672 \text{ W}$$

**(c)** When running, the power is much greater since $t$ is smaller:

$$P' = \frac{4704 \text{ J}}{2.0 \text{ s}} = 2352 \text{ W}$$

where the time taken has been assumed to be 2 s.

**6.3**   A 50-kg patient exercises on the treadmill, exerting a constant force of 500 N up the slope of the treadmill while running at a constant velocity of 4 m/s at an inclination angle of the treadmill $\theta = 30°$ for 5 min. The coefficient of kinetic friction of the treadmill is 0.45. Determine the work done by the patient.

**Solution**   The work done by the patient can be determined from the equation

$$W = Fd = (T - mg \sin \theta - \mu_k N)d \cos 60°$$
$$= (500 \text{ N} - 50 \text{ kg} \times 9.8 \text{ m/s}^2 \times \sin 30° - 0.45 \times 50 \text{ kg} \times 9.8 \text{ m/s}^2 \times \cos 30°)d \cos 60°$$
$$= 63.6 \, d \cos 60° \text{ N}$$

The displacement of the patient can be determined from the equation of motion:

$$d = vt = (4 \text{ m/s})(5 \text{ min})(60 \text{ s/min}) = 1200 \text{ m}$$

Thus, the work can now be determined:

$$W = (63.60 \text{ N} \times 1200 \text{ m}) \cos 60° = 3.8 \times 10^4 \text{ J}$$

**6.4**   A 70-kg person requires about 240 kcal of energy to walk 4000 m on level ground. How many more kilocalories are required if the walk is done at 30° to the horizontal? Assume a 25% efficiency of the body.

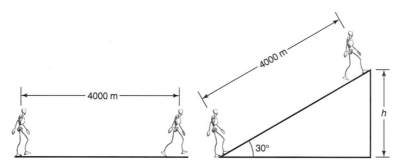

**Solution**   In terms of energy, the only difference between the level walk and the inclined walk is that there is no increase in gravitational potential energy for the level walk, whereas there is an increase for the inclined walk. The extra food energy required for the inclined walk is just the food energy required to increase the gravitational potential energy:

$$\text{Increase in gravitational PE} = mgh$$
$$= (70.0 \text{ kg})(9.8 \text{ m/s}^2)(4000 \text{ m}) \sin 30°$$
$$= 1.37 \times 10^6 \text{ J}$$
$$\text{Required food energy} = 4 \times 1.37 \times 10^6 \text{ J}$$
$$= 5.48 \times 10^6 \text{ J} \times 1 \text{ kcal}/4186 \text{ J}$$
$$= 1309 \text{ kcal}$$

# EXERCISES

**6.1**   The mechanical power $P$ delivered by a muscle is the amount of energy delivered per unit time. Power is the product of the load $W$ and the shortening velocity $v$.
**(a)** Find an expression for $P$ as a function of $W$ for a muscle obeying Hill's law.

**(b)** What is the maximum power the muscle can deliver?

**(c)** Determine the load $W$ for which this maximum power level is reached.

**6.2** Suppose you connect two identical muscles in series (i.e., one after the other). Each muscle obeys Hill's law separately and has certain values of $a$, $b$, and $F_{max}$. If you now measure the force–velocity curve of the two muscles in series, what are the new values of $a$, $b$, and $F_{max}$?

**6.3** Suppose you connect two identical muscles in parallel, i.e., next to each other. Each muscle obeys Hill's law separately. What are the resultant values of the parameters $a$, $b$, and $F_{max}$ for the system of two muscles in parallel? To achieve a high maximum force, is it better to make a muscle very long or very thick?

**6.4** A strong person can produce 100 W of power sustained over a period of several hours. An electric power plant generates on the order of 800 MW. What is the human equivalent of this amount of power generation?

**6.5** Estimate the time you can sustain your effort of cycling if you only ate an 800-kJ candy bar recently? The rate of energy consumption while cycling is 400 W.

**6.6** How much energy in kilocalories is expended against gravity alone by a 70-kg person in climbing 1800 m up a mountain?

**6.7** In a normal joint, the coefficient of friction is approximately 0.003. Assuming all other things equal, how much more energy is dissipated in an arthritic joint where $\mu = 0.03$ than in a normal joint?

# CHAPTER 7

# IMPULSE AND MOMENTUM

## PHYSICAL BACKGROUND

The impulse of a force is the product of the average force $\overline{F}$ and the time interval $\Delta t$ during which the force acts:

$$\text{Impulse} = \overline{F}\Delta t \tag{7.1}$$

Impulse is a vector that points in the same direction as the average force.

The linear momentum $\mathbf{p}$ of an object is the product of the object's mass and velocity $\mathbf{v}$:

$$\mathbf{p} = m\mathbf{v} \tag{7.2}$$

Linear momentum is a vector that points in the direction of the velocity. Momentum is given in units of kilogram meter per second. The total momentum of a system of objects is the vector sum of the momenta of the individual objects. Momentum can also be used to restate Newton's second law as

$$\mathbf{F} = m\mathbf{a} = \frac{d\mathbf{p}}{dt} = \frac{d(m\mathbf{v})}{dt} \tag{7.3}$$

where $\mathbf{F}$ is the net force exerted on a particle.

The impulse–momentum theorem states that an impulse produces a change in an object's momentum, according to

$$\overline{F}\Delta t = mv_f - mv_i$$

where $mv_f$ is the final momentum and $mv_i$ is the initial momentum.

The principle of conservation of linear momentum states that the total linear momentum of an isolated system remains constant. An isolated system is one for which the sum of the external forces acting on the system is zero. External forces are those that agents external to the system exert on objects within the system. In contrast, internal forces are those that the objects within the system exert on each other.

An elastic collision is one in which the total kinetic energy is the same before (B) and after (A) the collision. For a system of two bodies (labeled 1 and 2, respectively) which undergo an elastic collision, it can be stated that total linear momentum before collision is equal to the total linear momentum after collision. For two colliding objects we write

$$m_{1B}\mathbf{v}_{1B} + m_{2B}\mathbf{v}_{2B} = m_{1A}\mathbf{v}_{1A} + m_{2A}\mathbf{v}_{2A} \tag{7.5}$$

An inelastic collision is one in which the total kinetic energy is not the same before and after the collision. If the objects stick together after colliding, the collision is said to be completely inelastic. In an inelastic collision, only linear momentum is conserved while some kinetic energy is transformed to internal energy. For a system of two bodies which undergo a completely inelastic collision, the following can be stated: Total linear momentum before collision is equal to total linear momentum after collision:

$$m_{1B}\mathbf{v}_{1B} + m_{2B}\mathbf{v}_{2B} = (m_{1A} + m_{2A})\mathbf{v}_{12A} \tag{7.6}$$

The center of mass (CM) of an object is the fictitious point where the total mass is "thought" to be concentrated, and it can be determined from knowledge of the mass of the extended object and its distance from a central point defined as an origin. In two-dimensional coordinates, the center of mass is given by

$$x_{cm} = \frac{\sum xm}{\sum m} = \frac{m_1 x_1 + m_2 x_2 + m_3 x_3 + \cdots + m_n x_n}{m_1 + m_2 + m_3 + \cdots + m_n} \tag{7.7}$$

$$y_{cm} = \frac{\sum ym}{\sum m} = \frac{m_1 y_1 + m_2 y_2 + m_3 y_3 + \cdots + m_n y_n}{m_1 + m_2 + m_3 + \cdots + m_n} \tag{7.8}$$

Similar formulas may be obtained for the CM velocity. If the total linear momentum of a system of objects remains constant during an interaction such as a collision, the velocity of the CM also remains constant.

As an example consider the human body. In Table 7.1 we list the CM and hinge points or joints for the different components of a "representative" person (these are only rough averages as there are wide variations among people). The numbers in the figure represent a percentage of the total height which is regarded as 100 units. For example, if a person is 1.70 m tall, his or her shoulder joint would be $(1.70\text{ m})(81.2/100) = 1.38$ m above the floor. Knowing where the CM of a body is in various positions is of great use in studying body mechanics. For example, in Figure 7.1, if high jumpers can get into the position shown, their CM can actually pass below the bar which their bodies go over. This means that for a particular takeoff speed, they can clear a higher bar, and this is what they try to do.

The coefficient of restitution $q$ is the ratio between the relative velocities of the two bodies before and after the collision and is defined by

$$q = \frac{v_{2A} - v_{1A}}{v_{1B} - v_{2B}} \tag{7.9}$$

The coefficient of restitution is a pure number, and it reflects the ability of the two objects to physically respond following impact. It varies between 0 (no rebound, bodies stick together) and 1 (the relative velocities before and after impact are equal, or maximum rebound).

**TABLE 7.1   Center of Mass of Parts of a Typical Human Body**

| Distance of hinge points (%) | Hinge points (joints) | | Center of mass (% height above floor) | | Percent mass |
|---|---|---|---|---|---|
| 91.2 | Base of skull on spine | | Head | 93.5 | 6.9 |
| 81.2 | Shoulder joint | | Trunk and neck | 71.1 | 46.1 |
| 52.1 | Hip | | Upper arms | 71.7 | 6.6 |
| 76.3 | Knee | Elbow 62.2 | Lower arms | 55.3 | 4.2 |
| 4.0 | Ankle | Wrist 46.2 | Hands | 43.1 | 1.7 |
| | | | Upper legs (thighs) | 42.5 | 21.5 |
| | | | Lower legs | 18.2 | 9.6 |
| | | | Feet | 1.8 | 3.4 |
| | | | | 58.0 | 100.0 |

*Note:* Full height and mass = 100 units.

**FIGURE 7.1** A high jumper clearing a bar whose CM may be beneath the bar.

## TOPIC 7.1 Impulsive Force and Injury Due to a Fall

A person who falls from some height experiences a conversion from potential energy to kinetic energy prior to striking the ground. The resultant injury from the fall on impact depends on the impulsive force exerted on the person. If we neglect air resistance, we may use the equation

$$v_f^2 = v_i^2 + 2ay \tag{7.10}$$

where $v_f$ is the final velocity, $v_i$ the initial velocity, $a$ the acceleration, and $y$ the height through which the fall takes place. Thus in the case under consideration we assume the person began at rest, so $v_i = 0$, the acceleration is due to gravity, and suppose the height is $h$. Thus from equation (7.10)

$$v_f = \sqrt{2gh} \tag{7.11}$$

since $a = g$. The momentum of the person on impact is

$$p = mv = m\sqrt{2gh} = mg\sqrt{\frac{2h}{g}} \tag{7.12}$$

After impact the momentum is zero, so the change of momentum is

$$\Delta p = mv_i - mv_f = m\sqrt{2hg} \tag{7.13}$$

The impulsive force is then the change of momentum over the time taken. Taking, for example, $m = 80$ kg, $v_i = 0$, and $h = 32$ m, we obtain

$$\Delta p = 2005 \text{ kgm/s} \tag{7.14}$$

## TOPIC 7.2 Animal Propulsion

In Chapter 4 we have already mentioned the jet propulsion of the octopus, but this method of propulsion is common to the nautilus, the squid, and cuttlefish. It is also used by the frog fish, the dragonfly nymph, and microscopic organisms. In each case the animal absorbs water in a body cavity and expels this water through an orifice by contracting the cavity, and the animal thus moves in a direction opposite to that of the expelled water. Using conservation of momentum, we have that

$$MV = mv_0 \tag{7.15}$$

where $V$ is the velocity of the animal following a squirt of water, $M$ is its mass with an empty mantle or cavity, $m$ is the mass of the expelled water, and $v_0$ its speed. Consider as an example a squid which draws water through a slit in its belly and stores water inside its mantle. Suppose the squid has a mass $M = 0.4$ kg with its cavity empty and can store a mass $m = 0.1$ kg of water in its water. The velocity $v_0$ of the expelled water is then given, by rearranging equation (7.15), as

$$v_0 = \frac{MV}{m} \tag{7.16}$$

If, after such a squirt of water, the velocity of the squid is $V = 0.8$ m/s, we have that

$$v_0 = \frac{(0.4 \text{ kg})(0.8 \text{ m/s})}{(0.1 \text{ kg})} = 3.2 \text{ m/s} \tag{7.17}$$

What this is telling us is that the surrounding water exerts a significant drag force on the squid, slowing it down, and the squid exchanges momentum with the water. The law of conservation of momentum is exact only if all contributions to the total momentum (including the effects of drag) are taken into account, so our calculation above, neglecting drag, is not exact but only approximate.

## SOLVED PROBLEM

**7.1**    Estimate the force in the Achilles tendon in the human foot at the moment before a high jump. Assume the body weight to be $W = 800$ N, the cross-sectional area of the Achilles tendon to be $A = 1.61$ cm$^2$, and an additional force on the foot for a time interval of $t = 0.3$ s should be applied to a height $h = 1.5$ m.

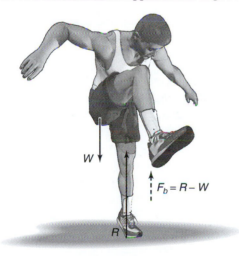

**Solution**    Just before the jump, the body weight should be poised on one foot ($R = 800$ N), the knee should be bent, and an additional force, $F_b$, should be generated. This extra force, which may be larger than $W$, acts for $\Delta t = 0.3$ s, and during that interval the impulse $F_b \Delta t$ provides an initial momentum $mv_i$ without a significant lift of the body. From Newton's second law, we have

$$F_b = \frac{mv_i}{\Delta t} = \frac{Wv_i}{g \, \Delta t}$$

Here, $v_i$ is the initial velocity, which can be determined from the height to which the mass is lifted against gravity. Since the initial kinetic energy $\frac{1}{2}mv_i^2$ is converted into the potential energy $mgh$ of the body at the top of the lift, we obtain

$$\tfrac{1}{2}mv_i^2 = mgh$$

where $g$ is the acceleration due to gravity. Numerically, the initial velocity is $v_i = \sqrt{2gh} = 5.42$ m/s from the above equation and the force can be calculated from the first equation as $F_b = 1473$ N. Thus, the reactive force acting on the sole of the jumping foot is $R = W + F_b = 800$ N + 1473 N.

## EXERCISES

**7.1** Suppose a fish with a known mass $M$ is initially at rest. After applying two kicks with its tail, it has a velocity $\mathbf{v}$ whose magnitude and direction are known. The velocity vectors of the two swirls are denoted by $\mathbf{v}_1$ and $\mathbf{v}_2$, where $v_{1x}$ and $v_{2x}$ are the magnitudes in the $x$ direction and $v_{1y}$ and $v_{2y}$ are the magnitudes in the $y$ direction. The swirls have unknown masses $m_1$ and $m_2$.

(a) Using the law of momentum conservation, obtain a set of two coupled equations for the unknown masses $m_1$ and $m_2$ in terms of the known quantities $M$, $v$, $v_{1x}$, $v_{2x}$, $v_{1y}$, and $v_{2y}$.

(b) Solve the two equations for $m_1$ and $m_2$ for the special case that $v_{1x} = v_{2x} = 0.10$ m/s, $v_{1y} = -v_{2y} = 0.03$ m/s, $M = 0.025$ kg, and $v = 0.05$ m/s.

**7.2** A 4-kg squid initially at rest ejects 0.3 kg of water at a velocity of 10 m/s. Calculate the squid's recoil velocity assuming that the process takes 0.1 s and that a friction force of 4.5 N opposes the squid's motion.

**7.3** Squid are known to travel up to 30 m in the horizontal direction when jumping out of the ocean. Assuming that a squid leaves the water at an angle of 25° to the horizontal, what is its initial velocity?

**7.4** A 1-kg duck is flying horizontally at 20 m/s when seized by an 0.8-kg hawk diving down at 30 m/s. The hawk is coming in from behind and makes an angle of 30° from the vertical just before contact. What is the velocity of the birds just after contact?

# CHAPTER 8

# ROTATIONAL KINEMATICS

## PHYSICAL BACKGROUND

When a rigid body rotates about a fixed axis, the angular displacement is the angle swept out by a line passing through any point on the body and intersecting the axis of rotation perpendicularly. The radian (rad) is the SI unit of angular displacement. In radians, the angle $\theta$ is defined as the circular arc length $s$ traveled by a point on the rotating body divided by the radial distance $r$ of the point from the axis:

$$\theta = \frac{s}{r} \qquad (8.1)$$

Measurement of angular displacement is given in degrees, revolutions, or radians, and they are related according to

$$360° = 1 \text{ rev} = 2\pi \text{ rad}$$

$$1 \text{ rad} = 57.3° = \frac{180°}{\pi \text{ rad}}$$

All units of angular displacement are pure numbers and have no physical dimension.

The average angular velocity $\overline{\omega}$ is the angular displacement $\Delta\theta$ divided by the elapsed time $\Delta t$:

$$\overline{\omega} = \frac{\Delta\theta}{\Delta t} \qquad (8.2)$$

As $\Delta t$ approaches zero, the average angular velocity becomes equal to the instantaneous angular velocity:

$$\omega = \frac{d\theta}{dt} = \frac{v}{r} \qquad (8.3)$$

where $v$ is the tangential velocity. The magnitude of the instantaneous angular velocity is called the instantaneous angular speed.

The average angular acceleration $\overline{\alpha}$ is the change $\Delta\omega$ in the angular velocity divided by the elapsed time $\Delta t$:

$$\overline{\alpha} = \frac{\Delta\omega}{\Delta t} \qquad (8.4)$$

As $\Delta t$ approaches zero, the average angular acceleration becomes equal to the instantaneous angular acceleration:

$$\alpha = \frac{d\omega}{dt} \qquad (8.5)$$

When a rigid body rotates with constant angular acceleration about a fixed axis, the angular displacement $\theta$, the finite angular velocity $\omega$, the initial angular velocity $\omega_0$, the angular acceleration $\alpha$, and the elapsed time $t$ are related as follows, assuming that $\theta_0 = 0$ at $t_0 = 0$:

$$\omega = \omega_0 + \alpha t \quad \text{and} \quad \theta = \tfrac{1}{2}(\omega_0 + \omega)t \tag{8.6}$$

The above equations of rotational kinematics, in the case of constant angular acceleration, are supplemented by

$$\theta = \theta_0 + \omega_0 t + \tfrac{1}{2}\alpha t^2 \tag{8.7}$$

and

$$\omega^2 = \omega_0^2 + 2\alpha\theta \tag{8.8}$$

There is a complete analogy between these equations and those of linear kinematics in Chapter 2 when we identify pairs of corresponding variables $x$ and $\theta$, $v$ and $\omega$, and $a$ and $\alpha$.

When a rigid body rotates through an angle $\theta$ about a fixed axis, any point on the body moves on a circular arc of length $s$ and radius $r$. Such a point has a tangential velocity $v_T$ and, possibly, a tangential acceleration $a_T$. The angular and tangential variables are related according to

$$s = r\theta \qquad v_T = r\omega \qquad a_T = r\alpha \tag{8.9}$$

where $\omega$ and $\alpha$ are, respectively, the magnitudes of the angular velocity and the angular acceleration. These equations refer to the magnitudes of the variables involved, without reference to positive or negative signs, and only radian measures can be used when applying them.

A point on an object rotating with nonuniform circular motion experiences a total acceleration that is the vector sum of two perpendicular acceleration components, the centripetal acceleration $\mathbf{a}_C$ and the tangential acceleration $\mathbf{a}_T$,

$$\mathbf{a} = \mathbf{a}_C + \mathbf{a}_T \tag{8.10}$$

The essence of rolling motion is that there is no slipping at the point where the object touches the surface upon which it is rolling. As a result, the tangential speed $v_T$ of a point on the outer edge of a rolling object, measured relative to the axis through the center of the object, is equal to the linear speed $v$ with which the object moves parallel to the surface. In other words, $v = v_T \ (=R\omega)$, where $R$ is the radial distance of the point from the axis. The magnitudes of the tangential acceleration $a_T$ and the linear acceleration $a$ of a rolling object are similarly related: $a = a_T \ (=R\alpha)$.

The moment of inertia $I$, or rotational inertia, represents the resistance of an object to rotational motion just as inertia represents the resistance of an object to translational or linear motion. The moment of inertia, $I$, for a small body of mass $m$ can be defined by

$$I = mr^2 \tag{8.11}$$

where $r$ is the distance from mass $m$ to the axis of rotation.

For large or irregular objects composed of several smaller, well-defined objects, the moment of inertia can be expressed as the total or sum of the moments of inertia for the smaller objects:

$$I = \sum mr^2 = m_1 r_1^2 + m_2 r_2^2 + m_3 r_3^2 + \cdots + m_n r_n^2 \tag{8.12}$$

The moment of inertia is a scalar quantity with units of kilogram meter squared (kg m$^2$).

The radius of gyration $k$ of a rigid body represents the distance between the axis of rotation and the point where the weight is considered to be concentrated without changing the moment of inertia of the body. The radius of gyration $k$ of a rigid body is related to its moment of inertia by

$$I = mk^2 \qquad (8.13)$$

## TOPIC 8.1   Physics of Basketball

One of the principles of basketball shooting is that the ball should be pushed by the fingertips rather than by the palm of the shooter's hand. First, the use of the fingertips gives better control over the ball's trajectory and, second, using a flick of the wrist when shooting from the fingertips gives the ball a backspin. In contrast to both the use of forward spin and no spin at all, shooting with backspin present, the friction force at the point of contact opposes both the translational and rotational motion of the ball. Consequently, the ball will lose considerable speed and may even bounce backward. Hence, a backspin shot seems "softer" and is more likely to drop in the basket. It can be shown that a back-spinning ball always experiences a greater decrease in translational energy and in total energy than a forward-spinning ball. The change in the linear momentum in the bounce is given as

$$F \, \Delta t = \Delta(mv_x) = m(v_f - v_i) \qquad (8.14)$$

and in the angular momentum as

$$Fr \, \Delta t = \Delta(I\omega) = \frac{2mr^2}{5}(\omega_i - \omega_f) \qquad (8.15)$$

where $I$ is the ball's moment of inertia, $r$ is its radius, and $m$ is its mass (see Figure 8.1). The no-skidding condition is

$$v_f = r\omega_f \qquad (8.16)$$

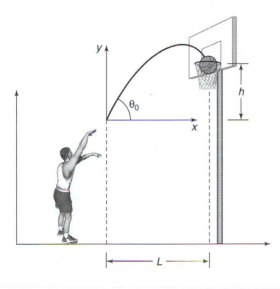

**FIGURE 8.1** Schematic of the process of shooting a ball into a basket.

What is the best trajectory for a basketball shot? Using projectile motion equations in the field of gravity yields for the coordinates (the origin being at the player's fingertips and $x = 0$ and $y = 0$ when $t = 0$)

$$x = v_0 t \cos \theta_0 \tag{8.17}$$

$$y = v_0 t \sin \theta_0 - \tfrac{1}{2} g t^2 \tag{8.18}$$

and for the velocity components

$$v_x = v_0 \cos \theta_0 \tag{8.19}$$

$$v_y = v_0 \sin \theta_0 - gt \tag{8.20}$$

Eliminating time $t$ from equations (8.17) and (8.18) gives

$$v_0 = \left[ \frac{gx}{2 \cos^2 \theta_0 (\tan \theta_0 - y/x)} \right]^{1/2} \tag{8.21}$$

Substituting $x = L$ and $y = h$ into equation (8.21) gives

$$v_0 = \left[ \frac{gL}{2 \cos^2 \theta_0 (\tan \theta_0 - h/L)} \right]^{1/2} \tag{8.22}$$

which gives a family of parabolas parameterized by $\theta_0$ and a corresponding $v_0$. It can be shown that a minimum launching velocity exists given by

$$v_0 = \{ g[h + (h^2 + L^2)^{1/2}] \}^{1/2} \tag{8.23}$$

A typical basketball shot has a speed of 6 to 9 m/s and a time of flight of the order of 1 s. This results in the trajectory of a basketball deviating from an ideal parabola by 10 to 20%.

## TOPIC 8.2   Ultracentrifuge

The centrifuge is an instrument that uses centripetal force to separate macromolecules of different mass suspended in solution. In a centrifuge, a test tube containing the sample solution is placed into a rotor and is subjected to a high rotational speed, typically from 5000 to 50,000 rotations per minute. The macromolecules in the sample are subjected to three forces, as shown in Figure 8.2:

1. A centrifugal force $F_c = m\omega^2 r$ acting to propel the macromolecule away from the axis of rotation toward the bottom of the test tube, where $m$ is the mass of the macromolecule, $\omega$ is the angular velocity, and $r$ is the radius of motion

2. A frictional force $F_f = \mu v$ due to motion of the particle through the fluid acting toward the axis of rotation, where $\mu$ is a frictional coefficient and $v$ is the velocity of the macromolecule in the medium (solution)

3. A buoyant force $F_b = (m/\rho)\rho_0 \omega^2 r$ produced by the weight of the fluid displaced by the macromolecule, where $\rho$ is the density of the macromolecule and $\rho_0$ is the density of the medium (see Chapter 11 for a more detailed description)

The centrifugal force is typically greater in magnitude than both the frictional and buoyant forces combined, causing the macromolecules in the solution to sediment toward the bottom of the test tube, in effect separating the solute from the solvent.

The next topic discusses the application of the centrifuge in biology.

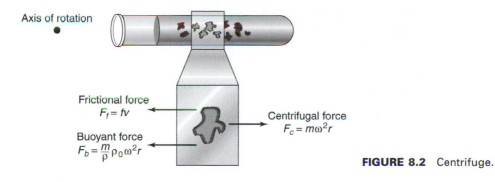

**FIGURE 8.2** Centrifuge.

## TOPIC 8.3   Sedimentation in Biology

Solutions and suspensions play an important role in biology, and in this section we discuss the phenomenon of sedimentation. When a centrifuge is being used to study a solution of biological interest, it is spun very rapidly and the processes of diffusion and sedimentation interact. One would expect that, after a sufficient time, all the particles would end up at the outside of the centrifuge. This is precisely what happens if the particles in solution are large enough and we say that the particles have been sedimented. However, if the particles are smaller, the process of diffusion tends to spread the particles back toward the axis of the centrifuge. These two conflicting tendencies result in an equilibrium gradient of concentration, as we show in Figure 8.3.

Let us suppose that the radial distance from the axis of rotation to the inner edge of the centrifuge is $r_0$. If the concentration at this latter position is $c_0$, then it may be shown that the concentration at a distance $r$ is

$$c = c_0 e^{A(r^2 - r_0^2)} \tag{8.24}$$

where $A$ is given by

$$A = \left(\frac{m\omega^2}{2kT}\right)\left(1 - \frac{\rho_0}{\rho}\right) \tag{8.25}$$

In equation (8.25), $\omega$ is the angular velocity of the centrifuge, $m$ is the mass of the particles, $k$ is Boltzmann's constant, $T$ is the temperature, and $\rho_0$ and $\rho$ are the densities of the solute and the solvent, respectively. For a detailed discussion concerning temperature and heat the reader is referred to Chapter 12. The concentration given in equation (8.24) can be measured optically, and hence the value of $A$ may be determined. By this method the mass of the particle can be found; in practice, the "particles" will be large molecules so $m$ will be the molecular mass. A key feature of this method is that the determination of

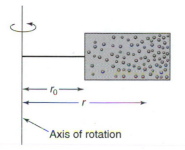

**FIGURE 8.3** Concentration of particles in suspension in a centrifuge as a function of distance from the axis of rotation.

*m* does not depend in any way on the shape of the "particles." This method is called the sedimentation equilibrium method.

## SOLVED PROBLEM

**8.1** In a particular experiment, an ultracentrifuge with a radial distance of 0.10 m is operated at 40,000 rpm. Determine the angular velocity of the centrifuge in radians per second and the centripetal force exerted on the macromolecule of mass *M* at a point 0.025, 0.05, 0.075, and 0.10 m from the center of rotation.

**Solution** The angular velocity of the centrifuge, given as 40,000 rev/min, can be converted to radians per second by

$$\omega = 40{,}000 \text{ rev/min} \times \frac{2\pi \text{ rad}}{1 \text{ rev}} \times \frac{1 \text{ min}}{60 \text{ s}} = 4188.8 \text{ rad/s}$$

The centripetal force is given by $F_c = M\omega^2 r$. By substituting the appropriate values, we get

$r = 0.025$ m $\qquad F_c = M\omega^2 r = M(4188.8 \text{ rad/s})^2(0.025 \text{ m}) = M \times 4.387 \times 10^5 \text{ m/s}^2$

$r = 0.05$ m $\qquad F_c = M\omega^2 r = M(4188.8 \text{ rad/s})^2(0.050 \text{ m}) = M \times 8.773 \times 10^5 \text{ m/s}^2$

$r = 0.075$ m $\qquad F_c = M\omega^2 r = M(4188.8 \text{ rad/s})^2(0.075 \text{ m}) = M \times 1.316 \times 10^6 \text{ m/s}^2$

$r = 0.10$ m $\qquad F_c = M\omega^2 r = M(4188.8 \text{ rad/s})^2(0.10 \text{ m}) = M \times 1.755 \times 10^6 \text{ m/s}^2$

## EXERCISES

**8.1** A baseball pitcher rotates his arm forward around his shoulder joint. Calculate the velocity of the ball in his hand if the length of the arm is 0.65 m and the angular velocity of the arm is 100 rad/s.

**8.2** Calculate the centripetal acceleration experienced by a jet fighter pilot flying at 300 m/s and making a turn of radius 1.5 km.

**8.3** A lab technician is centrifuging a blood sample at an angular speed of 3500 rpm. If the radius of the circular path followed by the sample is 0.25 m, find the speed of the sample in meters per second.

CHAPTER **9**

# ROTATIONAL DYNAMICS

## PHYSICAL BACKGROUND

The torque $\boldsymbol{\tau}$, or moment of force, represents the external rotational force acting perpendicular to an object with respect to a fixed point or pivot and is defined by

$$\boldsymbol{\tau} = \mathbf{r} \times \mathbf{F} = |r||F| \sin \theta \hat{\boldsymbol{n}} \tag{9.1}$$

where $r$ is the distance between the pivot and the point where the force is exerted, also known as the lever arm, $F$ is the external force, $\theta$ is the angle between $\mathbf{r}$ and $\mathbf{F}$, and $\hat{\boldsymbol{n}}$ is a unit vector perpendicular to both vectors $\mathbf{r}$ and $\mathbf{F}$ (see Figure 9.1). The direction of $\hat{\boldsymbol{n}}$ is that of a right-handed screw when turned from $\mathbf{r}$ to $\mathbf{F}$. Torque is a vector quantity with dimensions of force multiplied by distance, and units of newton-meter (Nm).

A rigid body is in equilibrium if it has zero translational acceleration and zero angular acceleration, so the net external force and the net external torque acting on the body are zero. For forces acting only in the $x$, $y$ plane, the conditions for equilibrium are

$$\sum F_x = 0 \tag{9.2}$$

$$\sum F_y = 0 \tag{9.3}$$

$$\sum \boldsymbol{\tau} = 0 \tag{9.4}$$

The center of gravity of a rigid object is the point where its entire weight can be considered to act when calculating the torque due to the weight of the object. For a symmetric body with uniformly distributed weight, the center of gravity is located at the geometric center of the body. When a number of objects whose weights $W_1, W_2, \ldots, W_n$ are distributed along the $x$ axis at locations $x_1, x_2, \ldots, x_n$, the center of gravity is located at

$$x_{\text{CG}} = \frac{W_1 x_1 + W_2 x_2 + \cdots + W_n x_n}{W_1 + W_2 + \cdots + W_n} \tag{9.5}$$

The center of mass is identical to the center of gravity, provided the acceleration due to gravity does not vary over the spatial extent of the objects.

For rigid bodies rotating about fixed axes, Newton's second law for rotational motion is

$$\sum \boldsymbol{\tau} = I\alpha \tag{9.6}$$

where $\sum \boldsymbol{\tau}$ is the net external torque applied to a body, $I$ is the moment of inertia of the body, and $\alpha$ is the angular acceleration (in radians per square second).

The rotational work $W_R$ done by a constant torque $\tau$ in turning a rigid body through an angular displacement $\theta$ (in radians) is expressed by

$$W_R = \tau\theta \tag{9.7}$$

**55**

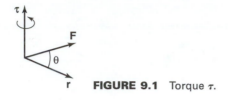

**FIGURE 9.1** Torque $\tau$.

The rotational kinetic energy of an object with angular speed $\omega$ (in radians per second) and moment of inertia $I$ is

$$KE_R = \tfrac{1}{2}I\omega^2 \tag{9.8}$$

The rotational kinetic energy of a large object composed of several smaller masses, each possessing its own moment of inertia, can be expressed as

$$KE_{rot} = \tfrac{1}{2}\left(\sum m_n r_n^2\right)\omega^2 = \tfrac{1}{2}(m_1 r_1^2 + m_2 r_2^2 + m_3 r_3^2 + \cdots + m_n r_n^2)\omega^2 = \tfrac{1}{2}I\omega^2 \tag{9.9}$$

The total kinetic energy of an object in rotational motion is the sum of (a) its rotational kinetic energy determined about an axis through its center of mass,

$$KE_{rot} = \tfrac{1}{2}I\omega^2 \tag{9.10}$$

and (b) its translational kinetic energy of the motion of the object's center of mass,

$$KE_{trans} = \tfrac{1}{2}mv_{cm}^2 \tag{9.11}$$

or

$$KE_{tot} = \tfrac{1}{2}I\omega^2 + \tfrac{1}{2}mv_{cm}^2 \tag{9.12}$$

The total mechanical energy is conserved, provided that the net work done by external nonconservative forces and torques is zero.

The rotational power $P_{rot}$ is the power output of an object in rotational motion performing work over a defined time interval and is given by

$$P_{rot} = \tau\omega \tag{9.13}$$

where $\omega$ is the angular velocity. Rotational power is expressed in units of watts.

The angular momentum $L$ is the product of the linear momentum $p$ and the radius $r$, or the distance from the pivot point where the momentum is applied.

$$L = mvr \tag{9.14}$$

The angular momentum $L$ of a body rotating with angular velocity $\omega$ about a fixed axis and having a moment of inertia $I$ is

$$L = I\omega \tag{9.15}$$

The principle of conservation of angular momentum states that the total angular momentum of a system remains constant if the net external torque acting on the system is zero.

## TOPIC 9.1 Levers and Biomechanics

Levers have an important role to play in biomechanics. An understanding of the vertebrate skeleton and its movement would be impossible without realizing that limbs in motion depend on lever action. In Figure 9.2 we show a hand holding a ball of weight $W$.

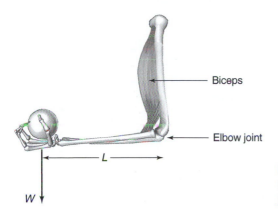

Biceps

Elbow joint

$L$

$W$

**FIGURE 9.2**   A ball being held in the hand at arm's length from the elbow joint.

The force which the biceps muscle creates to keep the ball stationary is affected not only by the ball's weight but also by the extension, $L$, of the arm. Clearly, the ball tries to rotate about the elbow joint, and the longer the arm, the greater the force, via the biceps, required to prevent this rotation. In Figure 9.3 we show a simple example of a lever that is a rod which can rotate about the fulcrum or pivot. A vertical force acts on the rod at a distance $L_F$ from the pivot. It is hoped that this will lift a weight $W$ located on the other side of the pivot, a distance $L_W$ from it. Archimedes determined that the minimum force, $F_{min}$, to lift the load is given by

$$F_{min} = W \frac{L_W}{L_F} \tag{9.16}$$

This is equivalent to balancing the moments of the forces about the pivot, that is, exactly in equilibrium we would have

$$L_F F = W L_W \tag{9.17}$$

If $F < F_{min}$, the lever rotates so as to lower the load. By making $L_F$ as large as possible, one can reduce $F_{min}$, making it possible to lift the load with a force much less than its weight $W$.

Figure 9.4$a$ shows the two bones of the forearm, the radius and ulna, and the humerus, which connects them via the elbow joint and itself is attached to the shoulder. We assume the person and ball are stationary, the biceps muscle is attached to the radius about 0.04 m from the elbow joint, and the mass $M = 4$ kg. We neglect the weight of the forearm itself and assume the humerus is vertical and that the force $F_b$ exerted by the biceps muscle is vertical. The gravitational force on the ball, $W = Mg$, acts vertically downward, at the end of the forearm of length $D = 0.24$ m. The humerus provides another force, $F_h$, on the elbow joint acting vertically downward. Figure 9.4$b$ shows what is called a free-force diagram with only the pivot, the forces and their line of action, and the distance of this line from the pivot. In addition, Figure 9.4$a$ labels the bones, the position of

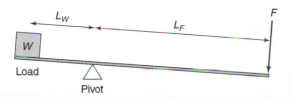

$L_W$

$L_F$

$F$

$W$

Load

Pivot

**FIGURE 9.3**   A lever subject to a load $W$ and force $F$ about a pivot.

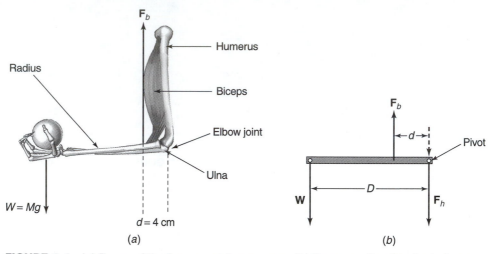

**FIGURE 9.4** (a) Bones of the forearm and upper arm. (b) Corresponding free-body force diagram.

the biceps muscle, the forces, and the elbow joint. As the arm is in static equilibrium, the horizontal and vertical forces must be zero, so that

$$F_b - W - F_h = 0 \qquad (9.18)$$

and there are no horizontal forces. We now apply the moment equation (9.17), noting that $F_h$ acts directly on the pivot and so exerts no torque or moment about it. Thus we have

$$F_b d = WD \qquad (9.19)$$

The force exerted by the biceps muscle is therefore

$$F_b = W\left(\frac{D}{d}\right) \qquad (9.20)$$

We can see that $D/d = 6$, so the biceps exert a force six times larger than the weight of the ball. The humerus force can now be found directly from equation (9.18) and

$$F_h = F_b - W \qquad (9.21)$$

which is about five times larger than the weight of the ball. The size of $F_b$ explains why it is tiring to hold a mass in our hands when the forearm is raised. We might surmise that the use of forearms for producing large torques was not a major adaptive advantage for our ancestors. However, other vertebrates have forepaws which are better designed in this respect; for example, the mole has a muscle attachment farther away from the elbow, which increases the torque about the elbow.

## TOPIC 9.2 Skeletal Mechanics of the Leg

Here, we apply the principles of mechanical equilibria to the mechanics of the leg. The thigh bone or femur rotates in a socket which forms part of the pelvis. Such movements are stabilized by three independent abductor muscles. Figure 9.5 illustrates these forces for a person standing and presents a simple example, using Newton's laws, of how the various forces are balanced. We denote the force exerted by the hip abductor muscle by

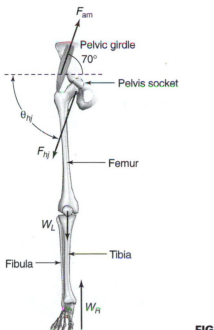

$F_{am}$
Pelvic girdle
70°
Pelvis socket
$\theta_{hj}$
$F_{hj}$
Femur
$W_L$
Tibia
Fibula
$W_R$

**FIGURE 9.5** Forces acting on a leg.

the vector $\mathbf{F}_{am}$, the force acting at the socket of the hip joint by $\mathbf{F}_{hj}$, $\mathbf{W}_R$ the vector describing the reaction force of the ground, and $W_L$ the weight of the leg acting downward through the center of gravity. The angles between $F_{am}$ and the horizontal may be written as $\theta_{am}$, and we choose it to be $\theta_{am} = 70°$, and between $F_{hj}$ and the horizontal as $\theta_{hj}$. Assuming that the person is in equilibrium, we balance the horizontal forces so that

$$F_{am} \cos \theta_{am} = F_{hj} \cos \theta_{hj} \tag{9.22}$$

and the vertical forces so that

$$F_{am} \sin \theta_{am} + W_R = W_L + F_{hj} \sin \theta_{hj} \tag{9.23}$$

We choose the hip joint or socket as the pivot point and approximate the weight of the leg, $W_L$, by $\frac{1}{6}W$. In addition, we assume the hip abductor muscle is attached to the femur 0.069 m from the pivot and also that the weight of the person $W = 690$ N. Hence,

$$W_L = \frac{1}{6}W = 0.167W \tag{9.24}$$

The leg must also be in rotational equilibrium so the moments of all the forces about the pivot point must be zero (see Figure 9.6). Therefore inserting the numbers, we obtain

$$F_{am} \sin 70°(0.069 \text{ m}) + W_L (0.03 \text{ m}) - W(0.11 \text{ m}) = 0 \tag{9.25}$$

where we have assumed that the line of action of $W_L$ is 0.03 m from the pivot and $W_R$ 0.11 m from the pivot. From equation (9.25) we therefore have

$$F_{am} = 1117 \text{ N} \tag{9.26}$$

Putting $F_{am}$ from equation (9.26) into equation (9.22) gives

$$F_{hj} \cos \theta_{hj} = (1117 \text{ N}) \times (0.3420) = 382.0 \text{ N} \tag{9.27}$$

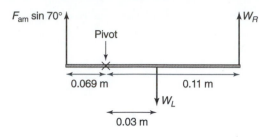

**FIGURE 9.6** Moments of forces about pivotal hip joint.

whereas equation (9.23) provides

$$F_{hj} \sin \theta_{hj} = 1624.4 \text{ N} \tag{9.28}$$

From equations (9.27) and (9.28) we find the result

$$\sqrt{F_{hj}^2 \cos^2 \theta_{hj} + F_{hj}^2 \sin^2 \theta_{hj}} = F_{hj} = \sqrt{(1624.4)^2 + (382.0)^2} = 1669 \text{ N} \tag{9.29}$$

and

$$\theta_{hj} = \tan^{-1}\left(\frac{F_{hj} \sin \theta_{hj}}{F_{hj} \cos \theta_{hj}}\right) = \tan^{-1}\left(\frac{1624.4}{382.0}\right) = 76.8° \tag{9.30}$$

## TOPIC 9.3 Forces at the Ankle Joint

We now turn to the forces acting on the foot. First, there is the upward force of the ground acting on the toes or the ball of the foot, equal and opposite in direction to the person's weight. Second, the Achilles tendon exerts a force at an angle to the vertical, and third, the weight of the leg acts downward on the ankle. Let us now use some reasonable magnitudes for the forces concerned to find the force exerted by the Achilles tendon. We assume that it is at 30° to the vertical and denote the magnitude of its vertical component by $F_{AT}$ as in Figure 9.7. We represent the vertical force on the ball of the foot by $F_G$ and its line of action is $d_2 = 0.10$ m from the ankle bone, which is acting as a pivot. We sup-

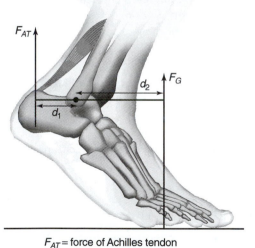

$F_{AT}$ = force of Achilles tendon
$F_G$ = force of ground (weight of body)

**FIGURE 9.7** Forces acting on ankle joint.

pose the line of action of $F_{AT}$ is $d_1 = 0.05$ m from the pivot. Taking moments about the ankle bone and assuming the ankle is in rotational equilibrium, we have that

$$F_{AT}d_1 = F_Gd_2 \tag{9.31}$$

Assuming the weight of a person is 700 N and substituting the values of $d_1$ and $d_2$, equation (9.31) becomes

$$F_{AT}\,(0.05\text{ m}) = 700\text{ N }(0.10\text{ cm}) \tag{9.32}$$

Hence

$$F_{AT} = 1400\text{ N} \tag{9.33}$$

which is a force twice as large as the person's weight.

## TOPIC 9.4  The Spinal Column

Here we consider a person bending forward to lift a box as in Figure 9.8. Two torques arise from the force of gravity. One is on the box and the other is on the upper body of the person. These forces together exert a force on the lower back, in particular on the fifth lumbar vertebra. We can consider the lower back as a natural pivot, the spinal column being considered as a lever. Bearing in mind that the lifting of heavy objects can damage the back, we calculate the muscle force exerted on the spinal column.

Note that the gravitational torque on the object as a whole is the same as the gravitational torque exerted on a point mass with the same mass as the object located at the CM. Now going back to the man with a box, we note that the CM of the torso lies near the lower part of the chest. We can now calculate the torque exerted by gravity on the torso assuming that this is bending forward, as in Figure 9.8a, at a 90° angle to the vertical and that the forearms are stretched downward. Drawing the free-body force diagram (see Figure 9.8b), we find that four forces are acting on the spinal column:

1. The gravitational force, $\mathbf{W}_t$, acting on the CM of the torso. We assume that this lies at a distance $L_{CG} = 0.42$ m from the lower back and that its magnitude is 523 N.

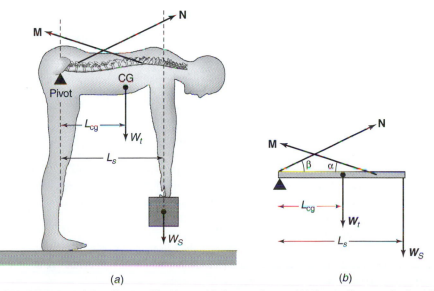

(a)  (b)

**FIGURE 9.8**  (a) Bending to lift a box with forces shown. (b) Force diagram for bending man.

2. The gravitational force on the box, $\mathbf{W}_s$, which is transmitted by the arms and shoulders. We assume $W_s = 140$ N and $L_S = 0.70$ m.

3. The unknown force $\mathbf{M}$ exerted on the spinal column by the back muscle responsible for the lifting. We assume this force is applied at a distance $d = 0.7L_S$ from the lower back at an angle $\alpha$ of about 12° above the horizontal. This is the force we wish to find.

4. An unknown compressive reaction force, $\mathbf{N}$, exerted on the pivot in the lower back. Let $\beta$ be the angle between $\mathbf{N}$ and the horizontal.

As the spinal column is stationary, the total vertical and horizontal forces must be zero. Therefore we have the following equations to balance force components: (a) horizontally,

$$N \cos \beta - M \cos \alpha = 0 \tag{9.34}$$

and (b) vertically,

$$N \sin \beta + M \sin \alpha - W_S - W_t = 0 \tag{9.35}$$

Similarly, taking moments about the pivot gives zero, so that

$$W_S L_S + W_t L_{CG} - 0.7 L_S M \sin \alpha = 0 \tag{9.36}$$

However, we also know that $L_{CG} = 0.6L_S$, and hence from equation (9.36)

$$W_S L_S + W_t 0.6 L_S - 0.7 L_S M \sin \alpha = 0 \tag{9.37}$$

From equation (9.37) we now evaluate $M = 3118$ N. Using this value of $M$ in equation (9.34), we find

$$N \cos \beta = 3050 \text{ N} \tag{9.38}$$

and in equation (9.35) we obtain

$$N \sin \beta = 14.71 \text{ N} \tag{9.39}$$

Hence from equation (9.38) and equation (9.39),

$$\tan \beta = 0.0048 \tag{9.40}$$

so $\beta = 0.27°$, which is very small, so the reaction force $\mathbf{N}$ points almost completely along the spinal column. Furthermore, it acts as a compressive force on the spine. It follows that $N$ is very close to $N = 3050$ N.

To emphasize, $\mathbf{N}$ is a very large force, and during lifting the fairly fragile spinal column experiences a compressive force comparable to a mass of about 311 kg. Our back muscles are sufficiently strong to provide the required forces, but our spinal column may not be able to support the huge compression and is certainly not arranged optimally from the point of view of lifting heavy weights.

It is well known that poor posture can produce back strain. Figure 9.9 illustrates why. When one stands erect, the weight of the upper body is directly over the legs and little force is exerted by back and leg muscles. Most of one's weight is supported by the skeletal system, and not by muscle action. If one slouches or leans forward, the center of gravity of the upper body is no longer directly over the pivot point. Hence, the back muscles must exert a torque about the pivot at the base of the spine to counteract the torque due to the weight of the upper body. Consequently, continual effort by the back muscles produces back strain.

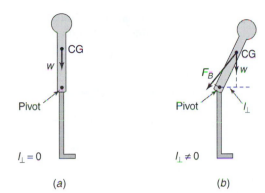

**FIGURE 9.9** (*a*) A person standing erect places the center of gravity of his upper body directly above the pivot in his lower back. (*b*) A person bending over allows the weight of his upper body to create a torque that must be compensated by a torque created by the back (erector spinae) muscles.

## SOLVED PROBLEMS

**9.1**  Estimate the force in the Achilles tendon in the human foot during a walk. The weight of a human is taken as $W = 800$ N, the cross-sectional area of the Achilles tendon is $A = 1.61$ cm$^2$, and the remaining parameters are given in Figure 9.9.

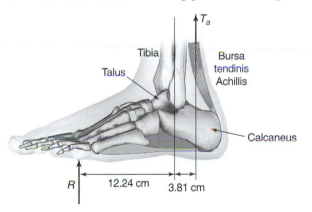

**Solution**  The joint between the tibia and talus may be regarded as a pivot. At equilibrium, the moments of the forces are equal; hence

$$T_a(0.0381 \text{ m}) = R(0.1224 \text{ m})$$

where $T_a$ is the tension in the Achilles tendon, $R$ is the reactive force of the ground exerted on the foot at the point of contact of the sole, and the numerical data are those in Figure 9.9. The magnitude of the reactive force $R$ depends on the state (position) of the foot. During a walk, the weight of the body is distributed between the two feet, and thus $R = \frac{1}{2}W = 400$ N, or loads one foot only, when $R = W = 800$ N. After rearrangement and substitution of the numerical values of $R$, the force exerted in the Achilles tendon can vary between $T_a = 1285$ N and $T_a = 2570$ N.

**9.2**  A ballet dancer puts all her weight on the toes of one foot. If her mass is 60 kg, what is the force that has to be exerted by her leg muscle to hold that pose?

**Solution**   To calculate the moments about the ankle bone, we have to take into account the normal force of the floor on the dancer equal to $mg$. We also assume that the force exerted by the leg muscle is perpendicular to the floor. We therefore have

$$F \times 0.05 \text{ m} = mg \times 0.2 \text{ m}$$
$$F = 60 \text{ kg} \times 9.81 \text{ m/s}^2 \times 0.2 \text{ m}/0.05 \text{ m} = 94.18 \text{ N}$$

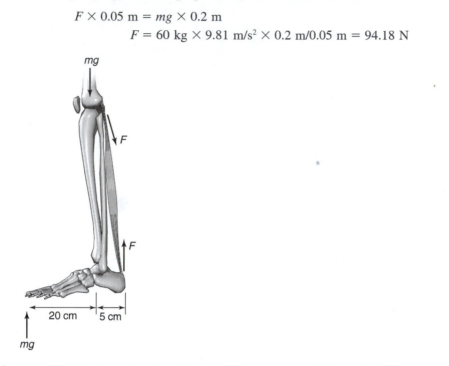

**9.3**   Calculate the force that the biceps muscle exerts to hold up the forearm and the book shown in Figure 9.10.

**Solution**   Four external forces act on the forearm: the weight of the forearm, the weight of the book, the force the biceps muscle exerts, and the force exerted at the elbow joint by the humerus. The second of the two conditions for equilibrium is sufficient to solve this problem. One simplification can be made by taking the pivot point to be the elbow joint. The torque exerted by $F_H$ is zero since it is applied at the pivot point. The weights

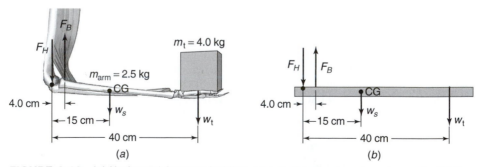

**FIGURE 9.10**   (a) Horizontal forearm holding a book showing bones and muscle. (b) Free force diagram corresponding to the situation in (a).

of the forearm and textbook produce clockwise torques, while $F_B$ exerted by the biceps muscle produces a counterclockwise torque. Hence

$$\tau_{cw,\,net} = \tau_{ccw,\,net}$$

$$(0.15\text{ m})w_{arm} + (0.40\text{ m})w_t = (0.04\text{ m})F_B,\text{ giving}$$

$$F_B = 484\text{ N}\quad\text{as the final result}$$

**9.4**  A woman whose upper body weight $W = 600$ N is leaning forward at an angle $\alpha = 60°$ to the vertical. What is the force $F$ exerted in the muscles of her back to maintain this position? The effective pivot point is in the pelvic cavity at a distance $\ell_W = 0.6$ m from the center of gravity and at a perpendicular distance $\ell_F = 0.06$ m from the muscles in the back.

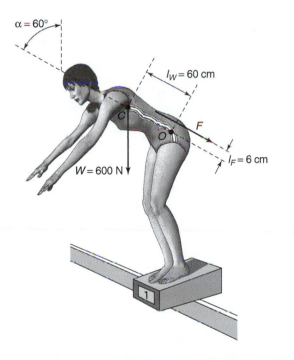

**Solution**  Since the body is in rotational equilibrium, the sum of the torques of the two forces tending to make the body rotate about the effective pivot point ($O$) must be zero. As the weight ($W$) tends to rotate the upper body counterclockwise and the force in the muscles ($F$) tends to rotate it clockwise, the balance of the torques can be given as

$$-W\ell_W\cos\alpha + F\ell_F = 0$$

Solving for $F$, the force in the muscles of the back, yields

$$F = W\frac{\ell_W}{\ell_F}\cos\alpha = 5196\text{ N}$$

**9.5**  The tension $T$ in the quadriceps tendon as it passes over the knee cap is 1500 N. What is the magnitude and direction of the resultant force $\mathbf{F}$ exerted on the femur by the cap for the configuration shown in the diagram in Figure 9.11?

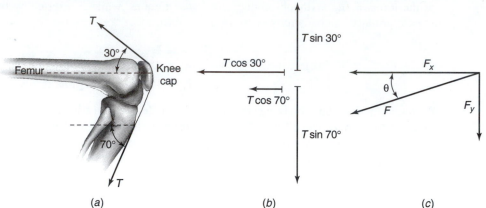

(a)                                    (b)                                    (c)

**FIGURE 9.11** (a) Quadriceps tendon passing over the kneecap. (b) Magnitude of horizontal and vertical forces. (c) Resultant force diagram.

**Solution**

$$\sum F_x = T \cos 30° + T \cos 70° = 1812 \text{ N}$$

$$\sum F_y = T \sin 70° - T \sin 30° = 660 \text{ N}$$

The resultant force has a magnitude

$$F = \sqrt{1812^2 + 660^2} = 1928 \text{ N}$$

and a direction given by

$$\tan \theta = \frac{660}{1812} = 0.36$$

Hence,

$$\theta = 20°$$

**9.6** Determine the position of the center of mass of the whole leg (a) when stretched out as in Figure 9.12a and (b) when bent at 90° as shown in Figure 9.12b. Assume the person is 1.70 m tall.

**TABLE 9.1  Masses of Leg Components in Percentage Units**

| | |
|---|---|
| Thigh | 21.5 |
| Knee | 9.6 |
| Ankle | 3.4 |

**Solution**

**(a)** Table 9.1 uses percentage units, meaning the person has a mass of 100 units and a height of 100 units. At the end we multiply by 1.70 m/100. We measure the distance from the hip joint in Table 9.1 and obtain the numbers shown in Figure 9.12a. We obtain

$$x_{CM} = \frac{(21.5)(9.6) + (9.6)(33.9) + (3.4)(50.3)}{21.5 + 9.6 + 3.4} = 20.4 \text{ units}$$

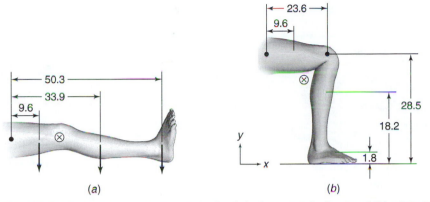

**FIGURE 9.12** Weight distribution in the leg (a) when stretched out and (b) when bent at 90°.

Thus the center of mass of the leg and foot is 20.4 units from the hip joint, or $52.1 - 20.4 = 31.7$ units from the base of the foot. Since the person is 1.70 m tall, this is $(1.70 \text{ m})(31.7/100) = 0.54$ m.

**(b)** In this part, we use an $x$–$y$ coordinate system, as shown in Figure 9.12b. First, we calculate how far right of the hip joint the CM lies:

$$x_{CM} = \frac{(21.5)(9.6) + (9.6)(23.6) + (3.4)(23.6)}{21.5 + 9.6 + 3.4} = 14.9 \text{ units}$$

For the 1.70-m-tall person, this is $(1.70 \text{ m})(14.9/100) = 0.25$ m. Next, we calculate the distance, $y_{CM}$, of the CM above the floor:

$$y_{CM} = \frac{(3.4)(1.8) + (9.6)(18.2) + (21.5)(28.5)}{21.5 + 9.6 + 3.4} = 23.1 \text{ units}$$

or $(1.70 \text{ m})(23.1/100) = 0.39$ m. Thus, the CM is located 39 cm above the floor and 25 cm to the right of the hip joint.

**9.7** Calculate the magnitude and direction of the force $\mathbf{F}_v$ acting on the fifth lumbar vertebra for the example shown in Figure 9.13.

**Solution** First we calculate $F_M$ using the torque equation, taking the axis at the base of the spine. To figure out the lever arms, we need to use trigonometric functions. For $F_M$, the lever arm will be the real distance to where the force acts (48 cm) times sin 12°, as shown in Figure 9.13c. The lever arms $w_1$, $w_2$, and $w_3$ can be seen from Figure 9.13b to be their respective distances times sin 60°. Thus, $\sum \tau = 0$ gives

$$(0.48 \text{ m})(\sin 12°)(F_M) - (0.72 \text{ m})(\sin 60°)w_1 - (0.48 \text{ m})(\sin 60°)w_2$$
$$- (0.36 \text{ m})(\sin 60°)w_3 = 0.$$

where we chose the positive sign for counterclockwise torque. Putting in the values for $w_1$, $w_2$, and $w_3$ given in the figure, we find $F_M = 2.2w$, where $w$ is the total weight of the body. To get the components of $F_v$ we use the $x$ components of the force equation:

$$\sum F_y = F_{vy} - F_M \sin 18° - w_1 - w_2 - w_3 = 0 \qquad F_{vy} = 1.3w$$
$$\sum F_x = F_{vx} - F_M \cos 18° = 0 \qquad F_{vx} = 2.1w$$

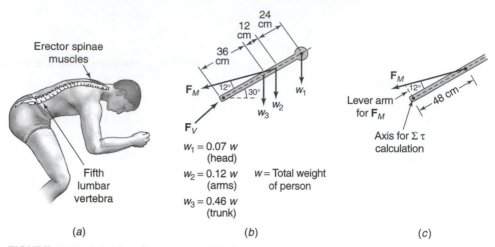

**FIGURE 9.13** (a) A bending man and (b) free-body diagram of spine.

and

$$F_v = \sqrt{F_{vx}^2 + F_{vy}^2} = 2.5w$$

The angle that $F_v$ makes with the horizontal is given by $\tan \theta = 0.625$, so $\theta = 32°$.

**9.8** How much force must the biceps muscle exert when a 5.0-kg mass is held in the hand (a) with the arm horizontal as in Figure 9.14a and (b) when the arm is at a 30° angle as in Figure 9.14b. Assume that the mass of the forearm and hand together is 2.0 kg and their CM is as shown.

**Solution**

**(a)** The forces acting on the forearm are shown in Figure 9.14a and include the upward force $F_M$ exerted by the muscle and the force $F_J$ exerted at the joint by the bone in the upper arm (both assumed to act vertically). To find $F_M$, we use the torque equation and choose the axis through the joint so that $F_J$ does not enter:

$$\sum \tau = (0.050 \text{ m})(F_M) - (0.15 \text{ m})(2.0 \text{ kg})(g) - (0.35 \text{ m})(5.0 \text{ kg})(g) = 0$$

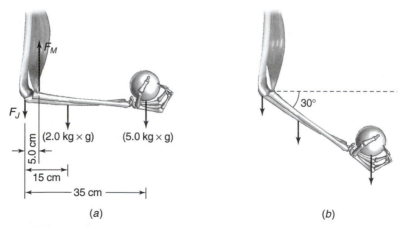

**FIGURE 9.14** Forces in the arm when it is (a) horizontal, and (b) at a 30° angle to the horizontal axis.

We solve this for $F_M$ and find $F_M = (41 \text{ kg})(g) = 400$ N.

**(b)** The lever arm, as calculated about the joint, is reduced by the factor cos 30° for all three forces. So our torque equation will look like the one just above, except that each term will have a "cos 30°." The latter will cancel out so the same result will be obtained, $F_M = 400$ N.

**9.9** A man of total weight 800 N stands on one leg. Assuming that each of his legs accounts for one-eighth of his body weight, calculate the force exerted on the head of his femur using the anatomical data shown.

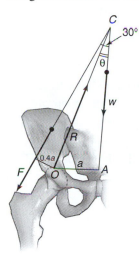

**Solution** We may solve this problem by calculating the total force exerted by gravity on all of the man's body except the leg on which he is standing. Since this leg weighs $\frac{1}{8} \times$ 800 N = 100 N, it follows that the weight of the rest of his body is $w = 700$ N. This weight acts vertically downward along the line $CA$ through the center of gravity of the rest of the body and exerts a counterclockwise torque about the head of the femur $O$ of magnitude $wa$, where $a$ is the perpendicular distance from $O$ onto the line $CA$. The gluteus medius and minimus muscles exert forces on the pelvis, which can be represented by a single force $F$ acting at an angle of 30° to the vertical, as shown in the figure. The perpendicular distance from $O$ to the line of action $F$ may be taken as $0.4a$, and hence this muscular force exerts on the rest of the body a counterclockwise torque, and so

$$0.4Fa = wa$$

or

$$F = \frac{w}{0.4} = 1750 \text{ N}$$

The third force which acts on the rest of the body is the force $R$ exerted on the hip socket by the head of the femur. Since the rest of the body is at rest, this force $R$ must balance the sum of $F$ and $w$. If $R$ acts at an angle $\theta$ to the vertical, then, resolving horizontally, we get

$$R \sin \theta = F \sin 30° = 875 \text{ N}$$

and, resolving vertically,

$$R \cos \theta = w + F \cos 30° = 2216 \text{ N}$$

Squaring and adding these two expressions gives

$$R^2 \sin^2 \theta + R^2 \cos^2 \theta = R^2(\sin^2 \theta + \cos^2 \theta) = R^2 = (875 \text{ N})^2 + (2216 \text{ N})^2$$
$$= 5.68 \times 10^6 \text{ N}^2$$

and so,

$$R = 2383 \text{ N}$$

The angle $\theta$ at which this force acts may be found from

$$\frac{R \sin \theta}{R \cos \theta} = \tan \theta = \frac{875 \text{ N}}{2216 \text{ N}} = 0.395$$

Therefore,

$$\theta = 21.5°$$

The force which the pelvis exerts on the head of the femur at $O$ is, by Newton's third law, equal and opposite to the force $R$ calculated above, that is, 2383 N downward at an angle of 21.5° to the vertical.

**9.10** An amateur weightlifter is trying to lift as much weight as possible, but he is not holding his arms very close to his body. If he has a mass of 100 kg, how much can he lift, given the conditions in the diagram?

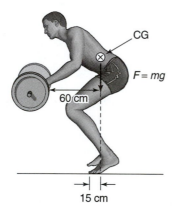

**Solution** It is quite simple to calculate the moments about his toes, and we obtain

$$mg \times 0.15 \text{ m} = W \times 0.45 \text{ m}$$
$$W = \frac{mg \times 0.15 \text{ m}}{0.45 \text{ m}} = 327 \text{ N}$$

This force is equivalent to the gravitational force due to a mass of 33.3 kg.

**9.11** Figure 9.15 shows a traction device used with a foot injury. The weight of the 2.2-kg object creates a tension in the rope that passes around the pulleys. Therefore, tension forces $T_1$ and $T_2$ are applied to the pulley on the foot. The foot pulley is kept in equilibrium. Ignoring the weight of the foot, find the magnitude of **F**.

**Solution** Since the sum of the $y$ components of the forces must be equal to zero, it follows that

$$\sum F_y = T_1 \sin 35° - T_2 \sin 35° = 0$$

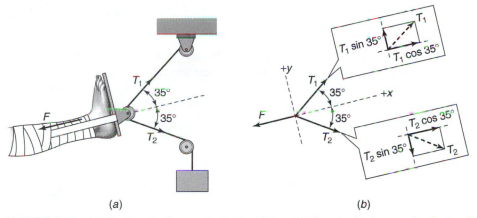

**FIGURE 9.15** (a) Traction device used with foot injury. (b) Free-body diagram for foot pulley.

or $T_1 = T_2$. In addition, the sum of the x components of the forces is zero, so we have that

$$\sum F_x = T_1 \cos 35° + T_2 \cos 35° - F = 0$$

Solving for F and letting $T_1 = T_2 = T$, we find that $F = 2T \cos 35°$. However, the tension T in the rope is determined by the weight of the 2.2-kg object: $T = mg = 2.2$ kg$(9.81$ m/s$^2) = 22$ N. Therefore,

$$F = 2(22 \text{ N}) \cos 35° = 36 \text{ N}$$

**9.12**   The masseter muscle is one of the muscles involved in the process of mastication, or chewing. Assume that the masseter muscle exerts an upward force of 250 N at a distance of 2.5 cm from the pivot or temporomandibular joint, as shown in the diagram. If the linear distance from the pivot to the molars is 4.8 cm, determine the force exerted on the molars during the chewing process.

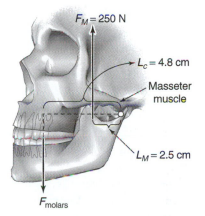

**Solution**   From the conservation of rotational equilibrium,

$$\tau_{cw,net} = \tau_{ccw,net}$$
$$(250 \text{ N})(0.025 \text{ m}) = F_{molars}(0.048 \text{ m})$$

Solving for $F_{\text{molars}}$ gives

$$F_{\text{molars}} = 130.2 \text{ N}$$

**9.13** What force is exerted in the back muscles and in the disk between the vertebrae if the man in the diagram is holding a loaded shovel weighing $W = 100$ N? The centers of mass of the loaded shovel and the back muscles are on opposite sides at distances $\ell_W = 1$ m and $\ell_b = 0.05$ m from the lumbar vertebrae of the backbone.

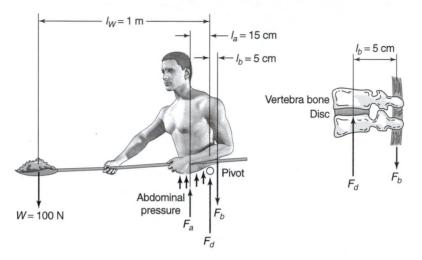

**Solution** The disk between the vertebrae serves as a pivot point of rotation. At equilibrium, the torque of the weight of the shovel is compensated by that of the back muscles. From the condition of equilibrium we express $F_b$, the force exerted in the back muscles, as

$$F_b = W \frac{\ell_W}{\ell_b} = 2000 \text{ N}$$

The load on the disk between the vertebrae is the sum of the two forces:

$$F_d = F_b + W = 2000 \text{ N} + 100 \text{ N} = 2100 \text{ N}$$

However, abdominal muscles also contribute if they are tensed up. Taking this effect into account, the balance of torques of the forces should be extended to give

$$F_a \ell_a + F_b \ell_b = W \ell_W$$

where the perpendicular distance of the line of force ($F_a$) of the abdominal muscles from the pivot of rotation is $\ell_a = 0.15$ m. If we suppose that $F_b = F_a$, that is, the tensions in the back muscles and the abdominal muscles are equal, we get $F_b = F_a = 500$ N.

**9.14** What force, $F$, is exerted in the deltoid muscle of a straight arm elevated at an angle of $\alpha = 60°$ to the vertical when a weight $W = 100$ N is held in the hand? The length of the arm is $\ell_W = 0.8$ m, the weight of the arm is $W_a = 50$ N, the center of mass of the arm is at a distance $\ell_a = 0.32$ m from the joint, and $\beta = 30°$ is the angle of insertion of the muscle at a distance $\ell_F = 0.16$ m from the joint, as shown in the diagram.

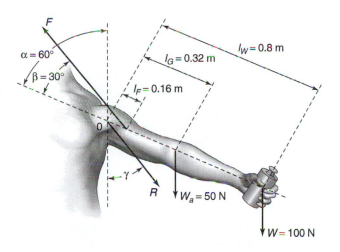

**Solution** The arm is at rest in an elevated position with a weight in the hand. Using the pivot point of the shoulder, the sum of the torques of the forces can be expressed in rotational equilibrium as

$$W_a\ell_a \sin \alpha + W\ell_w \sin \alpha - F\ell_F \sin \beta = 0$$

From this equation, the tension in the deltoid muscle can be calculated:

$$F = \frac{W_a\ell_a + W\ell_w}{\ell_F} \frac{\sin \alpha}{\sin \beta}$$

On substitution of the numerical data, we obtain $F = 1039$ N.

# EXERCISES

**9.1**   In the exercise situation shown in the figure the torque about the knee joint exerted by the 10-kg weight attached to the ankle varies with the elevation of the leg. Calculate the torque for the four positions shown.

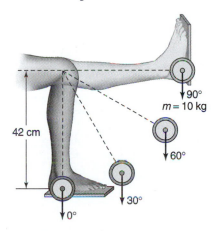

**9.2**   The accompanying figure shows an elastic cord attached to two back teeth and stretched across a front tooth to apply a force **F** to the front tooth. If the tension in the cord is 1 N, what is the magnitude and direction of the force **F** applied to the front tooth?

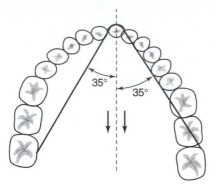

**9.3** The hip abduction muscle, which connects the hip to the femur, consists of three independent muscles acting at different angles. The figure shows the results of measurements of the force exerted separately by each muscle. Find the total force exerted by the three muscles together.

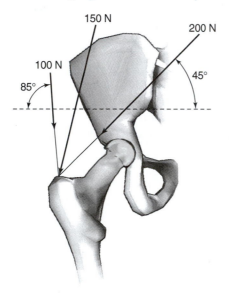

**9.4** Find the total force applied to the patient's head by the traction device shown in the figure.

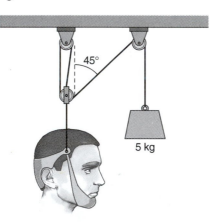

**9.5** Derive a formula for the tension in the deltoid muscle of the arm arising from elevation of the arm using figures (*a*) and (*b*). The arm is at rest in its elevated position so it is in a state of rotational equilibrium, where $\theta$ is the angle of elevation of the arm, $\Delta$ is the angle of insertion of the muscle, $C$ is the center of mass of the arm, $T$ is the tension in the muscle, $W$ is the weight of the arm considered to act at the CM, and $R$ is the reaction at the joint oriented at an angle $\Delta$ below the line $0B$.

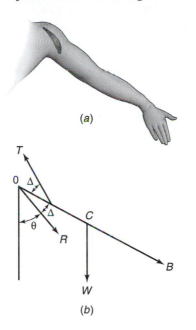

(*a*)

(*b*)

**9.6** The horizontal arm in the figure is composed of three parts: the upper arm (weight $W_1 = 20$ N), the lower arm ($W_2 = 10$ N), and the hand ($W_3 = 5$ N). Determine the center of gravity of the entire arm, relative to the shoulder joint.

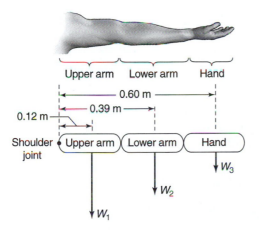

**9.7** A person uses the forearm to hold a ball of mass $M = 1.5$ kg. The length of the forearm is 0.40 m, as measured from the pivot point (i.e., the elbow joint) to the ball. The mass of the forearm itself is 1.2 kg. The center-of-gravity point of the forearm (without the ball) is 0.20 m from the elbow joint. The biceps muscle exerts an unknown force $F_b$ in the upward vertical direction a perpendicular distance of 0.07 m from the elbow joint. The humerus bone exerts an unknown force $F_h$ in the downward vertical direction on the elbow joint.

(a) What is the total torque exerted by gravity on the elbow joint?

(b) Find the magnitudes $F_b$ and $F_h$.

**9.8** The accompanying figure shows the ankle joint and the Achilles tendon attached to the heel. The tendon exerts a force of magnitude $F = 700$ N, as shown. Determine the torque (magnitude and direction) of this force about the ankle joint which is located 0.04 m away from point $P$.

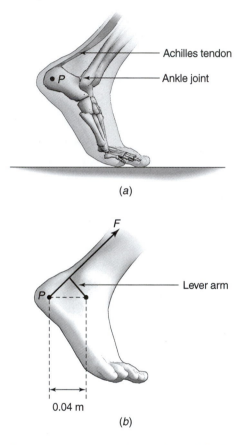

(a)

(b)

**9.9** A professional boxer is capable of dealing a blow with a net force of 2000 N generated by his forearm. Assume that the effective perpendicular lever arm is 0.05 m and the angular acceleration of the forearm is 100 rad/s$^2$. Calculate the moment of inertia of the boxer's forearm.

# CHAPTER *10*

# ELASTICITY AND SIMPLE HARMONIC MOTION

## PHYSICAL BACKGROUND

Elasticity is an inherent property of all solid materials describing their response to an external mechanical force. It is characterized by the presence and degree of deformation and depends primarily on the size and internal structure of the object and the magnitude of the force. The elasticity of a substance can be characterized by two physical parameters: stress and strain.

Stress $\sigma$ reflects the magnitude of an external force $F$ required to induce the given deformation of an object divided by the surface area $A$ upon which the force is exerted perpendicularly, or

$$\sigma = \frac{F}{A} \tag{10.1}$$

and is given in units of newtons per meter squared.

Strain $\varepsilon$ is a quantitative measure of the fractional extent of deformation of an object produced as a result of the applied stresses, and it is a unitless quantity. Strain is defined as the ratio of the increase in a particular dimension in the deformed state to that dimension in its initial, undeformed state and can be expressed as

$$\varepsilon = \frac{\ell_d - \ell_n}{\ell_n} \tag{10.2}$$

where $\ell_d$ and $\ell_n$ represent the dimensional quantities of the elastic object in the deformed and normal states, respectively.

Hooke's law of elasticity describes, for any material, a linear relation between tensile stress and tensile strain with the proportionality constant defined as an elastic modulus:

$$\text{Elastic modulus} = \frac{\text{stress}}{\text{strain}} \tag{10.3}$$

Since the elastic modulus is linearly proportional to stress, a large value for the elastic modulus implies that a large stress is required to produce a given strain.

The possible kinds of elastic deformation include stretch and compression, shear deformation, and volume deformation. The forces required to create them are given by

$$\frac{F}{A} = Y\left(\frac{\Delta L}{L_0}\right) \qquad \frac{F}{A} = S\left(\frac{\Delta x}{L_0}\right) \qquad \Delta P = -B\left(\frac{\Delta V}{V_0}\right) \tag{10.4}$$

where $\Delta x$ is the change of length of the object in the direction parallel to the force acting tangentially to the surface and the proportionality constants $Y$, $S$, and $B$ are called, respectively, Young's modulus, the shear modulus, and the bulk modulus. The original length is $L_0$, the volume $V_0$ and the corresponding deformations are given by $\Delta L$ and $\Delta V$, respectively.

Hooke's law is an approximate relationship whose validity is limited to moderate stresses. Every material has a characteristic breaking point at which fracture occurs. The corresponding value of stress is called the fracture stress. Table 10.1 summarizes the values of Young and shear moduli for a number of materials, including biological ones, and Table 10.2 lists values of fracture stress for various materials.

The pressure $P$ is the magnitude $F$ of the force acting perpendicular to a surface divided by the area $A$ over which the force acts:

$$P = \frac{F}{A} \tag{10.5}$$

The SI unit of pressure is the newton per square meter, a unit known as a pascal (Pa).

The force $F$ that must be applied to stretch or compress an ideal spring is $F = kx$, where $k$ is the spring constant and $x$ is the displacement of the spring from its unstrained length. A spring exerts a restoring force on an object attached to the spring. The restor-

**TABLE 10.1 Young and Shear Moduli for Selected Materials**

| Material | Young's modulus $Y$ (N/m²) |
|---|---|
| Aluminum | $6.9 \times 10^{10}$ |
| Bone | |
| Compression | $9.4 \times 10^{9}$ |
| Tension | $1.6 \times 10^{10}$ |
| Brass | $9.0 \times 10^{10}$ |
| Brick | $1.4 \times 10^{10}$ |
| Copper | $1.1 \times 10^{11}$ |
| Nylon | $3.7 \times 10^{9}$ |
| Pyrex glass | $6.2 \times 10^{10}$ |
| Steel | $2.0 \times 10^{11}$ |
| Teflon | $3.7 \times 10^{8}$ |
| Tungsten | $3.6 \times 10^{11}$ |
| Tendon | $2.0 \times 10^{7}$ |
| Rib cartilage | $1.2 \times 10^{7}$ |
| Blood vessels | $2 \times 10^{5}$ |
| Sapphire | $4.2 \times 10^{11}$ |
| Diamond | $1.20 \times 10^{12}$ |
| Jellyfish | $10^{3}$ |
| Elastin, rubber | $10^{6}$ |
| Collagen | $0.2 \times 10^{10}$ |
| Wood | $1.0 \times 10^{10}$ |

| Material | Shear modulus $S$ (N/m²) |
|---|---|
| Long bone | $0.8–1.5 \times 10^{10}$ |
| Steel | $8 \times 10^{10}$ |
| Aluminum | $2.6 \times 10^{10}$ |
| Rubber (soft, vulcanized) | $1.6 \times 10^{6}$ |

**TABLE 10.2  Fracture Stress for Selected Materials**

| Material | Fracture stress ($\times 10^6$ Pa) |
| --- | --- |
| Silk | 2000 |
| Iron | 400 |
| Bone | 200 |
| Wood | 100 |
| Collagen | 100 |
| Humerus | 5 |
| Arterial wall | 2 |

ing force produced by an ideal spring is $F = -kx$, where the minus sign indicates that the restoring force points in the opposite direction to the displacement of the spring.

Simple harmonic motion is the oscillatory motion that occurs when a restoring force of the form $F = -kx$ acts on an object. A graphical record of position versus time for an object in simple harmonic motion is sinusoidal (see Figure 10.1). The amplitude $A$ of the motion is the maximum distance that the object moves away from its equilibrium position. The period $T$ is the time required to complete one cycle of the motion, while the frequency $f$ is the number of cycles per second that occur. Frequency and period are related according to $f = 1/T$. The frequency $f$ (in hertz) is related to the angular frequency $\omega$ (in radians per second) according to

$$\omega = 2\pi f \tag{10.6}$$

The following equation describes the displacement of the oscillating object from its point of equilibrium:

$$x = A\cos(\omega t + \phi) \tag{10.7}$$

where $x$ is the position as a function of time $t$ in meters and $\phi$ is the phase angle. For an object of mass $m$ on a spring with spring constant $k$, the frequency is determined by

$$2\pi f = \sqrt{\frac{k}{m}} \tag{10.8}$$

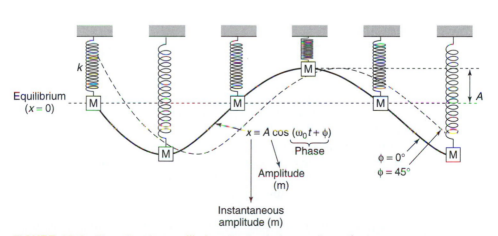

**FIGURE 10.1**  Plot of spring oscillations in simple harmonic motion.

The velocity and the acceleration in simple harmonic motion are continually changing with time. The elastic potential energy of an object attached to an ideal spring is

$$PE_{elastic} = \tfrac{1}{2}kx^2 \tag{10.9}$$

A simple pendulum is a point particle of mass $m$ attached to a frictionless pivot by a cable whose length is $L$ and whose mass is negligible. The small-angle back-and-forth swinging of a simple pendulum is simple harmonic motion, while large-angle motion is not. The frequency of small-angle motion is given by

$$2\pi f = \sqrt{\frac{g}{L}} \tag{10.10}$$

A physical pendulum consists of a rigid object, with moment of inertia $I$ and mass $m$, suspended from a frictionless pivot. For small-angle displacements, the frequency of simple harmonic motion for a physical pendulum is determined by

$$2\pi f = \sqrt{\frac{mgL}{I}} \tag{10.11}$$

where $L$ is the distance between the axis of rotation and the center of gravity of the rigid object.

Damped harmonic motion is motion in which the amplitude of oscillation decreases as time passes. Critical damping is the minimum degree of damping that eliminates any oscillations in the motion as the object returns to its equilibrium position. Driven harmonic motion occurs when an additional driving force is applied to an object along with the restoring force. Resonance is the condition under which a driving force can transmit large amounts of energy to an oscillating object, leading to large-amplitude motion. In the absence of damping, resonance occurs when the frequency of the driving force matches a natural frequency at which the object oscillates.

## TOPIC 10.1   Bone Stiffness and Strength

Animals develop a variety of solid materials such as bone, tooth, horn, shell, nail, and cartilage. Most of these organic compounds are complex, heterogeneous substances. The compact part of a bone, for example, consists of living cells embedded in a solid framework one-third of which is made up of collagen and the remainder, called bone salt, is a mixture of inorganic materials containing calcium, phosphorus, oxygen, and hydrogen. The presence of tiny crystallites attached to the collagen fibrils increases the bone's Young's modulus. The stress–strain curve for bone is shown in Figure 10.2 and includes both tensile and compressive applications of forces. It is worth noting that bone is constructed on the same principle as reinforced concrete. While concrete alone has great compressive strength, it lacks tensile strength. Steel rods are embedded in reinforced concrete to give it tensile as well as compressive strength. Similarly, collagen adds tensile strength to the compressive strength of hydroxyapatite that is the main building material for bone structures.

Bones not only support loads placed on them, in particular the weight of an animal or human, but also resist bending and twisting. Indeed, fractures are almost always caused by excessive bending and twisting torques. The ability to resist bending and twisting depends on its stiffness and hence, in turn, on the cross-sectional area and its shape. Hollow bones are stiffer than solid ones with the same weight (similar to the way pipes are stiffer than rods).

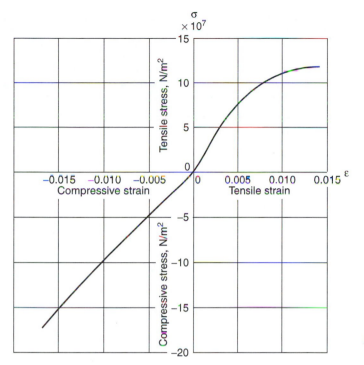

**FIGURE 10.2**   Stress–strain curve for bone.

A simple dimensional analysis demonstrates that the various monster creatures featured in horror and science fiction movies could not possibly exist due to the inability of their legs to carry them. For example, an ant 100 times larger than normal would be $100 \times 100 \times 100 = 1 \times 10^6$ times heavier while its legs would have a cross-sectional area $100 \times 100 = 10,000$ times larger, that is, the strength would have to increase by a factor of 100 also. Using the same bone material would therefore inevitably lead to the collapse of this hypothetical animal.

Biological materials lack the high degree of order characteristic of metals and all show anisotropy in their physical properties (e.g., wood). It is well known that the properties of wood parallel to the grain are quite different from those perpendicular to the grain. Collagen fibers are very markedly anisotropic since they contain the threadlike fibrils which lie in directions approximately parallel to the length of the fiber, but their alignment with the axis of the fiber is not perfect, and furthermore there are regions where the fibrils have a much more random arrangement.

When a tensile stress is applied to a fiber of collagen, the fibrils become more aligned parallel to the axis. This is the reason that the value of Young's modulus for collagen is only $10^9$ N/m$^2$. In the case of bone the tensile strength is due to the collagen and its compressive strength is due to hydroxyapatite. As a consequence, the Young's modulus of bone and other heterogeneous substances is different for tensile and compressive stress. Table 10.3 shows the tensile and compressive properties of some biological materials. From this table we see that bone has a Young's modulus which is nearly twice as large for tensile stress as for compressive stress; that is, a compressive stress produces twice as much strain as a tensile stress of equal magnitude. Surprisingly, the mechanical properties of bones from different animals are very similar when one considers how different they are in other respects. The tensile stress of bone is 1.4 that of steel and the compressive strength is close to that of granite. Bone is, of course, much lighter than steel or granite so as a

**TABLE 10.3   Tensile and Compressive Properties of Selected Biological Materials**

| Material | Tension | | | Compression | | |
|---|---|---|---|---|---|---|
| | Young's modulus, $10^9$ N/m² | Tensile strength, $10^7$ N/m² | Maximum strain | Young's modulus, $10^9$ N/m² | Compressive strength, $10^7$ N/m² | Maximum strain |
| Bone, | | | | | | |
| Human femur | 16.0 | 12.1 | 0.014 | 9.4 | 16.7 | 0.0185 |
| Horse femur | 23 | 11.5 | 0.0075 | 8.3 | 14.2 | 0.024 |
| Human vertebra | 0.17 | 0.12 | 0.0058 | 0.088 | 0.19 | 0.025 |
| Cartilage, | | 0.30 | 0.30 | | | |
| Human ear | | | | | | |
| Eggshell | 0.06 | 0.12 | 0.20 | | | |
| Tooth, human: | | | | | | |
| Crown | | | | | 14.6 | 0.023 |
| Dentin | | | | 6.8 | 18.2 | 0.042 |
| Nail, thumb | 0.15 | 1.8 | 0.16 | | | |
| Hair | | 19.6 | 0.40 | | | |

structural material it compares very favorably. Concrete has great compressive strength but it lacks tensile strength.

## TOPIC 10.2   Bone Fracture and Bone Design

As mentioned earlier, bone fractures are nearly always caused by excessive bending and twisting torques. The ability of bone to resist these phenomena depends on its stiffness. Channels for blood vessels and cavities for cells are in the hollow part of a limb bone. It has been suggested that their presence might help prevent any cracks from spreading.

Living bone is not a simple material like steel because it can increase its own strength in response to prolonged mechanical stress. This is known as Wolff's law and has a very important consequence in that the healing of a fracture is actually aided if the two portions of bone which must be rejoined are held together under stress during the healing process. It would seem a little puzzling, in view of Wolff's law, that the replacement of a living bone by a metal prosthesis (e.g., for the hip joint) often leads to a progressive destruction of the bone in contact with the metal and the loosening of the replacement part. This can be explained by observing that the elastic constants of bone are very much smaller than those of metals. The same stress will cause a much larger strain in a bone than metal and consequently repeated relative movements between bone and metal. These relative movements are known as fretting.

We must know the stress inside a bone to determine when it will break because a material fractures when the internal stress is larger than the fracture stress. At the surface of a curved bone the stress due to compression, $\sigma$, is proportional to the curvature of the bone, $1/R$. If we consider the bone to be a cylinder of radius $r$, then

$$\sigma = Y \frac{r}{R} \tag{10.12}$$

The relationship between the applied torque $\tau$ and the curvature of the bone is

$$\tau = \frac{Y\pi r^4}{4}\left(\frac{1}{R}\right) \tag{10.13}$$

By eliminating the curvature $(1/R)$ between equations (10.12) and (10.13), we find

$$\sigma = \frac{4\tau}{\pi r^3} \tag{10.14}$$

Thus the torque $\tau_F$ required to fracture a bone of radius $r$ by bending is found by replacing the stress in equation (10.14) with the fracture stress $\sigma_F$, that is,

$$\tau_F = \tfrac{1}{4}(\pi r^3 \sigma_F) \tag{10.15}$$

For a tibia bone of radius $r = 1$ cm, the fracture torque is approximately 160 Nm for a fracture stress of $\sigma_F = 2 \times 10^8$ Pa (or N/m$^2$) according to equation (10.15). We see from equation (10.15) that the fracture torque increases as the third power of the bone radius, so why do we not have heavier bones to avoid fractures completely? The disadvantage, of course, of heavy bones is the increased weight. The human skeleton is a compromise among conflicting requirements, that is, stiffness, so bones can act as a scaffold for our muscles; strength, so we can survive falls; and lightness, so we can move more rapidly to obtain prey and avoid being caught ourselves. According to equation (10.15), we can only increase the fracture torque by increasing the bone's diameter. However, it is possible to increase the fracture torque by making the bone hollow. Our own bones contain bone marrow but are nevertheless still effectively hollow.

Let us consider a quadruped as having four identical cylindrical columns of length $L$ and cross-sectional area $A$ as legs. If $W$ if the total weight of the animal, then each leg supports a weight of $\tfrac{1}{4}W$. The weight of an animal depends on its size as

$$W(L) = a\rho g L^3 \tag{10.16}$$

where $a$ is a constant, $\rho$ the animal's density, and $g$ the acceleration due to gravity. The applied stress on each leg is the force $\tfrac{1}{4}W$ divided by the cross-sectional area $A$. Thus

$$\sigma = \frac{a\rho g L^3}{4A} \tag{10.17}$$

A cylindrical sample of bone will break if we compress it with an applied stress which exceeds the fracture stress of bone, $\sigma_F$, which is about $2 \times 10^8$ Pa. The smallest possible cross-sectional area $A_{\min}$ of a leg bone which still prevents fracture is found by setting $\sigma = \sigma_F$ in equation (10.17) and solving for $A$, so that

$$A_{\min} = \frac{a\rho g L^3}{4\sigma_F} \tag{10.18}$$

We need a minimal cross section since this allows the minimum amount of material to be used for the bone.

## TOPIC 10.3   Bone Fracture from a Fall

Regardless of their composition and structure, many solid objects can absorb energy and quite safely avoid any permanent deformation. For bones this becomes of paramount importance since the amount of energy given to a bone dictates the severity of the injury inflicted on the limb. As a particular example, consider a bone of the lower leg. Let us suppose that its length $L = 0.5$ m with an average cross-sectional area $A = 0.0004$ m$^2$. The elastic modulus for bone is about $Y = 10^9$ N/m$^2$ and the breaking stress for bone is of order $S_b = 10^8$ N/m$^2$. The force required to fracture the bone is

$$F = S_b A = 4 \times 10^4 \text{ N} \tag{10.19}$$

The amount of energy capable of being stored within the bone is

$$E = \frac{1}{2} \frac{ALS_b^2}{Y} = 1 \times 10^3 \text{ J} \tag{10.20}$$

Assuming that the person lands on both legs, we must double the energy, giving

$$E = 2000 \text{ J} \tag{10.21}$$

Suppose we want to find out if this energy corresponds to the potential energy of a person who has fallen through a given height $h$. If the mass of the person is 70 kg, then

$$\text{Potential energy} = mgh \tag{10.22}$$

When we solve for $h$, we find

$$h = 2.9 \text{ m} \tag{10.23}$$

There are a number of other factors we should take into account, such as surface texture and weather conditions, because they would affect the energy transferred to the leg bones and hence alter the likelihood of fracture on impact.

## TOPIC 10.4   Stresses in the Leg during Movement

As a result of everyday activity, many bones in the human skeleton are continually subjected to stresses. Some bones receive more stress than others. As an example, suppose an 80-kg person is standing on the ball of one foot and we wish to calculate the stress exerted on the tibia in the lower leg. Clearly the weight of the person, that is, $W = 785$ N, acts downward on the leg and the Achilles tendon also pulls downward. We suppose this latter force to be 1400 N. The combined downward force is therefore $F = 2185$ N. An equal and opposite compressional force is exerted on the tibia. If the cross-sectional area $A$ of the tibia is 0.0005 m$^2$, the stress on the tibia, $S$, will be

$$S = \frac{F}{A} = \frac{2184 \text{ N}}{5 \times 10^{-4} \text{ m}^2} = 4.37 \times 10^6 \text{ N/m}^2 \tag{10.24}$$

## TOPIC 10.5   Physics of Karate: Breaking Wooden Blocks with Bare Hands

Karate was developed by the inhabitants of the island of Okinawa when, in the early seventeenth century, it was invaded by the Japanese, who confiscated all their weapons. An especially impressive application of the karate technique of focusing sheer physical strength is the ability to fracture a slab of material such as wood or brick using bare hands or feet. It is instructive to compare the various elastic parameters of wood, concrete, and bone. For example, the Young modulus in units of $10^8$ N/m² is (a) 1.4 ± 0.5 for wood, (b) 28 ± 9 for concrete, and (c) 180 for bone! The modulus of rupture $\sigma_0$ is in units of $10^6$ N/m²: (a) 3.6 ± 1.0 for wood, (b) 4.5 ± 0.5 for concrete, and (c) 210 for bone. The force required to break a block of wood with dimensions 28 × 15 × 1.9 cm³ is approximately 700 N, while for a block of concrete measuring 0.40 × 0.19 × 0.04 m³ it is roughly 3000 N. The force produced by a karate expert is given by

$$F = \frac{mv}{\Delta t} \tag{10.25}$$

where $m$ is the mass of the fist (approximately 0.7 kg), $v$ is the maximum speed attained by exerting a blow (10 m/s or more), and $\Delta t$ is the interaction time between the fist and the block (approximately 5 ms). Thus, the forces generated reach or even exceed 1400 N, sufficient to break wooden blocks. To break concrete blocks, one needs to reduce the contact time to 3 ms and increase the speed to 14 m/s.

## TOPIC 10.6   Elasticity of Ligaments

Many structural components of biological systems are assembled from a variety of substances. This can give rise to some rather unusual stress–strain diagrams. Consider the ligamentum nuchae, which is a very heavy "cable" along the top of the neck of cows. It must be strong enough to hold a very heavy head while grazing. It should yield readily at first if subjected to a sudden stress but should "tighten up" before the strain becomes excessive. It must be a biological shock absorber. The actual stress–strain diagram shows these desirable properties (see Figure 10.3, solid line). The two major structural components are the proteins elastin and collagen. If the collagen is removed in vitro, by the action of the enzyme collagenase, the resultant stress–strain relationship indicates that the elastin is principally responsible for the initial high-yield response of the intact ligament. On the other hand, removal of elastin by elastase leaves collagen, which is seen to be responsible for the final portion of the response of the ligament. Obviously the collagen has a greater Young's modulus than the elastin.

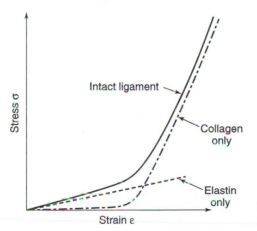

**FIGURE 10.3**  Stress–strain diagram for ligamentum nuchae.

## TOPIC 10.7   The Resilience of Tendons

For a tendon to be designed well, we require the material of the tendon to store a large amount of energy which is released when the applied force is removed. We call such materials resilient (e.g., tendons, rubber). In Figure 10.4 we have plotted a stress–strain curve for a piece of standard tendon and a piece of arterial wall. Clearly, neither obeys Hooke's law. The area under the stress–strain curve for the tendon is much larger than for the arterial wall material. The tensile strengths of the two materials are similar but the tendon has a greater extensibility. When you walk, elastic energy is stored in tendons but arterial tissue is not used for storing energy.

## TOPIC 10.8   Elasticity of Lungs

The concepts of elasticity can be applied to such organs as the lungs. If the lung of a cadaver is inflated with an excess pressure above that of the atmosphere, we can measure the lung volume $V$ as a function of lung pressure $P$ by determining how much air flows into the lung. The picture that emerges is displayed in Figure 10.5 in which the intercept $V_0$ is the volume corresponding to the uninflated lung. We see that for low applied pressures there is a regime in which the relationship between $V$ and $P$ is linear, reminiscent of Hooke's law. The slope of the curve in Figure 10.5 is called the compliance of the lung. High compliance means that we can inflate the lung easily, whereas low compliance at high applied pressures means it is harder to inflate the lung any further if it is already inflated.

## TOPIC 10.9   Elastic Properties of Blood Vessels

Three types of blood vessels make up the circulatory system (see Chapter 11 for more information): veins, arteries, and capillaries. Here we explore their elastic properties. Oxygen-rich blood is transported from the left side of the heart, under high pressure, via arteries and hence to a whole series of capillary vessels embedded in the organs and tissues of the human body. Veins transport oxygen-poor blood away from the capillaries and back to the right side of the heart under low pressure. The structural components which make up every blood vessel are the endothelial lining, smooth muscle, elastin fibers, and collagen fibers. The endothelial lining provides the smooth surface of the inner walls of

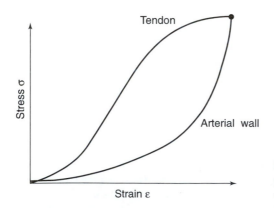

**FIGURE 10.4**   Stress–strain plot for a piece of tendon and a piece of arterial wall with the same maximum stress and strain.

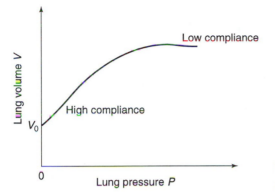

**FIGURE 10.5** Lung volume versus lung pressure.

the blood vessel and enables selective permeability to various substances being transported in the blood stream. However, this does not contribute to the elasticity of the vessel wall. The smooth muscle contracts and relaxes the vessel wall to control the volume of blood flowing through it. Its elastic modulus ranges from $6 \times 10^3$ to $6 \times 10^6$ N/m². Elastin and collagen are structural proteins that predominantly contribute to the elasticity of the blood vessel. The easily stretched elastin fibers possess an elastic modulus of $5 \times 10^5$ N/m², whereas collagen fibers are much stronger and more resistant to stretching, with an elastic modulus of $1 \times 10^7$ N/m².

## TOPIC 10.10   How Trees Bend

When a tree bends in a gust of wind, the force of gravity provides a torque which tries to bend the tree further. We say that the tree has buckled if it is unable to straighten up. We now try to determine the width of the tree required to avoid buckling. The tree will be described by a cylinder of height $H$ and radius $r$ with its base fixed (see Figure 10.6a). The bending of the tree is characterized by the curvature $1/R$, where $R$ denotes its radius of curvature. The elastic torque exerted on a cylindrical tree, $\tau_{el}$ (elastic because an elastic force tries to keep the tree upright), is given by a relationship in equation (10.13) which is written here as

$$\tau_{el} = Y \frac{\pi r^4}{4}\left(\frac{1}{R}\right)$$

(10.26)

where $Y$ is Young's modulus. The gravitational torque which tries to buckle the tree, $\tau_g$, is given by

$$\tau_g = Wd$$

(10.27)

where $d$ is the distance between the pivot and the line of action of the weight of the tree, $W$, through its center of gravity. From the geometry of the bent tree in Figure 10.6b, we see that

$$d = R[1 - \cos(\tfrac{1}{2}\theta)]$$

(10.28)

where $\theta$ is the total angle of the circular section subtended by the tree. For small bending angles $\theta$, expanding $\cos\left(\tfrac{1}{2}\theta\right)$ yields

$$d \cong R\left\{1 - \left[1 - \frac{1}{2}\left(\frac{\theta}{2}\right)^2\right]\right\} = \frac{1}{2}R\left(\frac{\theta}{2}\right)^2$$

(10.29)

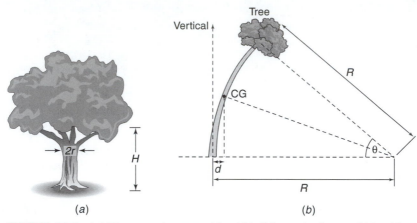

**FIGURE 10.6** (a) Diagram of a tree with width ($H$) required to avoid buckling. (b) The geometry of a bent tree.

If $H$ is the height of the tree, then

$$H = R\theta \tag{10.30}$$

so that

$$d \cong \frac{H^2}{8R} \tag{10.31}$$

The weight of the tree is

$$W = \rho g V = \rho g \pi r^2 H \tag{10.32}$$

where $V$ is the volume of the tree and $\rho$ the average density of wood.

When the magnitude of the elastic torque is greater than the magnitude of the gravitational torque, from equations (10.26), (10.31), and (10.32), we see that

$$\frac{Y\pi r^4}{4R} > \rho g r^2 H \frac{H^2}{8R} \tag{10.33}$$

As both the elastic and gravitational torque are proportional to the curvature $1/R$, from equation (10.33) we obtain

$$Yr^2 > \tfrac{1}{2}\rho g H^3 \tag{10.34}$$

We can find the limit on the maximum tree height, $H_{\text{max}}$, whereby trees taller than $H_{\text{max}}$ will buckle, by equating both sides of equation (10.34):

$$H_{\text{max}} = \left[\frac{2Yr^2}{\rho g}\right]^{1/3} \tag{10.35}$$

Thus we observe that the maximum height of the tree increases with the radius of the tree and is also increased if $g$ is reduced.

## TOPIC 10.11 The Human Leg as Physical Pendulum during Walking

We can view walking at a steady pace as a pendulum motion where the leg is the length of the pendulum and the hip joint is the pivot. Let us describe the leg by a thin rod of length $L$ rotating about an axis at one end. The moment of inertia of such an object would be

$$I = \tfrac{1}{3}mL^2 \tag{10.36}$$

where we take $m$ to be the mass of the person, say 70 kg, and the leg length to be 0.90 m. The associated time period $T$ is given by

$$T = 2\pi\sqrt{\frac{I}{mgd}} \tag{10.37}$$

where $d$ is the distance between the pivot point and the center of gravity. Here we suppose $d = \tfrac{1}{2}L$, so

$$I = \tfrac{1}{3}[(70\text{ kg})(0.90\text{ m})^2] = 18.90\text{ kg} \cdot \text{m}^2 \tag{10.38}$$

Thus the time period becomes

$$T = 2\pi\sqrt{\frac{18.90\text{ kg} \cdot \text{m}^2}{(70\text{ kg})(9.8\text{ m/s}^2)(0.90\text{ m}/2)}} = 1.55\text{ s} \tag{10.39}$$

and the corresponding frequency $f = 1/T = 0.6$ Hz.

## TOPIC 10.12   Insect Flight and Mechanism of Resonance

Insect wings oscillate up and down with a frequency in the range 100 to 1000 Hz, each wing being attached to the thorax in an elastic assembly. The whole system forms a kind of harmonic oscillator with its own frequency. For any mechanical oscillator a power drive is required to supplement the energy by dissipation. A vibrating muscle called the fibrillar flight muscle drives the wings of the insect. Interestingly, the vibrational frequency of this muscle is not fixed. When the flight muscle of a fly is stimulated with a tunable electronic oscillator, the muscle oscillates with the frequency of the oscillator over a whole range of frequencies so the flight muscle does not choose the particular wing oscillation frequency. It actually chooses its drive frequency to match the natural frequency of the whole wing–thorax system to obtain a maximum amplitude and to ensure the whole system is in resonance. This behavior is called mode locking and has a clever design strategy. To see this, note that the mass of the wings and elasticity of the thorax must change during the lifetime of the insect and the natural frequency thus changes as well. The flight muscle is designed to lock in to the natural frequency to ensure its metabolic energy is optimally used.

The mode-locking effect can be studied experimentally by cutting off a part of the wing of a fly, which reduces the mass $m$ of the air moved up and down by the insect wings. As the natural angular frequency is given by

$$\omega = \frac{2\pi}{T} = \sqrt{\frac{k}{m}} \tag{10.40}$$

this has the effect of increasing the natural frequency. The fly accommodates to this new situation by buzzing at a higher frequency so the fibrillar muscle has adjusted itself to this new natural frequency. This particular mechanism makes sure the only work done by the flight muscle is the minimum to make up for energy dissipation. It makes sure the fly does not have to do work to accelerate and decelerate the wings every cycle of the oscillation, and permits very high frequencies—such resonance systems are not used in the flight of birds or bats.

## SOLVED PROBLEMS

**10.1** A strip of tissue 5 cm long with a cross-sectional area $A$ of 0.10 cm² is cut from the wall of the aorta. Such material has a Young's modulus of approximately $2 \times 10^5$ N/m². What mass suspended from the strip hung vertically will cause a 0.5-cm elongation if $\Delta A$ can be assumed to be negligible?

**Solution** The stress is given by

$$\sigma = \frac{F}{A} = \frac{mg}{10^{-5}} \text{ N/m}^2$$

where $mg$ is the weight of the suspended mass. We then use

$$\varepsilon = \frac{\Delta \ell}{\ell} = \frac{0.5}{5} = 0.1$$

From Hooke's law

$$\sigma = Y\varepsilon$$

thus

$$m = 0.02 \text{ kg}$$

**10.2** Show that the maximum work which can be done by an animal in a single slow contraction of its muscle is proportional to the animal's total mass. Assume that the animal's muscle volume is roughly proportional to its total volume.

**Solution** Let $F$ be the force applied by the muscle and $\Delta L$ the distance over which the muscle contracts. The work done by the muscle is then

$$W = F\Delta L$$

For slow contraction, the maximum muscle force is given approximately by

$$F_{\text{max}} = \sigma A$$

with $A$ the cross-sectional area of the muscle and $\sigma$ a constant. Therefore, the maximum work that can be done is proportional to the volume $A\,\Delta L$ of the muscle, which is assumed proportional to the animal's total volume. Since all organisms have approximately equal densities, it follows that the maximum work done by the muscle is proportional to the animal's total mass.

**10.3** A sample of collagen fiber with a cross-sectional area of $1.2 \times 10^{-7}$ m² is hung from a fixed support, and a gradually increasing load is applied at the lower end. Estimate the load in kilograms that can be applied before the fiber breaks. The tensile strength of collagen is $0.6 \times 10^8$ N/m².

**Solution** Since the cross-sectional area of the fiber is $1.2 \times 10^{-7}$ m², this corresponds to a breaking force of $0.6 \times 10^8$ N/m² $\times 1.2 \cdot 10^{-7}$ m² = 7.2 N, and the corresponding mass is

$$m = \frac{7.2 \text{ N}}{9.8 \text{ m/s}^2} = 0.73 \text{ kg}$$

**10.4**  In a certain experiment, a sample of bone in the form of a cylinder of cross-sectional area 1.5 cm² is loaded on its upper end by a mass of 10 kg. The length of the cylindrical sample is observed to decrease by 0.0065%. Estimate the value of Young's modulus for the specimen.

**Solution**  A decrease in length of 0.0065% is a linear compressive strain of $6.5 \times 10^{-5}$. The compressive stress exerted on the sample is

$$\sigma = \frac{mg}{A} = \frac{10 \text{ kg} \times 9.8 \text{ m/s}^2}{1.5 \times 10^{-4} \text{ m}^2} = 6.53 \times 10^5 \text{ N/m}^2$$

Hence Young's modulus is

$$Y = \frac{\text{stress}}{\text{strain}} = \frac{6.53 \times 10^5 \text{ N/m}^2}{6.5 \times 10^{-5}} = 1.0 \times 10^{10} \text{ N/m}^2$$

**10.5**  Assume that Gulliver was 12 times as tall as the Lilliputians and therefore 12 times as wide and 12 times as deep. What would have to be the radius of his bones to sustain the weight?

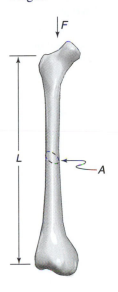

**Solution**  The maximum allowed stress $\sigma_{max}$ is a characteristic of the bone material, and we can assume it is the same for all animals and humans. We assume that the applied force on the leg bone is proportional to the weight of the person so that $\sigma_{max}$ can be written as

$$\sigma_{max} = \frac{mg}{A} = \frac{V}{A}$$

The volume of a person is proportional to $L^3$ and the cross-sectional area of the bone is proportional to $r^2$. So, if $\sigma_{max}$ is the same for both Gulliver and a Lilliputian soldier, we have

$$\frac{L_1^3}{r_1^2} = \frac{L_2^3}{r_2^2}$$

or

$$\frac{r_1}{r_2} = \left(\frac{L_1}{L_2}\right)^{3/2}$$

So, if Gulliver is 12 times as "large," the radius of his leg bones should be $12^{3/2} = 41.5$ times as big.

**10.6**  In a circus act, a performer supports the combined weight (1640 N) of a number of his colleagues. Each thighbone (femur) of this performer has a length of 0.55 m and an effective cross-sectional area of $7.7 \times 10^{-4}$ m². Determine the amount by which each thighbone compresses under the extra weight.

**Solution**  The amount of compression $\Delta L$ of each thighbone is

$$\Delta L = \frac{FL_0}{YA} = \frac{(820 \text{ N})(0.55 \text{ m})}{(9.4 \times 10^9 \text{ N/m}^2)(7.7 \times 10^{-4} \text{ m}^2)} = 6.2 \times 10^{-5} \text{ m}$$

This is a very small change, the fractional decrease being $\Delta L/L_0 = 0.00011$.

**10.7**  A sample of muscle in a relaxed state elongates 4.5 cm when subjected to a force of 40 N. The same muscle sample under conditions of maximum tension requires a force of 675 N to achieve the same elongation. Assuming that the muscle sample can be approximated as a uniform cylinder of 0.35 m length and 80 cm² cross-sectional area, determine the stress, the strain, and Young's modulus for the muscle sample in both cases.

**Solution**  For the muscle in the relaxed state

$$\text{Stress} = \frac{F}{A} = \frac{40 \text{ N}}{80 \times 10^{-4} \text{ m}^2} = 5.0 \times 10^3 \text{ N/m}^2$$

$$\text{Strain} = \frac{\Delta L}{L} = \frac{0.045 \text{ m}}{0.35 \text{ m}} = 0.13$$

$$Y = \frac{\text{stress}}{\text{strain}} = \frac{5000 \text{ N/m}^2}{0.13} = 3.85 \times 10^4 \text{ N/m}^2$$

For the muscle under maximum tension

$$\text{Stress} = \frac{F}{A} = \frac{675 \text{ N}}{80 \times 10^{-4} \text{ m}^2} = 8.4 \times 10^4 \text{ N/m}^2$$

$$\text{Strain} = \frac{\Delta L}{L} = \frac{0.045 \text{ m}}{0.35 \text{ m}} = 0.13$$

$$Y = \frac{\text{stress}}{\text{strain}} = \frac{8.4 \times 10^4 \text{ N/m}^2}{0.13} = 6.46 \times 10^5 \text{ N/m}^2$$

**10.8**  What is the compression in the tibia of length $\ell = 30$ cm of a person of mass m = 80 kg standing erect? The bone is regarded as a hollow circular tube with internal diameter $d_i = 2.4$ cm and external diameter $d_e = 3.5$ cm, and Young's modulus along the axis is $Y = 2 \times 10^{10}$ N/m². What are the strain and compression at the point of fracture? The maximum tensile strength $\sigma_f = 1.4 \times 10^8$ N/m².

**Solution**  The person's weight, $mg = 785$ N, is supported by two legs, and thus the load on one bone is $F = \frac{1}{2}mg = 392$ N. The cross-sectional area of the tibia is $A = \pi(r_e^2 - r_i^2) = 5.09$ cm². When the person is standing erect, the stress in the leg bone is elastic and obeys Hooke's law: $\sigma = Y\varepsilon$; that is, within the elastic limit of the bone, the stress is proportional to the strain. Substituting the numerical values, we obtain $\varepsilon = 4.2 \times 10^{-5}$, which means that the length decreases by 0.004%. Taking into account the fact that the bone length $\ell = 30$ cm, the compression under the weight of the person is equal to $\Delta\ell = \varepsilon\ell = 1.3 \times 10^{-3}$ cm. If we assume that the leg bone remains elastic under compression until fracture, then the ultimate strain will be $\varepsilon_f = \sigma_f/Y = 0.7\%$. The compression of the bone just before the fracture is $\Delta\ell_f = \varepsilon_f\ell = 2.1$ mm.

**10.9** The cylindrical cell of *Nitella* is $L = 10$ cm long, the radius $r = 0.5$ mm, and the thickness of the cell wall is $h = 5$ $\mu$m. What are the longitudinal ($\sigma_L$) and the tangential ($\sigma_T$) stresses in the cell wall due to an internal excess pressure $P = 600$ kPa? What is the relative extension of the linear dimensions of the cell? Young's modulus for the cell wall is $Y = 7 \times 10^8$ N/m$^2$.

**Solution**  The geometric data on the cylindrical cell, together with the directions of the longitudinal and the tangential tensions in the cell wall and the internal excess pressure are indicated in the accompanying figure.

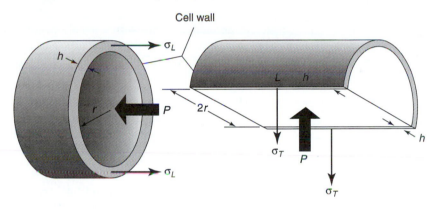

Cell wall

The area of the circular end plate of the cylinder is $\pi r^2$, and the pressing force attributed to the internal excess pressure is $P\pi r^2$. This force is compensated by the longitudinal component of the tension:

$$Pr^2\pi = \sigma_L(2\pi r)h$$

The longitudinal stress can be expressed as

$$\sigma_L = \frac{rP}{2h}$$

Substituting the numerical values, we obtain $\sigma_L = 3 \times 10^7$ N/m$^2$. Whereas the longitudinal tension rules out the deformation of the cell along its longitudinal axis, the tangential tension acts against the radial expansion of the cell. The tangential tension can be calculated from the conditions of statics applied to one-half of the cell. The longitudinal cross-sectional area of the inside of the cell is $2(r - h)L$, which is exposed to the internal pressure, yielding a force of $2P(r - h)L$. This force is counterbalanced by the tangential force, which can be determined from the product of the cross-sectional area of the cell wall, $2hL$, and the tangential stress, $\sigma_T 2hL$. From this condition

$$\sigma_T = P\left(\frac{r}{h} - 1\right)$$

Numerically, we obtain $\sigma_T = 6 \times 10^7$ N/m$^2$. This is twice as large as the longitudinal stress. Within the limit of elasticity, Hooke's law relates the longitudinal stress and strain:

$$\frac{\Delta L}{L} = \frac{\sigma_L}{Y}$$

Substitution of the numerical data yields $\Delta L/L = 0.04 = 4\%$.

# EXERCISES

**10.1**  A tree is approximated as a 60-m-tall cylinder with a diameter of 3 m.
(a) What is the weight of the tree?
(b) Calculate the amount of compression of the tree due to gravity.

**10.2**  A force of 60 pN is required to stretch a sample of DNA of length 45 nm and Young's modulus $Y = 1 \times 10^8$ Nm$^{-2}$ by 12% of its original length. Determine the cross-sectional area of the DNA molecule.

**10.3**  An aneurysm can be approximated as an elastic sphere with a small opening by which blood circulates and exerts pressure against the inner wall. Determine the force, in newtons, exerted by the blood on an aneurysm, given a blood pressure of 150 mm Hg and a cross-sectional area of 0.0025 m$^2$.

**10.4**  Assume that the human femur is a cylinder of inner radius 0.0250 m and of outer radius 0.0350 m. With no weight on the leg, the length of the femur is 0.250 m. How much shorter is the femur in a person of mass 72 kg when the mass is distributed evenly on both legs? Assume $Y = 16.0 \times 10^9$ N/m$^2$. The tensile stress at fracture is $12.0 \times 10^7$ Nm$^2$ in bone. Calculate the force in newtons required to pull the femur apart.

**10.5**  Between each pair of vertebrae of the spine is a disc of cartilage of thickness 0.5 cm. Assume the disc has a radius of 0.04 m. The shear modulus of cartilage is $10^7$ Nm$^{-2}$. A shearing force of 10 N is applied to one end of the disc while the other end is held fixed. What is the resulting shear strain? How far has one end of the disc moved with respect to the other end?

**10.6**  A mass is hung on a tendon of length 0.20 m and cross-sectional area $2.0 \times 10^{-6}$ m$^2$. The tendon stretches 1 cm. If Young's modulus for the tendon is $2 \times 10^7$ Nm$^{-2}$, what mass was hung on the tendon?

**10.7**  Consider two identical cylindrical bones of individual length $L$, cross-sectional area $S$, and Young's modulus $Y$. Each bone obeys Hooke's law.
(a) Connect one end of the first bone to one end of the second bone in a series arrangement. Find the effective spring constant of the connected bones in terms of $L$, $S$, and $Y$.
(b) Suppose you can glue the two bones next to each other in a parallel arrangement. Find the effective spring constant of the combined bones in terms of $L$, $S$, and $Y$.

**10.8**  The cross-sectional area of the tendon is $90 \times 10^{-6}$ m$^2$, and its length is 0.25 m. When subjected to a force of 7000 N, it is stretched by 3 mm. From this information, calculate Young's modulus of the tendon.

**10.9**  A 60-$\mu$m strand of DNA is stretched by an optical trap from equilibrium until its ends are separated by an additional 25 $\mu$m. Assume that the spring constant $k$ is approximately $10^{-10}$ N/m.

**(a)** What is the work done by the trap on the DNA strand?

**(b)** How much force must the trap exert on the DNA strand to maintain the 20-$\mu$m spacing?

**10.10** **(a)** Give an argument why, for a given curvature, the stress should not depend on the length of the bone.

**(b)** From the equation

$$\sigma = Y\frac{r}{R}$$

estimate the strain inside a bone with radius $r$ and curvature $1/R$.

**10.11** A vertebra experiences a shearing force $F_s = 600$ N. What is the magnitude of shear deformation if the vertebra is a cylinder 0.03 m high with 0.02 m radius?

**10.12** For the human tibia bone the ultimate tensile strength is $1.4 \times 10^8$ N/m$^2$. What is the strain at the point of fracture? Assume that Hooke's law is valid up to the point of fracture.

# CHAPTER 11

# FLUIDS

## PHYSICAL BACKGROUND

The mass density $\rho$ of any substance is its mass divided by its volume $V$:

$$\rho = \frac{m}{V} \tag{11.1}$$

Table 11.1 lists the density, molecular weight, and concentration in blood plasma of the most important proteins (see Chapter 14 for the definition of molecular weight).

Fluids are materials that can flow. In the presence of gravity, the upper layers of a fluid push downward on the layers beneath, with the result that fluid pressure is related to depth. In an incompressible static fluid whose density is $\rho$,

$$P_2 = P_1 + \rho g h \tag{11.2}$$

where $P_1$ is the pressure at one level and $P_2$ is the pressure at a level that is $h$ meters deeper.

According to Pascal's principle, any change in the pressure applied to a completely enclosed fluid is transmitted undiminished to all parts of the fluid and the enclosing walls.

The buoyant force is the net upward force that a fluid applies to any object that is immersed partially or completely in it. Archimedes' principle states that the magnitude of the buoyant force equals the weight of the fluid that the immersed object displaces.

The equation of continuity states that (a) the total volume of an incompressible fluid, that is, fluid that maintains constant density regardless of changes in pressure and temperature, entering a tube will be equal to that exiting the tube and (b) flow measured at one point along the tube will be equal to the flow at another point along the tube, regardless of the cross-sectional area of the tube at each point. This can be expressed as

$$Q = A_1 v_1 = A_2 v_2 = \text{const} \tag{11.3}$$

where $Q$ is the fluid flow rate. Table 11.2 summarizes data for fluid flows in various organs of the body.

Bernoulli's principle, which is the fluid equivalent of conservation of energy, states that the energy of fluid flow through a rigid vessel by a pressure gradient is equal to the sum of the kinetic, gravitational potential, and pressure energy densities:

$$E_{\text{tot}} = P + \tfrac{1}{2}\rho v^2 + \rho g h = \text{const} \tag{11.4}$$

Torricelli's theorem is a special case of Bernoulli's principle and describes the speed of a liquid flowing from an opening in a tank filled with liquid to a height $h$. The outward velocity $v$ of a liquid from an opening a distance $h$ from the surface level of the liquid is given by

$$v = \sqrt{2gh} \tag{11.5}$$

The coefficient of viscosity, $\eta$, is the proportionality constant which determines how much tangential force is required to move a fluid layer at a constant speed $v$ past an im-

**TABLE 11.1  Physical Characteristics of Selected Blood Proteins**

| Protein in plasma | Molecular weight (kDa) | Density ($10^3$ kg/m³) | Concentration (mg/mL) |
|---|---|---|---|
| Immunoglobulin M | 1000 | 1.38 | 0.8–0.9 |
| $\alpha_2$-Macroglobulin | 820 | 1.36 | 2.65 (men), 3.35 (women) |
| High-density lipoprotein | 435 | 0.91 | 0.37–1.17 |
| Factor I (fibrinogen) | 341 | 1.38 | 2–6 |
| Factor XIII (fibrinase) | 320 | 1.37 | — |
| Immunoglobulin G | 153 | 1.35 | 12–18 |
| Ceruloplasmin | 134 | 1.40 | 0.27–0.39 |
| Plasminogen | 81 | 1.40 | 0.48 |
| Lactoferrin | 77 | 1.40 | 0.004 |
| Transferrin | 76 | 1.38 | 2–4 |
| Factor II (prothrombin) | 70.2 | 1.39 | 0.11–0.23 |
| Serum albumin | 69 | 1.36 | 35–45 |
| Transcortin | 51.7 | 1.41 | 0.041 |
| Orosomucoid | 44.1 | 1.48 | 0.75–1.0 |
| Retinol-binding protein | 21 | 1.39 | 0.04–0.06 |
| Lysozome | 15 | 1.39 | 0.005 |
| $\beta_2$-Microglobulin | 11.5 | 1.38 | 0.0013 |

mobile surface. When the layer has an area $A$ and is located a perpendicular distance $y$ from the surface, the magnitude of the tangential force is

$$F = \eta \frac{Av}{y} \tag{11.6}$$

Viscosity has SI units of pascals times seconds. A fluid with viscosity $\eta$ which flows through a pipe of radius $R$ and length $L$ has a volume flow rate $Q$ given by

$$Q = \frac{\pi R^4 (P_2 - P_1)}{8\eta L} \tag{11.7}$$

where $P_1$ and $P_2$ are the pressures at the ends of the pipe. Equation (11.7) is known as Poiseuille's law. Applying a greater and greater pressure difference to a tube through which

**TABLE 11.2  Current Flow in Different Organs of the Body**

| Organ | Current $Q$ (L/min) |
|---|---|
| Kidney | 1.2 |
| Liver | 1.4 |
| Heart | 0.25 |
| Brain | 0.75 |
| Skin | 0.2 |
| Muscles | 0.9 |
| Other organs | 0.9 |

a fluid is flowing, the flow will increase linearly, in accordance with Poiseuille's law, but only up to a certain point, beyond which it takes significantly greater increments of pressure to produce flow increases. Beyond the Poiseuille law region, the flow will be noisy and the tube may even vibrate. Ink injected would be observed to swirl around rather than advance in streamlines. The flow is said to be turbulent.

The Reynolds number, Re, characterizes fluid flows and is defined by

$$Re = \frac{\rho v D}{\eta} \tag{11.8}$$

where $D$ is the diameter of the tube. Large values of Re correspond to turbulent flow, small ones to laminar flow.

Surface tension, $\gamma$, is the tension, or force per unit length, created by the cohesive forces of molecules on the surface of a liquid acting toward the interior (see Figure 11.1). Surface tension is given as force per unit length and defined as the ratio of the surface force to the length $d$ along which the force acts:

$$\gamma = \frac{F}{d} \tag{11.9}$$

Surface tension is given in units of newtons per meter. See Table 11.3 for related values of $\gamma$.

Capillary action refers to the rise or fall of a liquid in a narrow tube or capillary, owing to the cohesion of the liquid to itself and its adhesion to the walls of the tube, causing the formation of a curved surface or the meniscus, at the wall of the tube, with the height $h$ given by

$$h = \frac{2\gamma \cos \theta}{r \rho g} \tag{11.10}$$

where $\gamma$ is the surface tension, $\theta$ is the angle of the contact point between the capillary wall and the tangent to the liquid surface, and $r$ is the radius of the capillary tube (see Figure 11.2). For water at room temperature one finds that

$$\rho = 1.00 \times 10^3 \text{ kg/m}^3 \qquad \eta = 0.0010 \text{ Ns/m}^2 \qquad \gamma = 7.0 \times 10^{-2} \text{ N/m} \tag{11.11}$$

The potential energy $U$ of a fluid surface is proportional to the area of the surface, $A$, with the surface tension $\gamma$ as the proportionality constant:

$$U = \gamma A \tag{11.12}$$

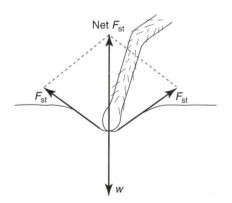

**FIGURE 11.1**  Surface tension supporting the weight of an insect leg.

**TABLE 11.3  Surface Tension of Some Liquids**

| Liquid | Surface tension $\gamma$ (N/m) |
| --- | --- |
| Blood plasma | 0.073 |
| Blood, whole | 0.058 |
| Ethyl alcohol | 0.023 |
| Mercury | 0.4355 |
| Tissue fluids | 0.050 |
| Water | |
|   At 0°C | 0.076 |
|   At 20°C | 0.072 |
|   At 100°C | 0.059 |
| Soapy water (surfactant) | 0.037 |
| Lung surfactant | 0.001 |
| Benzene | 0.0289 |

Laplace's law describes the relation between the circumferential tension $\gamma$ and the radius $r$ for any curved elastic surface. In effect, it is based on Newton's third law and equates a force $F_P$ produced by the transmural fluid (e.g. blood) pressure over the cross-sectional area of the structure to a circumferential force (tension) $T$ that compensates for the distension and is required to maintain equilibrium,

$$\Delta P = \frac{2\gamma}{r} \tag{11.13}$$

where $\Delta P$ is the difference in pressure between the inside and the outside of a bubble.

## TOPIC 11.1  Capillary Rise in Plants

Capillarity plays an important role in the transport of water in plants, but can it explain the rise of water in trees? Taking the radius of the xylem (vascular system of a plant or tree) as approximately 20 $\mu$m and a contact angle $\theta = 0°$ with the surface tension of water $\gamma = 73 \times 10^{-3}$ N/m, by equating the vertical component of the force of tension,

$$F_T = 2\pi r \gamma \cos \theta \tag{11.14}$$

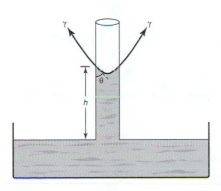

**FIGURE 11.2**  Capillary action and the meniscus in a narrow cylindrical tube.

to the force of gravity pulling down a cylinder of water in a tube (xylem),

$$F_g = \pi r^2 \rho g y \qquad (11.15)$$

where $\rho$ is the density of water, we find that the maximum height $y$ is

$$y = \frac{2\gamma \cos \theta}{\rho g r} \qquad (11.16)$$

which upon substitution of the numbers given above results in $y \cong 0.75$ m, not sufficient to explain the height of trees. How can this be explained? Two main factors should be included in this discussion: (a) the pores through which water rises may become much smaller than the estimate of 20 $\mu$m given above and (b) the tensile stress of a column of water or tree sap is not the same as that of free water.

To explain this paradox, we explore the phenomena of the transport of water in plants below. Plants generate their own food from water containing nutrients and air. Inside they have sophisticated pipelines taking water from the soil to the cells to be used and then excess water is emitted through their leaves. In the reverse direction the leaves take in carbon dioxide and produce food by photosynthesis. Figure 11.3 shows a cross section through a tree. Water and mineral nutrients move upward through the xylem and food moves downward through the phloem. The spongy pithy areas store food and water. The transport of water occurs through very thin capillary vessels within the xylem which are elongated tubules of length 100 to 500 $\mu$m joined end to end with diameters of the order of 1 to 20 $\mu$m. One finds that, to obtain a rate of flow which is known to exist in living trees, a pressure of about 0.15 atmospheres per meter (atm/m) is required. Thus for a 100-m tree we need 15 atm to overcome resistance in the capillaries plus approximately 10 atm due to the height of the tree, that is, amounting to some 25 atm.

We now try to determine the physical processes which might give rise to such large pressures. Water in root hairs moves into the xylem and is pushed upward by the excess osmotic pressure (see Chapter 12). However, the maximum osmotic pressure is only about 3 atm and so cannot account for the movement of water to the top of a tall tree. Capil-

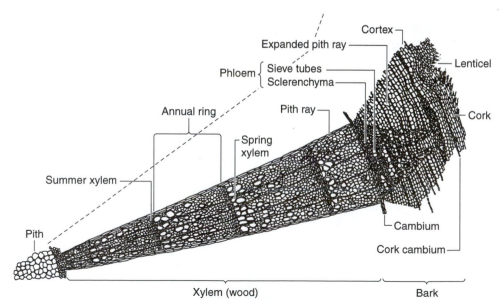

**FIGURE 11.3** Cross section of the stem of a tree.

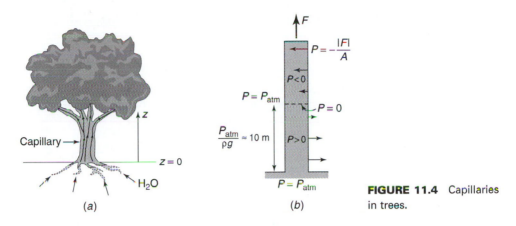

**FIGURE 11.4**  Capillaries in trees.

larity is another process which has been suggested for the movement of water in plants. Unfortunately, the diameters of the conduits in the xylem are too large. The narrower of these was found to only give a capillary rise of 3 m. Water, it has been suggested, might rise to great heights by capillarity if it was conducted through the narrow spaces between the cellulose fibers in the cell wall, much like the action of a wick.

The force $F$ (illustrated in Figure 11.4) applies a tensile stress on the column of water. Liquids, like solids, have internal cohesion trying to prevent fracture. This breakup can be expressed as a fracture stress $\sigma_f$—just as for solids. If the excess pressure $P - P_{atm}$ at the top of the capillary exceeds the fracture stress, then the capillary column will break. However, the fracture stress of water is close to that of aluminum, and if we equate the fracture stress with $\rho gh$, $h$ being the height of the tree, we find that this limits trees to heights of about 2 km. However, the maximum height of trees is not set by the fracture of the capillary column since the maximum height is determined by the buckling mechanism we have already discussed in Chapter 10.

Since water transport through xylem tissue is similar to flow through a narrow vessel, to determine the hydrostatic pressure gradient, we use Poiseuille's law. To do this, we must know the volume of fluid flowing during a given time interval, the viscosity of fluid (assumed to be water with $\eta = 10^{-3}$ Pa · s at 20°C), and the length and radius of the capillary vessels in the xylem. With the exception of fluid flow, every parameter is available. Flow volume is difficult to determine because not all the vessels measured within a given cross section of a stem contribute to water transport. Thus the problem can be circumvented by expressing Poiseuille's law in terms of velocity, which can be measured easily directly from plants. Two expressions for the volume flow rate are

$$Q = \frac{\pi}{8} \frac{\Delta P r^4}{L \eta} \tag{11.17}$$

and

$$Q = Av \tag{11.18}$$

where $v$ is the velocity of the fluid, $A = \pi r^2$ is the cross-sectional area, $L$ is the length of the vessel where the cross section has a radius $r$, and $\eta$ is the viscosity.

Solving equations (11.17) and (11.18) for $v$, we find

$$v = \frac{\Delta P}{L} \frac{r^2}{8\eta} \tag{11.19}$$

Thus we may solve for the hydrostatic pressure gradient $\Delta P$ to give

$$\Delta P = \frac{8\eta vL}{r^2} \tag{11.20}$$

As an example, suppose that a fluid with viscosity $\eta = 1.5 \times 10^{-3}$ Pa $\cdot$ s is driven at a velocity of $9.03 \times 10^{-4}$ m/s through xylem vessels of radius $2.4 \times 10^{-6}$ m and length $5.0 \times 10^{-6}$ m. Then from equation (11.20) we obtain

$$\Delta P = \frac{8 \times (1.5 \times 10^{-3} \text{ Pa} \cdot \text{s})(9.03 \times 10^{-4} \text{ m/s})(5.0 \times 10^{-6} \text{ m})}{(2.4 \times 10^{-6} \text{ m})^2} = 9.40 \text{ N/m}^2 \tag{11.21}$$

which is very small.

As the water moves through the stomata to reach the air outside, it literally pulls up the water below it, and this in turn pulls up the water below *it* until the whole column of water in the xylem is stretched like an elastic wire. The molecules of water are linked together so strongly by hydrogen bonds that pure water has a tensile strength of about $3 \times 10^7$ N/m$^2$, which is about the same as that of collagen. Water under tension has a negative gauge pressure which at the leaves of the tallest trees is 25 atm. This corresponds to a tensile stress of $2.5 \times 10^6$ N/m$^2$, which is much less than the tensile stress of either water or sap.

However, the limiting factor to the height of trees is not the tensile strength of sap, because this is large enough to support a column of sap of height 300 m or more. The height of trees is limited by the maximum pressure difference that can be maintained across the cells of the leaf by the process of transpiration, which is of the order of 30 atm.

## TOPIC 11.2   Examples of Pressure in Human Organs

Since most of the human body's composition is water, one expects fluids to play a major role in the functioning of the organism. Various types of fluid flow in separate networks can be characterized by their typical pressure values. We have compiled some of these values in Table 11.4.

Below we discuss the mechanisms involved in more detail.

### Bladder Pressure

Bladder pressure varies over quite a large range. It is zero when the bladder is empty and climbs steadily to about 25 mm Hg when the bladder reaches its normal capacity of some 500 cm$^3$. The micturition reflex is triggered by a bladder pressure of about 25 mm Hg. That reflex stimulates the feeling of needing to urinate, and it further triggers muscle contractions around the bladder that can raise bladder pressure to 110 mm Hg. Bladder pressure while urinating is normally 15 to 30 mm Hg, but an obstruction of the urinary tract, such as from a swollen prostate gland, can necessitate pressures as large as 70 mm Hg.

### Cerebrospinal Pressure

The skull and spinal column contain cerebrospinal fluid (CSF), as shown in Figure 11.5. The CSF supports the weight of the brain with buoyant force, acts as a protective cushion, and supplies nutrients filtered from the blood. Cerebrospinal fluid is generated in the skull and circulates around the brain through cavities in the brain called ventricles and down the central canal of the spinal cord. Normally, CSF is absorbed in the spinal col-

**Table 11.4  Typical Fluid Pressures in Human Organs**

| Organ | $P$ (mm Hg) |
|---|---|
| Arterial blood pressures | |
|   Maximum (systolic) | |
|     Adult | 100–140 |
|     Infant | 60–70 |
|   Minimum (diastolic) | |
|     Adult | 60–90 |
|     Infant | 30–40 |
| Venous blood pressure | |
|   Venules | 8–15 |
|   Veins | 4–8 |
|   Major veins (central venous pressure) | 4 |
| Capillary blood pressure | |
|   Arteriole end | 35 |
|   Venule end | 15 |
| Bladder | |
|   Average | 0–25 |
|   During micturition | 110 |
| Brain, lying down (cerebrospinal fluid) | 5–12 |
| Eye, aqueous humor | 12–24 |
| Gastrointestinal | 10–20 |
| Intrathoracic | −4 to −8 |
| Middle ear | <1 |

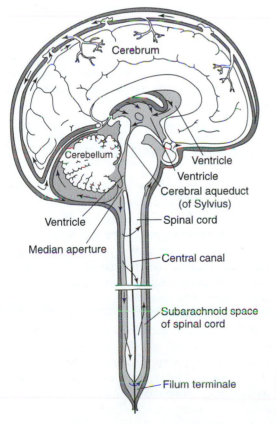

**FIGURE 11.5**  Brain, spinal cord, and CSF. Black arrows indicate the circulation of CSF.

Cerebrum

Cerebellum

Ventricle

Ventricle

Cerebral aqueduct (of Sylvius)

Spinal cord

Ventricle

Central canal

Median aperture

Subarachnoid space of spinal cord

Filum terminale

umn as fast as it is generated in the skull. The skull and spinal column contain only about 125 cm³ of CSF, so even a small loss is significant. The CSF is slowly replenished over a period of four or five days, and in the meantime lying down distributes the remaining CSF evenly.

### Pressure in the Gastrointestinal System

Food, drink, and waste products moving through the 6-m-long digestive tract or gastrointestinal (GI) system are fluid or fluidlike in character. Their flow is regulated by pressure and especially by valves and sphincter muscles in the system. As shown in Table 11.4, pressures in the GI system are usually positive. The esophagus is an exception; its pressure is directly related to thoracic (chest) cavity pressure and is negative. The stomach is elastic, so pressure in it increases gradually, becoming large only when the stomach is overfilled. The sensation of hunger occurs when stomach pressure is low.

### Pressure in the Eye

Part of the human eye is filled with a fluid called the vitreous humor which in a normal eye is continuously produced and drained from the eye. (See Chapter 26 for more information regarding the structure of the eye.) The shape of the eye is maintained by fluid pressure called the intraocular pressure; whose normal range is 12 to 24 mm Hg. The eye has two fluid-filled chambers. The front chamber contains aqueous humor and the rear chamber contains vitreous humor. About 5 cm³ of aqueous humor is produced daily in the eye. Excess fluid flows out through a canal and is absorbed into the bloodstream. Aqueous humor is similar in character to cerebrospinal fluid and carries nutrients to the lens and the cornea and the eye, neither of which has blood vessels. Pressure created by the aqueous humor is transmitted through the vitreous humor, which is a jellylike substance that does not circulate and is not replenished. If the drainage of the aqueous humor becomes obstructed, the pressure can increase to between 25 and 50 mm Hg and the clinical onset of glaucoma occurs. This additional pressure increases wall stresses in the eyeball and compresses the optic nerve, adversely affecting vision and ultimately damaging the eyeball. The treatment of glaucoma may be a surgical procedure to remove the drainage obstruction and reestablish drainage flow.

## TOPIC 11.3    The Circulation of Blood around the Body

It was Harvey, the English physician and physiologist, who established how blood circulates. The circulation supplies, via the blood, food and oxygen to tissues, carries away waste products from the cells, and distributes heat to equalize body temperature. It also carries hormones which stimulate the activity of organs, distributes antibodies to fight infection, and possesses many other functions. The circulation of the blood is therefore not only a complex process but of first importance for the good health of the body.

Blood is made up of two main components. First, its basis is the intercellular fluid. This plasma is constituted of approximately 90% water, 9% proteins, 0.9% salt, and the remaining 0.1% in the form of sugars, and traces of other components. Second, there are cells suspended in the plasma, namely white and red blood cells. The red cells are biconcave in shape and have a diameter of about 7.5 $\mu$m and a density of $5 \times 10^3$ kg/m³. The average concentration of white blood cells is approximately 8000 per mL³ and can vary between about 4500 and 11,000 per mL³. The viscosity of blood varies between 2.5 and 4.0 times that of water and its density is about 1050 kg/m³.

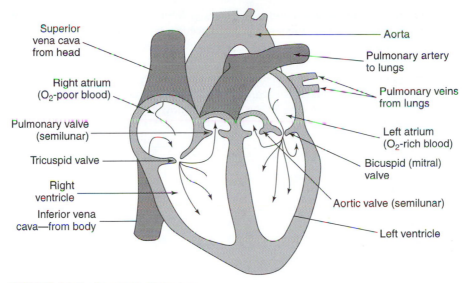

**FIGURE 11.6** Structure of the heart.

## Structure of the Heart and Its Action as a Double Pump

Each side of the heart is made up of an upper atrium and lower ventricle. Blood depleted of oxygen enters the right atrium and flows into the right ventricle, which then pumps it to the lungs. It then returns through the left atrium into the left ventricle, which pumps the now-oxygenated blood out through the aorta to the rest of the body. The heart has a system of valves to ensure the blood flows in the right direction, without backflow. The two ventricles pump at the same time, but the pressure in the right ventricle is quite low, of order 25 mm Hg, compared to the left ventricle, where the pressure can be greater than 120 mm Hg at the peak (systolic) pressure. During the resting stage of the heartbeat (diastole), the pressure is typically about 80 mm Hg. Figure 11.6 shows the structure of the heart and Figure 11.7 shows its pumping action.

The blood volume for an average person is about 5 L. At rest, the cardiac output is approximately 5 L/min. Hence the average time it takes for the entire volume of blood to

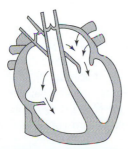

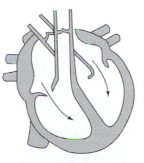

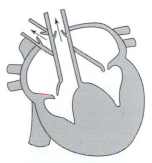

1. Blood fills both artria, some blood flows into ventricles—diastole phase of atria.

2. Atria contract, squeezing blood into ventricles— ventricular diastole.

3. Ventricles contract, squeezing blood into aorta and pulmonary arteries—ventricular asystole phase.

**FIGURE 11.7** Pumping action of the heart.

make one full circuit and return to the heart is 1 min! The mean circulation time during strenuous exercise of an athlete may be reduced to just 12 s. All of this work is done by the heart against the resistance to the viscous blood flow given by the combined network of interconnected blood vessels.

We define the power output $P$ of the heart as the work done by the heart per second in pumping the blood. It is equal to the average force $F$ exerted by the heart on the blood times the distance $d$ by which the blood moves in 1 s:

$$P = Fd \tag{11.22}$$

The force is equal to the pressure $p$ exerted by the heart on the aorta times the cross-sectional area $A$ of the aorta:

$$F = pA \tag{11.23}$$

The blood flow $Q$ is the volume of blood that passes through the aorta in 1 s, so in 1 s the volume of blood moves the distance

$$d = \frac{Q}{A} \tag{11.24}$$

In a normal adult, $Q = 0.83 \times 10^{-4}$ m$^3$/s and the average blood pressure is $p = 100$ mm Hg $= 1.3 \times 10^4$ N/m$^2$. Therefore, the power output of the heart is

$$P = Fd = pQ = 1.1 \text{ W} \tag{11.25}$$

which amounts to only 1% of the total power dissipated by the body. Furthermore, equation (11.25) shows that the heart of a person with abnormally high blood pressure does more work per second to maintain the same blood flow.

It is found by measuring oxygen consumption that a 70–kg man consumes approximately 10 W, so from equation (11.25) the heart is about 10% efficient. However, the blood pressure may increase considerably and the volume of blood pumped by as much as a factor of 5. Thus there will be an increase of 7.5 in the power generated by the left ventricle. The power required by the right ventricle, since its systolic pressure is roughly one-fifth of the left ventricle, is about a fifth of the left ventricle.

## TOPIC 11.4    More About Blood Pressure

The beginning of circulation of the blood occurs when the left ventricle ejects blood from the heart into the arterial system under a pressure of about 120 mm Hg. This is the systolic phase of the cardiac cycle. The blood pressure begins to decline gradually as it penetrates further into the system of arteries. It flows from the aorta through the larger arteries at 110 mm Hg, then through the medium-sized arteries at 75 mm Hg to the smaller arteries at 40 mm Hg. Under a pressure of 30 mm Hg it enters the capillary bed and exits under a pressure of 16 mm Hg. Under this low pressure the blood is then able to sustain the cellular metabolism and drains into the smallest veins at 16 mm Hg. It then continues through the medium-sized veins under a pressure of 12 mm Hg and thence into the large veins at 4 mm Hg. The blood then enters the heart for another cardiac cycle. Thus blood pressure ranges from 120 to 4 mm Hg and maintains this range on a continual basis.

The systolic and diastolic pressures for a normal person are about 120 and 80 mm Hg, respectively. The systemic blood pressure during a physical examination is defined by

$$\text{Systemic blood pressure} = \frac{\text{systolic blood pressure}}{\text{diastolic blood pressure}} = \frac{120}{80} \tag{11.26}$$

The pulse pressure is defined as the difference between systolic and diastolic pressures, and this is about 40 mm Hg. It is more convenient in most cases to condense these two blood pressures into one called the mean blood pressure, $BP_{mean}$, defined by

$$BP_{mean} = \tfrac{1}{3}(\text{systolic BP} + 2 \times \text{diastolic BP}) \qquad (11.27)$$

This mean pressure is the average pressure throughout the cardiac cycle and as the systole is shorter than the diastole, so the mean is slightly less than the value halfway between systolic and diastolic pressures. The mean blood pressure can only be determined precisely by integrating the area under a blood pressure–time curve.

The heart generates extra pressure when the ventricles contract. To consider this system, we assume the ventricles can be treated as hollow balls. Thus, when the ventricles contract, the tension $\gamma$ of the walls increases. The excess pressure inside the balls generated by contraction can be obtained from Laplace's law as

$$\Delta P_{ventricle} = \frac{2\gamma}{R} \qquad (11.28)$$

where $R$ is the radius of the ball describing the ventricle. When contracting, the tension and pressure increase since $R$ decreases according to equation (11.28) (systolic pressure). The pressure difference in equation (11.28) above is approximately the net overall pressure difference between the left ventricle and the right atrium and drives the flow of blood through the body. Since the heart contracts periodically, the blood pressure continues to vary periodically with time, so if we plot pressure felt by, say, a red blood cell as a function of distance from the heart, we must draw two curves. One of these corresponds to the maximum systolic pressure and the other to the minimum diastolic pressure. Clearly, the pressure gradually decreases, the largest pressure drop being in the capillaries. When the blood enters the veins, the pressure continues to drop until it re-enters the heart (see Figure 11.8).

The measurement of blood pressure uses a number of gauges, one of the simplest being the mercury-filled manometer. It is usually calibrated in millimeters of mercury. It is attached to a closed air-filled jacket which is wrapped around the upper arm at the approximate level of the heart. Two values of pressure are measured: the systolic pressure, which is the maximum pressure when the heart is pumping, and the diastolic pressure, the pressure in the resting part of the cycle. At the beginning of the procedure the air pres-

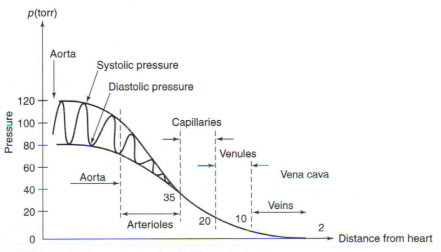

**FIGURE 11.8**   Pressure vs. distance from the heart.

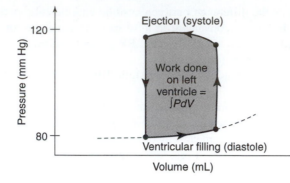

**FIGURE 11.9**   Pressure vs. volume curve and the work done on the left ventricle.

sure in the jacket is increased well above the systolic pressure using a hand pump. This compresses the brachial artery in the arm and cuts off the supply of blood. Slowly, this air pressure is reduced until the blood again begins to flow into the arm, detected using a stethoscope at the elbow, listening to one of the sharp sounds. At this point the systolic pressure is just the pressure in the air jacket and can be read off the gauge. When the pressure is further reduced, the sharp sound disappears when the blood at low pressure can enter the artery. The gauge then records the diastolic pressure. Normally the systolic pressure is about 120 mm Hg and the diastolic 80 mm Hg. In Figure 11.9 the work done by the left ventricle is shown to be the area between the systolic pressure–volume curve and the corresponding curve during the diastolic phase. This may be expressed by the integral

$$W = \int_{V_f}^{V_i} PV \, dV \qquad (11.29)$$

where $V_i$ and $V_f$ are the initial and final volumes and $V$ denotes the volume and $P$ the pressure. In a diseased heart, there is hardening of the arteries, or atherosclerosis, and calcified fatty or lipid deposits are formed on the inside walls of the aorta and major arteries. This not only reduces the lumen or inner diameter of the blood vessels but diminishes their elasticity. Their ability to expand and accommodate large volumes of blood upon ejection from the left ventricle can be severely affected. Consequently, internal regulatory mechanisms tell the heart to exert more pressure and hence perform more work upon each heartbeat. Hence, curves like that in Figure 11.9, for a patient, can yield useful information concerning the health or otherwise of the circulatory system.

As we have seen earlier, the blood pressure produced by the heart, $\Delta P$, depends on the surface tension $\gamma$ according to equation (11.28). The maximum stress produced by muscles is roughly constant. This means that the tension of the heart wall, $\gamma$, is proportional to its thickness $D$. If we treat the radius of a ventricle of the heart as isometric, then $R$ should be proportional to the body size. Suppose now that we wish to achieve the same blood pressure in two animals of different size. Since $\Delta P_{\text{ventricle}} = 2\gamma/R$, as in equation (11.28), this means that we need to increase the wall thickness $D$ in proportion to the size of the animal: that is, larger animals require larger heart muscles to produce the same blood pressure as smaller animals. This is indeed found to be the case.

## TOPIC 11.5   Cardiovascular System

We examine in this section the mammalian blood circulatory system in more detail and note that this is a good example of a biological transport system based on fluid flows. Another is air circulation through the lungs, which will be discussed later. In each there are a number of similarities. For example, each possesses wide pipes which rapidly transport large volumes of liquid or gas. These divide again and again into narrower and narrower pipes with slower and slower flow velocities. The pipe walls become so very thin that molecular exchange can take place across the walls.

The cardiovascular system is made up of the heart, the arteries, and the veins. The arteries carry blood to the organs, muscles, and skin of the body and the veins return it. The oxygenated blood finally reaches the very small capillaries, which are so small that single red cells can only pass through one at a time.

We have already seen when we discussed the structure of the heart that it is essentially two pumps connected by elastic pipes. The pressure generated in the heart forces the blood into the arteries. The right side pumps oxygen-deficient blood to the lungs via the pulmonary trunk. It then passes through pulmonary veins to the left side of the heart, which then pumps it through the aorta into the arteries, which branch first into arterioles and then into the very narrow capillaries where the blood gives up its oxygen. Most capillaries are 50 to 100 times longer than their diameter, typically 1000 $\mu$m long (1.00 mm), compared to 5 to 20 $\mu$m in diameter. The flow rate of the blood changes as the blood goes through the cardiovascular system. The effective cross-sectional area increases as it goes through the regime with very small capillaries, that is, the cross-sectional area $A$, and the number of capillaries is much greater than the cross-sectional area of the aorta. The flow rate through the aorta and the capillaries must be the same, so we have

$$A_{\text{aorta}} v_{\text{aorta}} = A_{\text{cap.eff.}} v_{\text{cap}} \qquad (11.30)$$

In Figure 11.10 we give diagrammatic representations of this circulatory system for mammals and birds.

Interestingly, although the cross section for each capillary is very small, the total cross section of all the capillaries together is on the order of 400 times that of the aorta. Figure 11.11 shows the total cross-sectional area $S$ across a set of parallel branches, taking into account the cross sections separately of the aorta, the arteries, the arterioles, the capillaries, the venules, and the veins. One sees immediately that $S$ has a pronounced maximum for the cross section of capillaries, but the mean flow velocity $U$ has a pronounced minimum for the capillaries. However, the product $Q = SU$ is the same wherever you take the cross section. This is an example of the generalized continuity principle. From this principle it follows that the flow rate through the pulmonary trunk and the aorta is the same, even though the pulmonary trunk only has to transport blood to the lungs whereas the aorta provides blood for the rest of the body.

The aorta is so large ($r = 9$ mm) that a pressure difference of only 3 mm Hg is required to maintain normal blood flow through it. Thus, if the pressure of the blood is 100 mm Hg when the blood enters the aorta, it is reduced to 97 mm Hg when the blood enters the major arteries. Because these vessels have much smaller radii than the aorta, a pressure drop of 17 mm Hg is required to maintain the flow through them. Therefore, the pressure is only 85 mm Hg when the blood enters the small arteries. These latter vessels have still smaller radii, so that a pressure drop of 55 mm Hg is required to maintain blood flow through them. Finally, there is a further drop of 20 mm Hg when the blood passes through the capillaries, in which the pressure drop is less than in the small arteries even though the capillaries have much smaller radii, because the number of capillaries is very large. The blood pressure drops to only 10 mm Hg by the time it finally reaches the veins.

The Bernoulli equation

$$p + \tfrac{1}{2}\rho v^2 + y\rho g = \text{const} \qquad (11.31)$$

applies to the blood flow in the aorta with its large radius. For example, it is possible to calculate the percentage of the pressure drop in the aorta due to the presence of an ath-

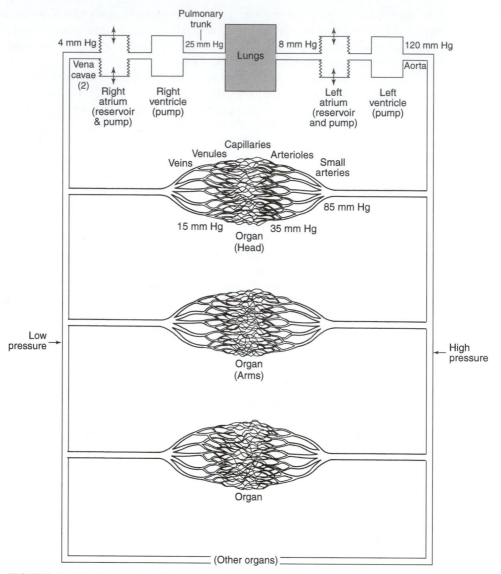

**FIGURE 11.10** Schematic of the major features of the circulatory system with representative blood pressures.

erosclerotic plaque which locally reduces the cross-sectional area to only one-fifth of the normal value. We take $P_1 = 100$ mm Hg $= 13,600$ Pa, a typical value of $v_1 = 0.12$ m/s, and $\rho \cong 100$ kg/m³ and use the Bernoulli equation to obtain

$$\Delta P = P_1 - P_2 = \tfrac{1}{2}\rho(v_2^2 - v_1^2) \tag{11.32}$$

with the continuity equation $A_1 v_1 = A_2 v_2$ and the assumption that $y_1 = y_2$ and we find that $\Delta P/P = 1.2\%$ decrease in the constricted area. However, the typical diameter of a small artery is only 25 $\mu$m with a blood velocity of approximately $2.8 \times 10^{-3}$ m/s. Therefore, to describe the blood flow in the aorta, one needs to consider the Poiseuille equation where the flow rate $Q$ is given by

$$Q = \frac{\pi r^4 \Delta P}{8\eta\ell} \tag{11.33}$$

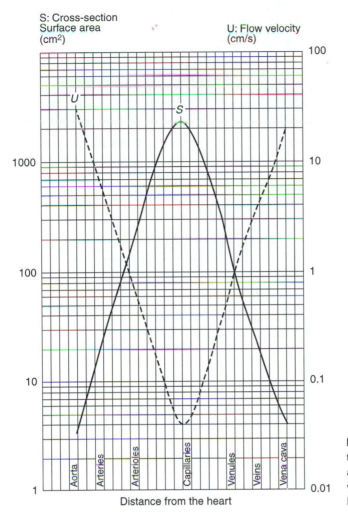

S: Cross-section
Surface area
(cm²)

U: Flow velocity
(cm/s)

**FIGURE 11.11** Plot of the total cross-sectional area and the mean flow velocity versus distance from the heart.

where $\eta = 4.5 \times 10^{-3}$ Ns/m² is the typical viscosity of blood. Hence, it is easy to calculate the pressure drop $\Delta P$ in a small artery whose length is $\ell = 5 \times 10^{-3}$ m as $\Delta P = 3200$ Pa, which, compared to the total pressure drop of 13,600 Pa, accounts for approximately 24% of the total drop in circulation.

In the larger arteries and the thoracic veins, the beating action of the heart creates a pulsatile flow. The blood still advances in streamlines, but the shape of the velocity profile is not parabolic but rather square across most of the area. If the velocity profile is examined over several cycles, one gets the impression of a slug of fluid which advances four steps and then backs up one and repeats this over and over. As we proceed away from the heart, the pulses are gradually damped out due to several factors, including the viscosity of the blood and the extensibility of the blood vessel. Many capillaries have a diameter of only 5 or 6 $\mu$m, whereas the erythrocytes have a diameter of about 8 $\mu$m. Consequently, the erythrocyte must deform to pass through the capillary. The cells occupy the entire cross section of the interior of the vessel and pass through as a series of moving plugs with short sections of plasma trapped between them. This is known as bolus flow and is represented schematically in Figure 11.12. The flow in the trapped plasma sections becomes a specialized form of streamline flow. Rapidly moving streamlines down the center catch up with an erythrocyte and are deflected to the outside where their velocity, relative

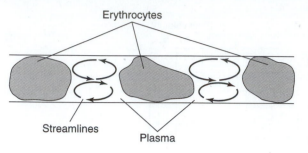

Erythrocytes

Streamlines

Plasma

**FIGURE 11.12**   Bolus flow.

to the wall, becomes essentially zero. The following erythrocyte picks up these layers and forces them once more down the center. This results in the plasma being well stirred up.

The decrease in velocity going from the aorta to the capillaries is not caused by resistance to flow. Resistance causes pressure to drop but does not affect velocity. This can be seen from the fact that blood velocity increases again in the veins while pressure continues to drop in the veins leading back to the heart.

### Blood Flow Waveforms, Velocity Profiles, and Flow in Curved Vessels

The heart sends blood into the circulatory system in a periodic manner, and a complete cycle lasts approximately 1.1 s. The oscillatory motion of the blood may be represented by plotting vertically the velocity of the blood flow against time (see Figure 11.13). We see that initially there is an impulsive rise in blood flow velocity from the heart, corresponding to its amplitude at the onset of systole, followed by a gradual decrease in velocity while the heart is filling for its rest heartbeat.

When flowing through a rigid vessel, blood flow exhibits a characteristic parabolic profile along the cross section of the vessel, where

$$v = \frac{\Delta P}{4L\eta}(R^2 - r^2) \tag{11.34}$$

where $v$ denotes the fluid velocity, $\Delta P$ the pressure gradient applied to both ends of the vessel to initiate blood flow, $L$ the length of the vessel, $\eta$ the viscosity of the blood, $R$ the radius of the vessel, and $r$ the distance from the center. Figure 11.14 shows this dia-

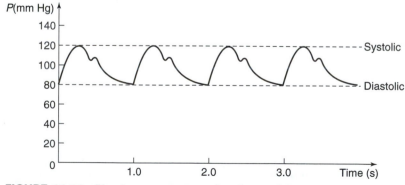

**FIGURE 11.13**   Blood pressure versus time in a major artery.

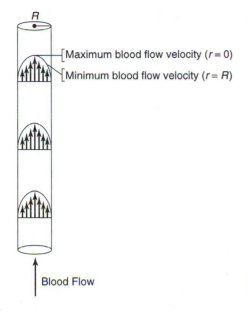

Maximum blood flow velocity ($r = 0$)

Minimum blood flow velocity ($r = R$)

Blood Flow

**FIGURE 11.14**  Velocity profile across a blood vessel's diameter.

grammatically, where it is clear that the flow velocity is largest at the center of the vessel and a minimum at the edges.

Often blood vessels are curved. A good example of this is the carotid artery, the major blood vessel supplying blood to the brain. It originates from the aorta, passes through the neck, and then bifurcates into two other major arteries which supply the front and back circulation of the brain. The points of the velocity profile in contact with the vessel wall exhibit the maximum wall stress (see Figure 11.15). As the blood flows through a curved section of a blood vessel, the velocity profile becomes bent away from the radius of vessel curvature (see Figure 11.15). A blood constituent experiences circular motion which is maintained by a centripetal or "center-seeking" force $F_C$ given by

$$F_C = \frac{mv^2}{r} \tag{11.35}$$

where $m$ is the mass of the object, $v$ its velocity, and r the radius of the circle. As $r$ decreases, the bend becomes sharper and the centripetal force increases. The centripetal force is counteracted by a centrifugal force of the same magnitude, but it is always directed outward from the center of the circle.

**FIGURE 11.15**  Blood flow in a curved vessel.

## Poiseuille's Law, Blood Flow, and Viscosity of Blood

When a pure liquid flows through a rigid tube, there is a relationship between the volumetric flow rate (or volume of liquid per unit time) $Q$, the pressure gradient $\Delta P$, the length of the tube $L$, its radius $r$, and the fluid viscosity $\eta$. This is known as Poiseuille's law and can be expressed by

$$Q = \frac{\pi \Delta P r^4}{8 \eta \ell} = \frac{\Delta P}{R_F} \tag{11.36}$$

where $R_F$ denotes the resistance to flow. However, applicability of Poiseuille's law to blood flow gives the correct sense of the dependence of resistance on radius and viscosity. That is, resistance always decreases when radius increases and increases when viscosity increases. It is not surprising that the body adjusts blood flow by changing vessel radii, since resistance to flow is more sensitive to this parameter than to any other. Most vessel dilation (vasodilation) and constriction (vasoconstriction) take place in the small arteries and arterioles, with some occurring in those capillaries that have sphincters.

## Equation of Continuity and Blood Flow

For blood to flow through an artery or a vein, an external force must be applied, that is, the pressure gradient between the ends of the vessel exists, and obviously blood flows from a high pressure to a low pressure. This motion will continue unless another force acts, for example, a vessel obstruction (stenosis or artherosclerotic blockage). The velocity of blood is much greater in the major arteries (about 0.3 m/s) than in the capillaries (about $3 \times 10^{-4}$ m/s). Surprisingly, the blood velocity increases again when the capillaries rejoin to form veins. Most changes in average blood velocity other than those produced by the heart are due to changes in the total cross-sectional area of the system during branching. For example, the aorta branches into the major arteries, each of which has a smaller cross-sectional area but whose combined area is larger than the area of the aorta. The total flow rate in the major arteries is the same as in the aorta since all the blood which passes through the aorta must also pass through the major arteries. Therefore, $Q_{\text{aorta}} = Q_{\text{major arteries}}$.

Above we have considered blood vessels to be approximated by circular cylinders, although in practice they may steadily decrease in size. The equation of continuity assures us that flow is equal at any point whatever the degree of tapering. If the cross sections and corresponding velocities at two points are respectively $A_1$, $A_2$ and $v_1$, $v_2$, then from the equation of continuity we have

$$A_1 v_1 = A_2 v_2 \quad \text{and thus} \quad v_1 = \frac{A_2}{A_1} v_2 \tag{11.37}$$

This, of course, may be generalized to several points in a vessel, as we see in Figure 11.16.

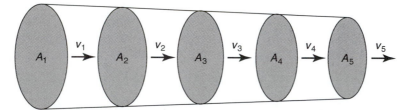

**FIGURE 11.16** Fluid flow in a tapered vessel and the equation of continuity.

Another situation which often arises is when a blood vessel bifurcates into two or more daughter vessels (labeled $d_1$ and $d_2$). Clearly, from the equation of continuity we have that the flow in the parent artery (labeled $p$) is equal to the flow in the branching daughter vessels, or

$$A_p \mathbf{v}_p = A_{d_1} \mathbf{v}_{d_1} + A_{d_2} \mathbf{v}_{d_2} \tag{11.38}$$

when we have labeled the velocities with vectors to take into account the relative orientation of the bifurcating vessels.

## TOPIC 11.6  Vascular Turbulence

A turbulent flow of a fluid can be predicted based on the value of the so-called Reynolds number, given by

$$\mathrm{Re} = \frac{\rho v D}{\eta} \tag{11.39}$$

where $D$ is the diameter of the tube, $v$ the flow velocity, $\rho$ the density of the fluid, and $\eta$ its viscosity. When Re exceeds 3000, the flow will be turbulent, below 2000 it is laminar, while in the range 2000 to 3000 it is metastable, that is, it can be either turbulent or laminar depending on the surface roughness and the presence of obstacles. The question then arises: Can the flow of blood ever be turbulent? Direct observation suggests that little if any turbulence occurs in the aorta of a human even when the flow rate increases dramatically during heavy exercise. Let us calculate the value of Re for humans assuming $\rho \cong 10^3$ kg $\cdot$ m$^{-3}$, $\eta = 0.004$ Nm$^{-2}$s, peak velocity of blood $v = 0.4$ ms$^{-1}$, and the diameter of the aorta $D = 0.02$ m. This gives

$$\mathrm{Re} = \frac{(10^3 \text{ kg} \cdot \text{m}^{-3})(0.4 \text{ ms}^{-1})(0.02 \text{ m})}{(0.004 \text{ Nm}^{-2} \text{ s})} = 2000 \tag{11.40}$$

which is right at the borderline. However, even if any turbulence is temporarily created, it dies down away from the valves. When the person engages in strenuous exercise, the velocity of flow increases severalfold, which would indicate that Re may exceed the critical value. However, the aorta and all blood vessels are not rigid tubes but rather quite elastic, and the pulsatile flow of blood is considered viscoelastic.

Turbulence can sometimes be detected by the sound it makes. The sound of turbulence is a valuable diagnostic indicator in the circulatory system. The turbulent sounds made by blood flow in this measurement are called the Korotkoff sounds.

## TOPIC 11.7  Diseases Related to Fluid Flow or Abnormal Blood Vessels and Bernoulli's Principle

The first disease we discuss is the arteriovenus fistula (an abnormal opening), a direct connection between an artery and a vein which results in low peripheral resistance and low venous pressure. The former represents the cumulative effect of resistance from all vessels in the human circulation. The high arterial pressure forces a volume of blood through the fistula. The pressure of such a fistula might be translated into a serious overloading of the heart and a decrease of blood flow to adjacent organs and tissue. If left untreated, it results in lack of oxygen or ischemia and ultimately tissue death. The treatment of such a malfunction involves the surgical closure of the fistula and a redirection of blood through the normal circulation channels.

Difficulty of breathing characterizes respiratory distress syndrome (RDS) and is due to underdeveloped lungs typically found in premature newborn babies. Normally air enters the windpipe and passes through the bronchi, bronchioles, and eventually the alveoli. These latter are tiny bubblelike sacs where gaseous exchange of $CO_2$ and $O_2$ takes place. Primarily due to a layer of water mixed with lung surfactant which lines the inner surface of the alveoli, they tend naturally to contract, or become smaller in size. Lung surfactant possesses a low surface tension ($\gamma = 25$ mN/m), which is extremely important in preventing collapse of the alveoli and making sure blood flows to the capillary vessels. In newborn babies with RDS a clear membrane can be found which corresponds to a region of reduced amounts of surfactant. If the lung surfactant is reduced or even missing, the alveoli become coated with a higher proportion of water, exhibiting a much larger surface tension ($\gamma = 75$ mN/m), contracting the alveoli to a deleterious degree, sometimes to the point of collapse.

In the diagnosis of vessel disease a qualitative assessment and quantitative measurement of blood flow through diseased vessels becomes of great importance. Standard X-ray images are required or X-ray angiography following an injection of an opaque contrast agent. Blood flow measurements can be made by employing equations of motion to describe the movement of the opaque dye along a linear portion of the blood vessel. This method is rather crude, but nevertheless useful, due partly to (a) the pulsing nature of the blood flow, (b) the elasticity of the vessel walls, and (c) the gradual narrowing and curvature of blood vessels.

The arteries, which transport blood through the circulatory system, can suffer through vessel disease in two main ways. The first is the development of an aneurysm or ballooning of the vessel wall (see Topic 11.11). The second problem which may arise is a constriction in an obstructed or clogged vessel due primarily to atherosclerosis, commonly called hardening of the arteries. This is a development of calcified, fibrotic growths which tend to accumulate along the inside of the vessel wall.

The idea behind Bernoulli's principle, when applied to an expansion or partially constricted vessel, is that the conservation of mass holds. One point in a vessel is normal and neither constricted nor expanded and a second is either a region where the flow is restricted or a region that is enlarged. Assuming that there is no gravitational potential energies in both regions, the pressure, density, and velocity in the first region are $P_1$, $\rho_1$, and $v_1$, and the corresponding quantities in region 2 are $P_2$, $\rho_2$, $v_2$, Bernoulli's principle states that

$$P_1 + \tfrac{1}{2}\rho_1 v_1^2 = P_2 + \tfrac{1}{2}\rho_2 v_2^2 \tag{11.41}$$

We will discuss forces acting on blood vessels in Topics 11.10 and 11.11.

## TOPIC 11.8   Intravenous Supply of Nutrients, Fluids, Blood, and Drugs

In the treatment of a patient it may become necessary to supply fluids, nutrients, or drugs intravenously (see Figure 11.17). The infusion is placed at a distance $h$ above the vein into which it will be administered. A thin, flexible tube connects the container to a metal cannula inserted into the vein. There is a mechanism on the tube to control the rate of

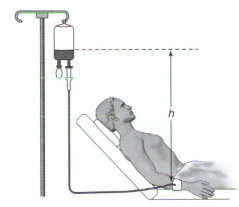

**FIGURE 11.17**    Intravenous supply of blood.

flow as prescribed by the physician. The venous pressure is of order 3 mm Hg. The pressure of the intravenous fluid is proportional to the height of the surface of the liquid in the container above the needle in the vein. As an example, let us determine the pressure required for a blood transfusion. The density of blood is $1.04 \times 10^3$ kg/m³, and we calculate the net pressure acting to transfer blood into the vein. Assuming that $h = 1$ m, the pressure of the blood is given by

$$P_{\text{blood}} = \rho gh = (1.04 \times 10^3 \text{ kg/m}^3)(9.80 \text{ m/s})(1 \text{ m}) = 10.19 \times 10^3 \text{ Pa} \quad (11.42)$$

However, atmospheric pressure is approximately $1.01 \times 10^5$ Pa, which clearly corresponds to 760 mm Hg. Hence, $P_{\text{blood}} = 76.7$ mm Hg and

$$P_{\text{net}} = (76.7 - 3) \text{ mm Hg} = 73.7 \text{ mm Hg} \quad (11.43)$$

which we interpret as the pressure exerted by the blood on the vein.

## TOPIC 11.9   Physiological Effects of Hydrostatic Pressure

The change of hydrostatic pressure of a fluid with density $\rho$ due to a difference in height $h$ is given by the expression

$$\Delta P = P_B - P_A = \rho gh \quad (11.44)$$

This means, for example, that the pressure $P_B$ in an artery in the foot is greater than the pressure $P_A$ in the aorta. Taking the density of blood as $\rho = 1.05 \times 10^3$ kg/m³ and the distance from the aorta to the foot as $h = 1.35$ m, we obtain $\Delta P = 103$ mm Hg. Since the pressure in the aorta is typically $P_A = 100$ mm Hg, the pressure in the foot, $P_B = 203$ mm Hg, is about twice as large. This leads to swelling of the legs of some people who must stand all day at work.

Similarly, we may compare a blood pressure difference in the brain when standing and when lying down. We find that in this case blood pressure drops by 35 mm Hg when a person stands up. To compensate for this effect, the arteries in the head expand. However, the result is not instantaneous—hence the common experience of dizziness when a person sits up or stands up rapidly.

Another experience associated with hydrostatic pressure effects is that of diving. Being submerged by 1 m results in an additional pressure of $10^4$ Pa or one-tenth of the normal atmospheric pressure. A dive to 2 m below the surface may already cause problems as the pressure difference of $0.2 \times 10^5$ Pa exerts a force on the eardrum whose cross-sectional area can be estimated as $0.6 \times 10^{-4}$ m$^2$ such that

$$F = PA = 1.2 \text{ N} \tag{11.45}$$

exceeding an amount necessary to rupture the eardrum. The latter usually does not happen since muscle contraction in the respiratory system and increased internal pressure have a compensatory effect.

In scuba diving at depths below 10 m additional hazards arise in the form of a medical condition called "the bends." The problem relates to the fact that under sufficiently high pressures nitrogen gas dissolves in the blood. When the diver returns after the dive to the water surface, the nitrogen diffuses out of the blood if the return journey is gradual enough. However, a rapid decompression results in nitrogen forming bubbles in the blood with painful or even deadly results depending on the depth and the time of diving. For example, at the depth of 20 m the maximum safe time is approximately 40 min. Note that the extra hydrostatic pressure at this depth is approximately 1500 mm Hg, or twice the atmospheric pressure.

The pressure in different parts of the circulatory system is due in part to the pressure drop when fluids are flowing. Another reason why the blood pressure varies can be explained using Pascal's law. In words, this states that any change in the pressure applied to a completely enclosed fluid is transmitted undiminished to all parts of the fluid and the enclosing walls. Suppose $P_{heart}$ is the blood pressure inside the heart at a given instant. Using Pascal's law, we calculate the blood pressure at any other point in the body within the arteries and take $h$ to be the vertical distance between the heart and this other point. Hence the pressure at this point, P(h), is given by

$$P(h) = P_{heart} + \rho g h \tag{11.46}$$

where $\rho$ is the density of blood and $h$ can be positive or negative depending on whether the corresponding point is below or above the heart, respectively. To calculate the pressure in the veins is more complicated because there are valves in the leg veins, capillaries have resistance, and a number of other factors enter into consideration.

If we use equation (11.46) to calculate the pressure in the head, as this lies above the heart, $h$ must be negative. That is, there must be a pressure deficit in the arteries of the head compared to that in the heart (see Figure 11.18). As an example, suppose the head is 0.3 m above the heart; then the pressure reduction $\Delta P$ is given as

$$\Delta P = (0.3 \text{ m})(9.8 \text{ m/s}^2)(10^3 \text{ kg/m}^3) = 2.9 \times 10^3 \text{ Pa} \tag{11.47}$$

Thus if the systolic pressure were less than $2.9 \times 10^3$ Pa, blood could not reach the head from the heart, and this becomes even more significant if we take into account pressure loss due to flow. Thus biologists use the following rule: The systolic pressure of an animal can never be less than $\rho g h$, where again $\rho$ is the density of blood, $g$ the acceleration due to gravity, and $h$ the height of the animal.

Consider flow from the heart to the legs and back to the heart in a standing person. Gravity is in the same direction as downward flow, and in the opposite direction to upward flow, so it has no net effect. However, gravity can have an effect on fluid balance. The pressure due to the weight of the fluid is

$$P = h \rho g \tag{11.48}$$

where $h$ is the depth in the fluid. For a standing person, pressure in the major arteries increases below the heart and decreases above the heart by an amount $h \rho g$, where $h$ is pos-

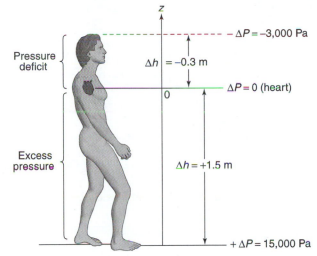

$z$

$-\Delta P = -3{,}000$ Pa

Pressure deficit

$\Delta h = -0.3$ m

$\Delta P = 0$ (heart)

0

Excess pressure

$\Delta h = +1.5$ m

$+\Delta P = 15{,}000$ Pa

**FIGURE 11.18**   Blood pressure and Pascal's law.

itive for any point below the heart and negative for any point above the heart. The blood pressure in the major arteries in the head is then

$$P_{\text{head}} = P_{\text{heart}} - h_{\text{head}}\,\rho g \qquad (11.49)$$

This can cause fainting if $P_{\text{head}}$ is too low as we mentioned earlier and calculated in equation (11.47). Pressure in the major arteries in the legs is

$$P_{\text{legs}} = P_{\text{heart}} + h_{\text{legs}}\,\rho g \qquad (11.50)$$

Let us suppose that the vertical distance between the heart and the feet in a human is $H$. Then the excess blood pressure, $\Delta P$, in the arteries of the feet compared with that at the heart is $\rho g H$. For a human with $H = 1.5$ m, $\Delta P = 15 \times 10^3$ Pa (see Figure 11.18). When this excess pressure is large enough, blood can be forced into the surrounding tissue, causing swelling. Elderly patients may require firm elastic bandages around their ankles to prevent this swelling. The larger pressure in the legs can cause fluid buildup (edema) by reverse osmosis through capillary walls, especially in people who are on their feet many hours a day. One remedy is to sit down and elevate the legs. Giraffes, with $H = 2.5$ m and pressure in the feet that is correspondingly greater, have a very tight elastic skin around their ankles to prevent such swelling.

## TOPIC 11.10   Forces Acting on a Blood Vessel

The blood circulates in complex elastic tubes and may be approximated by long circular cylinders. In Figure 11.19b we present a diagram of such a cylindrical vessel of radius $R$ and length $L$. Two forces acting on it keep it in static equilibrium: circumferential tension and internal pressure. As shown in Fig. 11.19 the circumferential tension $T$ acts along both sides of the curved section, hence the total force downward on the shaded section of the blood vessel due to the tension $T$ is

$$F_{\text{down}} = (T \sin \theta)2L \qquad (11.51)$$

On the other hand, the total force upward due to internal pressure $P$ is

$$F_{\text{up}} = PL(2R\theta) \qquad (11.52)$$

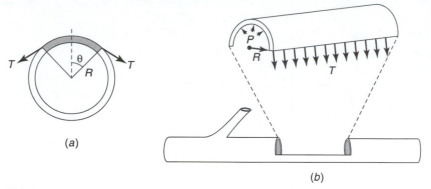

**FIGURE 11.19**   Section through a blood vessel represented by a circular cylinder.

Thus, in equilibrium, the two forces are balanced

$$F_{down} = F_{up} \qquad (11.53)$$

so

$$(T \sin \theta)2L = PL(2R\theta) \qquad (11.54)$$

Since for small angles, $\theta$, $\sin \theta \cong \theta$, equation (11.52) reduces to a simple relationship between internal pressure and circumferential tension:

$$T = PR \qquad (11.55)$$

## TOPIC 11.11   Forces Acting within a Brain Aneurysm

A ballooning of a diseased region of a blood vessel wall is called an aneurysm. A common site for such a swelling is where the arteries branch at the base of the brain and is caused by congenital weakness. We approximate this as a sphere. In Figure 11.20 we present a free-body diagram of the cross section through a sphere of radius $R$. Two forces act on the spherical surface. One of these, acting outward, against the spherical surface is produced by the pressure, which we denote by $F_{out}$.

The pressure is a radial force per unit area and acts perpendicular to the vessel surface at every point along the inner surface of the sphere. The force due to pressure, $F_{out}$, will simply be the product of pressure and cross-sectional area of the sphere, namely

$$F_{out} = P\pi R^2 \qquad (11.56)$$

This latter force, $F_{out}$, is countered by the tension force $F_T$, defined as the tension force due to the walls of the aneurysm which keeps the upper half of the elastic sphere attached to the lower. It is defined by

$$F_T = 2\pi RT \qquad (11.57)$$

where $T$ is the average tension per unit length. Equating these two forces from equations (11.56) and (11.57) yields

$$P\pi R^2 = 2\pi RT \qquad (11.58)$$

so the tension is

$$T = \tfrac{1}{2}PR \qquad (11.59)$$

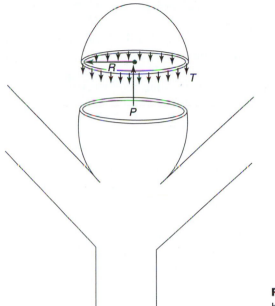

**FIGURE 11.20** Diagram of forces for a brain aneurysm.

For a cylindrical artery of radius $R$, the tension is proportional to $R^2$, as shown in Figure 11.21. At the same time the relationship with tension for pressure difference is

$$\Delta P = \frac{T}{R} \tag{11.60}$$

or

$$T = R\,\Delta P \tag{11.61}$$

Consequently, with a given elastic tension and applied pressure $\Delta P$, one finds a stable radius $R_1$, shown in Figure 11.21. The aorta, which withstands 100 mm Hg pressure, is about 0.002 m thick, whereas the capillaries experience blood pressures of 25 mm Hg magnitude but are only $5 \times 10^{-6}$ m thick. The discrepancy is explained by the huge difference in the diameters and hence cross-sectional areas.

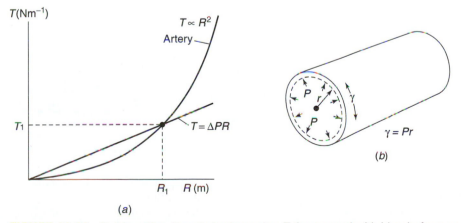

**FIGURE 11.21** Stable radius $R_1$ and elastic tension $T_1$ for a vessel with blood of pressure $\Delta P$.

## TOPIC 11.12 Fluid Dynamics of Respiration

The most important function of the respiratory system is to oxygenate blood and remove carbon dioxide from it. Oxygen is needed to combine chemically with food, and carbon dioxide is the primary gaseous waste product of cells. Air flows into the lungs when the pressure in them is less than atmospheric and out of the lungs when the pressure is greater than atmospheric. The flow of air obeys the familiar equation

$$Q = \frac{P_1 - P_2}{R_F} \qquad (11.62)$$

where $P_1$ is the pressure in the lungs, $P_2$ is atmospheric pressure, and $R_F$ is the resistance of the breathing passages to air flow. If gauge pressures are used, $P_2$ is zero; then air flows out of the lungs when $P_1$ is positive (greater than atmospheric) and into the lungs when $P_1$ is negative (less than atmospheric).

The body changes pressure in the lungs by increasing or decreasing their volume. During inhalation the lungs are expanded by muscle action in the diaphragm and rib cage (see Figure 11.22). The diaphragm is a sheet of muscle lying below the lungs. When relaxed, it has an upward curvature, and when contracted, it moves downward, expanding the lungs.

The interior surface of the lungs is the largest surface of the human body in contact with the environment. The exchange of $CO_2$ and $O_2$ molecules taking place between blood and the atmosphere requires about 1 m² of lung surface per kilogram of body weight. This is achieved by compartmentalizing the lungs into the tiny air sacs called alveoli. The alveoli are covered by a multitude of tiny blood capillaries with a very large total surface area (about 100 m²). Across the surface of the alveoli, oxygen and carbon dioxide are exchanged between the blood and lungs. The surface of an alveolus is covered by a thin film of water (about 0.5 $\mu$m thick) and surrounded by lung surfactant molecules.

During inhalation the radius of the alveoli expands from about $0.5 \times 10^{-4}$ to $1.0 \times 10^{-4}$ m. They are lined with mucous tissue fluid, which normally has a surface tension of 0.050 N/m. With this surface tension, the pressure difference required to inflate an alveoli would be given by the formula

$$\Delta p = \bar{p}_i - \bar{p}_o = \frac{2\gamma}{r} \qquad (11.63)$$

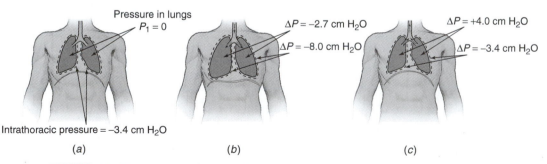

**FIGURE 11.22** Three stages in the breathing cycle: (*a*) At rest, the pressure in the lungs is zero. The lungs are open. (*b*) During inhalation the chest expands, the diaphragm moves downward, and pressures drop. (*c*) During exhalation, the chest contracts, the diaphragm rises, and pressures rise.

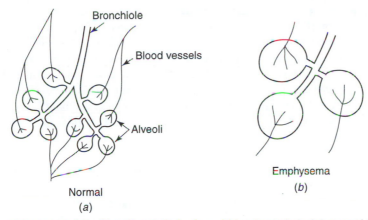

**FIGURE 11.23** Alveoli and their air and blood supply in (*a*) normal lungs and (*b*) an emphysema victim, where alveoli have joined to form fewer and larger sacs.

where $r$ denotes the radius and $\gamma$ the surface tension. Taking $r = 0.5 \times 10^{-4}$ m and a value of $\gamma$ typical for water, that is, 0.070 N/m, one obtains $\Delta p = 2.8 \times 10^3$ N/m$^2$ = 21 mm Hg, meaning that the gauge pressure $\overline{p}_0$ outside the alveolus would have to be $-24$ mm Hg since the internal pressure $\overline{p}_i = -3$ mm Hg. The outside pressure in this case is the pressure in the space between the lungs and the pleural cavity that holds the lungs. The gauge pressure in this space is in fact negative but only about $-4$ mm Hg, so that the actual pressure difference $\overline{p}_i - \overline{p}_0$ is only about 1 mm Hg. To overcome this difficulty, the walls of the alveoli secrete a surfactant that reduces the surface tension by a factor of 15 or so.

The surfactant in the lungs is a lipoprotein, that is, a molecule composed of protein and lipid constituents. The polar head group of the lipid is hydrophilic while its long fatty acid chains are hydrophobic. There appears to be a fixed amount of this surfactant in each alveolus, and its ability to reduce surface tension depends on its concentration. Therefore, when the alveolus is deflated, the concentration of the surfactant is high and the surface tension is very low, so that the alveolus is expanded without difficulty. However, as it expands, the concentration of the surfactant decreases and the surface tension increases, until a point of equilibrium is reached at maximum expansion.

The importance of surface tension in the lungs is seen in victims of emphysema. Many of the alveoli in a person suffering from emphysema join together to form fewer and larger alveoli, as shown in Figure 11.23. One effect of the larger sacs is a reduction in pressure because of their larger radii. When a person with emphysema attempts to exhale, the alveoli create a small pressure and air flow is less than normal. The lungs have effectively lost some of their elasticity.

## TOPIC 11.13  Buoyancy and Drag in Animals and Fishes

According to Archimedes, the buoyancy force on a submerged or partly submerged object in a liquid is equal to the weight of the fluid displaced. Thus, by altering an object's density, when it is submerged, it can be made to rise or sink. For a fish the optimal situation would be to maintain neutral buoyancy so it would then not need to expend effort to prevent sinking or floating. This condition requires that the density of fish is exactly the same as water. The densities of marine animals are found to be very close to water but always a little greater (e.g., muscle has a density of about 1060 kg/m³ whereas sea-water has a density of 1030 kg/m³). A typical excess density is 50 kg/m³. A large shark with a mass of $10^3$ kg on land has an effective mass of about 50 kg in water.

In nature there are a number of ways to control buoyancy to compensate for excess weight. To maintain constant swimming is one method of doing this because this produces an upward lift force (e.g., sharks, rays, and squids are constantly swimming). Some fish (e.g., herring and salmon) have a swim bladder filled with gases like $O_2$, $N_2$, and $CO_2$ and have densities which are negligible compared to water. This float or swim bladder needs to only occupy roughly one-twentieth of the volume of the fish, independent of the size of the fish.

If one had to design a fish, one would have to decide where to place the swim bladder. Putting it near the fish's tail or head would generate a force which would create a torque and rotate the fish from its preferred direction. The correct place would be just above the fish's center-of-gravity (CG), where these torques, about the CG, would not arise or would be much smaller depending on its actual position. In reality, fishes have an in-built compensating mechanism which modifies the volume of the float chamber. Suppose a fish is neutrally buoyant at depth $h$ below the surface of the water. If we displace the fish downward by a small amount, the ambient pressure rises due to Pascal's law and compresses the swim bladder, reducing its volume, which in turn decreases the buoyant force. The fish starts to sink since the float no longer compensates the weight, and the more it sinks, the more the weight dominates the buoyancy force. On the other hand, if the fish is displaced upward, the pressure decreases slightly, the float will expand, and the fish rises faster and faster. This instability is overcome by a feedback control mechanism which regularly checks the volume of the fish's float. In fact, by releasing gas into or absorbing gas from the float chamber, the fish can control its instability.

The swimming mode of fishes is not a form of jet propulsion as, for example, in the squid. The momentum exchange between the fish and the surrounding water now becomes important. A fish swims forward by pushing the surrounding water backward by giving sideways kicks to the water with its tail. In Figure 11.24 a fish has just given two kicks with its tail, and it can be seen that two swirls of water seem to move sideways and partially backward. The mass of the water on each swirl can be determined by the law of conservation of momentum; that is, the total momentum of the fish plus the swirls, after the kicks, must be the same as before kicks were applied. The fish, of course, needs to overcome the hydrodynamic drag.

If we assume that during a power stroke of the fish's tail a mass $m$ of swirling water is pushed directly backward with speed $v$ and the fish delivers a power stroke every $\tau$

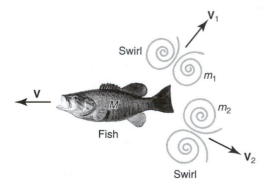

**FIGURE 11.24**  Two swirls from a fish's tail.

seconds, then the rate of change of the water's momentum is $mv/\tau$. The force $F_d$ on the fish is

$$F_d = \frac{mv}{\tau} \tag{11.64}$$

which is the drag force.

When a fish swims, it performs work in two ways: by overcoming the drag $F_d$ of the surrounding water and by setting up swirls in motion. The kinetic energy of a swirl of mass $m$ and velocity $v$ is

$$\text{KE} = \tfrac{1}{2}mv^2 \tag{11.65}$$

Thus the average amount of work done every second must be

$$W = \frac{1}{2}\frac{mv^2}{\tau} \tag{11.66}$$

The work done by the fish against the drag is equal to $F_d$ multiplied by the distance $d$ moved per stroke. The total work done by the fish per second or the power utilized, $P$, is then given by

$$P = \frac{(1/2)mv^2}{\tau} + F_d V = \frac{(1/2)mv^2 + mvV}{\tau} \tag{11.67}$$

where $V = d/\tau$ is the fish's velocity and we have inserted the drag force $F_d$ from equation (11.64).

Optimal design for a fish demands maximum speed with minimum power. The propeller efficiency, eff, is defined as the work done per second against the drag divided by the total work done per second, that is, if eff = 1, the fish performs work only against the drag. Thus

$$\text{eff} = \frac{mvV/\tau}{\dfrac{(1/2)mv^2 + mvV}{\tau}} = \frac{1}{1 + v/2V} \tag{11.68}$$

We can clearly attain a high efficiency by making the swirl velocity $v$ as low as possible, although the swirl momentum should be as high as possible. Thus, efficient swimming demands that swirls of water are set in motion and have the highest possible mass and move as slowly as possible. It is for this reason that mammals such as seals, dolphins, and whales have evolved large flippers and tails.

## TOPIC 11.14  Pressure Vessels in Cells

In this section we apply the concept of surface tension to cells. A double layer of phospholipid molecules with a variety of imbedded proteins make up the plasma membrane or the outer surface of a cell. This plasma membrane does not resemble the surface of a fluid or even the interface between two fluids. The reason for this is that the plasma membrane has an essentially fixed surface area; that is, there are only a fixed number of phospholipid molecules and proteins which when packed together make up the membrane. Each lipid molecule or protein has a preferred surface area so, unlike the surface of a fluid, the plasma membrane is for practical purposes inextensible.

Suppose we increase the osmotic pressure of a cell. The cell will try to swell, but this is prevented because the surface area of the plasma membrane is nearly fixed; that is, some sort of elastic stress will be built up inside the membrane, and if this is too great, the cell will burst—a condition called lysis. A cell under osmotic pressure is a special example of a mechanical system called a pressure vessel (e.g., a pressure cooker or the boiler of a steam engine). Another example is the human heart. To predict the point where a boiler will explode, engineers need to know the elastic stress inside the wall of the boiler as a function of boiler pressure.

To find this stress, suppose we cut the wall of a pressure vessel along some line of length $\ell$. To prevent the two sides of the cut from separating, a force $F$ must be present which is proportional to $\ell$. Writing $F = \gamma\ell$, then the proportionality constant $\gamma$ is called the elastic tension in the wall. This tension is most definitely not the surface tension $T$ of a fluid but it is playing a similar role. The excess pressure inside a bubble of radius $R$, over and above atmospheric pressure, is $2T/R$ (see Topics 11.10 and 11.11). It turns out that a similar relation to this may be used to compute the excess pressure $\Delta P$ inside a spherical pressure vessel if we replace the surface tension $T$ by the elastic tension $\gamma$. Thus we have

$$\Delta P = \frac{2\gamma}{R} \tag{11.69}$$

The elastic stress $\sigma$ inside the vessel wall is related to the elastic tension by

$$\gamma = D\sigma \tag{11.70}$$

where $D$ is the wall thickness. The reason for this is that the surface area of the cut is $\ell D$, and hence the force per unit area on the surface of the cut is $F/\ell D = \gamma/D$, and this is the stress of the vessel wall. From equations (11.69) and (11.70) the elastic stress $\sigma$ is given by

$$\sigma = \frac{R\,\Delta P}{2D} \tag{11.71}$$

so that the vessel will burst when this stress $\sigma$ exceeds the fracture stress $\sigma_{\max}$ of the material from which the vessel was made (see Chapter 10).

## SOLVED PROBLEMS

**11.1**   If the pressure in the cerebrospinal fluid (CSF) is measured as shown in the figure, using a spinal tap with the patient sitting erect, then the pressure due to the weight of the CSF in the spinal column increases the pressure.

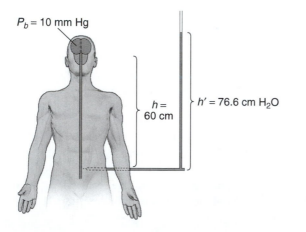

$P_b = 10$ mm Hg

$h = 60$ cm

$h' = 76.6$ cm $H_2O$

**(a)** What pressure is measured in centimeters of water if the pressure around the brain is 10 mm Hg and the tap is at a point 60 cm lower than the brain?

**(b)** What pressure in centimeters of water is measured if the patient lies down? The density of CSF is 1.05 g/cm$^3$.

### Solution

**(a)** Since the bottom of the manometer is at the same vertical height as the needle, the pressure measured in the manometer is

$$P = \rho g h + P_b$$

where $P_b$ is the pressure in the brain and $\rho g h$ is the pressure due to the 60-cm column of CSF. Expressed in SI units, this yields

$$P = (0.60 \text{ m})(1.05 \times 10^3 \text{ kg/m}^3)(9.8 \text{ m/s}^2) + (0.01 \text{ m})(13.6 \times 10^3 \text{ kg/m}^3)(9.8 \text{ m/s})$$
$$= 7.51 \times 10^3 \text{ N/m}^2$$

To convert this to centimeters of water, we use $P = \rho_w g h_w$, or

$$h_w = \frac{P}{\rho_w g} = \frac{7.51 \times 10^3}{(10^3 \text{ kg/m}^3)(9.8 \text{ m/s}^2)} = 76.6 \text{ cm H}_2\text{O}$$

**(b)** If the patient lies down, then there is no excess pressure due to gravity and the pressure measured will be equivalent to 10 mm Hg. That is,

$$P = P_b$$
$$\rho_w g h_w = \rho_{Hg} g h_{Hg}$$

or

$$h_w = h_{Hg} \frac{\rho_{Hg}}{\rho_w} = \frac{(10 \text{ mm})(13.6 \times 10^3 \text{ kg/m}^3)}{(1000 \text{ kg/m}^3)} = 13.6 \text{ cm H}_2\text{O}$$

so there is an extra pressure of 63 cm H$_2$O when the person sits erect due to the weight of the fluid in the spinal column.

**11.2** **(a)** Calculate the average blood velocity in the major arteries given that the aorta has a radius of 1.0 cm, the blood velocity is 30 cm/s in the aorta, and the total cross-sectional area of the major arteries is 20 cm$^2$.

**(b)** What is the total flow rate?

**(c)** On the assumption that all the blood in the circulatory system goes through the capillaries, and given that the average velocity of the blood in the capillaries is 0.03 cm/s, what is the total cross-sectional area of the capillaries?

### Solution

**(a)** If the vessels have a circular cross section, then $A = \pi r^2$. Now, the continuity equation gives

$$\bar{v}' = \frac{A}{A'} \bar{v} = \frac{\pi (1 \text{ cm})^2}{20 \text{ cm}^2} (30 \text{ cm/s}) = 4.7 \text{ cm/s}$$

**(b)** The total flow rate is the same in each system. Calculating it in the aorta gives

$$Q = A\bar{v} = [\pi(1.0 \text{ cm})^2](30 \text{ cm/s}) = 94 \text{ cm}^3/\text{s} = 5.7 \text{ L/min}$$

**(c)** Because $Q = A_{cap}\bar{v}_{cap}$, we get

$$A_{cap} = \frac{Q}{\bar{v}_{cap}} = \frac{94 \text{ cm}^3/\text{s}}{0.03 \text{ cm/s}} = 3133 \text{ cm}^2$$

where $A_{cap}$ is the total area of the capillary system.

**11.3**  When a person eats a large meal, blood flow in the digestive system is increased by dilating blood vessels supplying that system. By what factor must the flow-regulating vessels dilate to increase flow from 1.0 to 5.0 L/min?

**Solution**  If all other factors remain constant (blood viscosity and the pressures at the inlet and outlet of the system), then the $\eta$'s, $P$'s, and $L$'s in Poiseuille's equation cancel since their primed and unprimed values are the same, leaving

$$\frac{Q'}{Q} = \frac{r'^4}{r^4}$$

Hence

$$\left(\frac{Q'}{Q}\right)^{1/4} = \frac{r'}{r} = (5.0)^{1/4} = 1.495$$

So,

$$r' = 1.495r$$

and a 500% increase in flow rate is accomplished by a 49.5% increase in the radii of the blood vessels.

**11.4**  The rate of fluid flow in an intravenous (IV) needle is observed to be 2.0 cm³/min for a glucose solution of density 1.05 g/cm³. The surface of the solution is a height 1.0 m above the entrance of the needle. If the height is increased to 1.5 m, what is the new flow rate? Note that the blood pressure ($P_2$) remains at 8.0 mm Hg.

**Solution**  The only variable that changes is the pressure at the entrance of the needle, $P_1$. Therefore,

$$\frac{Q'}{Q} = \frac{P_1' - P_2'}{P_1 - P_2} \quad \text{or} \quad Q' = \frac{P_1' - P_2}{P_1 - P_2}Q$$

since $P_2 = P_2'$. The pressures $P_1$, $P_1'$, and $P_2$ can be expressed in terms of $h$ only. The pressures $P_1$ and $P_1'$ are 100 and 150 cm of glucose, respectively, and $P_2$ is 8.0 mm Hg, which is converted to the same units as the other pressures as follows:

$$P_2 = \rho_{Hg}gh_{Hg} = \rho_{gl}gh_{gl}$$

$$h_{gl} = h_{Hg}\frac{\rho_{Hg}}{\rho_{gl}} = (0.80 \text{ cm})\frac{13.6 \text{ g/cm}^3}{1.05 \text{ g/cm}^3} = 10.4 \text{ cm}$$

Therefore, $P_2 = 10.4$ cm of glucose. Now $Q'$ can be calculated as

$$Q' = \frac{150 \text{ cm glucose} - 10.4 \text{ cm glucose}}{100 \text{ cm glucose} - 10.4 \text{ cm glucose}}(2.0 \text{ cm}^3/\text{min}) = 3.1 \text{ cm}^3/\text{min}$$

**11.5**  Calculate the wall tension $\gamma$ created by blood pressure in the aorta and a capillary under the following circumstances: Average blood pressure in the aorta is 100 mm Hg, and the radius of the aorta is 1.0 cm.

**Solution**   We use Laplace's law and convert the pressure to newtons per square meter using $P = \rho gh$:

$$\gamma = Pr = (\rho gh)r = (0.10 \text{ m})(13.6 \times 10^3 \text{ kg/m}^3)(9.8 \text{ m/s}^2)(0.01 \text{ m}) = 133 \text{ N/m}$$

**11.6**   Calculate the kinetic power, pressure power, and total power generated by the left ventricle of the heart for a typical resting adult. The speed of the blood emerging from the left ventricle is 30 cm/s, the flow rate is 83 cm³/s, the density of blood is 1.05 g/cm³, and the pressure is 120 mm Hg ($1.60 \times 10^4$ N/m²).

**Solution**   The kinetic power output is

$$\text{Kinetic power} = (\tfrac{1}{2}\rho v^2)Q = [\tfrac{1}{2}(1050 \text{ kg/m}^2)(0.30 \text{ m/s})^2] \, (83 \times 10^{-6} \text{ m}^3/\text{s}) = 4 \times 10^{-3} \text{ W}$$

The pressure power output is

$$\text{Pressure power} = PQ = (1.60 \times 10^4 \text{ N/m}^2)(83 \times 10^{-6} \text{ m}^3/\text{s}) = 1.3 \text{ W}$$

The total useful power output is the sum of these two results, or

$$\text{Total power} = 1.3 \text{ W}$$

**11.7**   What is the peak Reynolds number in the abdominal aorta of the dog where the diameter $D = 0.3$ cm and the peak velocity $v = 0.60$ m/s? If a disturbance is initiated, will it propagate or die out?

**Solution**

$$\text{Re} = \frac{\rho Dv}{\eta} = \frac{1000 \times 3 \times 10^{-3} \times 0.60}{0.004} = 450$$

Since Re $< 2000$, the disturbance will probably die out.

**11.8**   Would turbulence be expected in the human aorta? Assume the diameter is $D = 0.02$ m, $\rho = 1000$ kg/m³, peak velocity $= 0.4$ m/s (for a man at rest), and $\eta = 0.004$ Ns/m².

**Solution**

$$\text{Re} = \frac{\rho Dv}{\eta} = \frac{1000 \times 0.02 \times 0.40}{0.004} = 2000$$

This is so close to the traditional dividing point that you cannot answer the question in a decisive manner.

**11.9**   The total flow rate $Q$ of the adult circulatory system is about $10^{-4}$ m³/s. This is thus the flow that passes through either the right or left side of the heart. The aorta, which carries the blood from the left side of the heart, has a diameter of 2.5 cm. What is the average flow velocity in the aorta?

**Solution**   We use

$$v = \frac{Q}{A}$$

The cross-sectional area $A$ is $5 \times 10^{-4}$ m². The resulting flow velocity $v$ is then 0.5 m/s.

**11.10**   Capillaries have a cross section of $3 \times 10^{-11}$ m² and a flow velocity of $10^{-3}$ m/s. Estimate the total number of capillaries in the body.

**Solution**   The flow through one capillary is the product of the area and flow velocity: $3 \times 10^{-14}$ m³/s. Let the number of capillaries be $N$. The generalized continuity principle states that the total flow through the capillaries must equal the flow through the aorta ($10^{-4}$ m³/s), or

$$10^{-4} = (3 \times 10^{-14} \text{ m}^3/\text{s})N$$

Therefore, $N = 3 \times 10^9$.

**11.11**   Water-soluble molecules such as water, glucose, albumin, and hemoglobin encountered inside a cell can be modeled as little spheres with diameter $D$. For a water molecule $D = 3$ Å, for glucose $D = 8$ Å, and for hemoglobin $D = 60$ Å. What is the dependence of the molar weight $M$ on the radius $R = \frac{1}{2}D$?

**Solution**   If $\rho$ is the density of the molecule, then $\rho V$ is its mass, and the molar weight $M$ of the substance is

$$M = \tfrac{4}{3}\pi R^3 \rho N_A$$

Assuming that water-soluble molecules in cells or in blood generally have a density close to that of water, we can treat $\rho$ as a constant, so $M$ is proportional to the third power of the radius. Thus, we should be able to plot the radius $R$ of a molecule as a function of its molecular weight and find that $R$ is approximately proportional to $M^{1/3}$.

**11.12**   The mass concentration of hemoglobin inside a red blood cell is 0.34 g/cm³. The volume of a red blood cell is 87 $\mu$m³. A hemoglobin protein has a molecular weight of 68 kD and a volume of $45 \times 35 \times 25$ Å³. What is the total hemoglobin mass inside a red blood cell?

**Solution**   The total hemoglobin mass inside a cell is the mass concentration multiplied by the volume. Converting mass concentration and volume into SI units, we find

$$M = [(0.34 \text{ g/cm}^3)(1 \text{ kg}/1000 \text{ g})(10^6 \text{ cm}^3/1 \text{ m}^3)][(87 \ \mu\text{m}^3)(1 \text{ m}/10^6 \ \mu\text{m})^3] = 29.6 \times 10^{-15} \text{ kg}$$

**11.13**   Water walkers (*Halobates*) are insects that can walk on water. The lower parts of their legs can be treated as thin cylinders, each of length $L = 2$ mm, lying horizontally on the air–water surface. What is the maximum weight of *Halobates* which can be supported by the surface tension of water?

**Solution**   Each of the six legs of *Halobates* must support one-sixth of the total weight $W$. As $W$ increases, each leg makes a deeper dimple. The surface tension force is directed upward. The maximum upward force is equal to $2\gamma L$ because two air–water surfaces are attached to the leg. The maximum weight is equal to 6 times this force (i.e., $12\gamma L$), which is equivalent to a mass of about 0.18 g.

**11.14**   Estimate the amount by which the blood pressure $P_2$ in the anterior tibial artery at the foot exceeds the blood pressure $P_1$ in the aorta at the heart when the body is (a) reclining horizontally or (b) standing, as in figures (*a*) and (*b*), respectively.

Anterior tibial artery          Aorta

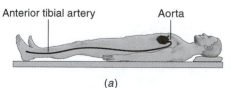

(*a*)

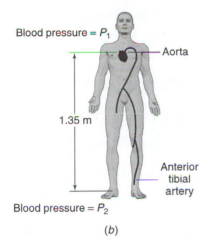

Blood pressure = $P_1$

Aorta

1.35 m

Anterior
tibial
artery

Blood pressure = $P_2$

(b)

**Solution** When the body is horizontal, there is little or no vertical separation between the foot and the heart. Since $h = 0$,

$$P_2 - P_1 = \rho g h = 0$$

When an adult is standing erect, the vertical separation between the feet and the heart is 1.35 m, as the figure indicates. The density of blood is 1060 kg/m³, so that

$$P_2 - P_1 = \rho g h = (1060 \text{ kg/m}^3)(9.8 \text{ m/s}^2)(1.35 \text{ m}) = 14024 \text{ N/m}^2$$

**11.15** In the condition known as atherosclerosis, a deposit, or atheroma, forms on the arterial wall and reduces the opening through which blood can flow. In the carotid artery in the neck, blood flows three times faster through a partially blocked region than it does through an unobstructed region. Determine the ratio of the effective radii of the artery at the two places.

**Solution** Blood, like most liquids, is incompressible, and the equation of continuity can be applied. Since the area of a circle is $\pi r^2$, it follows that

$$\underbrace{(\pi r_U^2)v_U}_{\substack{\text{Unobstructed} \\ \text{volume flow} \\ \text{rate}}} = \underbrace{(\pi r_O^2)v_O}_{\substack{\text{Obstructed} \\ \text{volume flow} \\ \text{rate}}}$$

The ratio of the radii is

$$\frac{r_U}{r_O} = \sqrt{\frac{v_O}{v_U}} = \sqrt{3} = 1.7$$

The unobstructed artery has an effective radius that is 70% larger than the radius of the obstructed region.

**11.16** During extreme exercise, the blood output from each ventricle increases fourfold, and the pressure exerted by each ventricle increases by 50%. Calculate the work done by the heart after exercise.

**Solution** The work done by the heart during extreme exercise can be determined from the work done by the left and right ventricles at the increased values of pressure and volume:

Left ventricle:      $P_{LV} = 2 \times 10^4$ N/m²      $V_{LV} = 3.32 \times 10^{-4}$ m³
Right ventricle:    $P_{RV} = 0.4 \times 10^4$ N/m²    $V_{RV} = 3.32 \times 10^{-4}$ m³

Therefore, the work of the left and right ventricles can be determined as follows:

Left ventricle: $W_{LV} = P_{LV} \Delta V_{LV} = (2 \times 10^4 \text{ N/m}^2)(3.32 \times 10^{-4} \text{ m}^3) = 6.64 \text{ J}$

Right ventricle: $W_{RV} = P_{RV} \Delta V_{RV} = (0.4 \times 10^4 \text{ N/m}^2)(3.32 \times 10^{-4} \text{ m}^3) = 1.33 \text{ J}$

The total average work done by the heart, $W_T$, during a single contraction under exercise conditions is

$$W_T = W_{LV} + W_{RV} = 6.64 \text{ J} + 1.33 \text{ J} = 7.97 \text{ J}$$

**11.17** How much work is done by the resting heart in an average lifetime of 80 years?

**Solution** The volume of blood $\Delta V$ or cardiac output ejected from the heart at each heartbeat is 83 cm³. Assuming each heartbeat occurs every 1.0 s, the total volume of blood ejected from the heart over a lifetime of 80 years is

$$\Delta V = (83 \text{ cm}^3\text{/s})(3600 \text{ s/h})(24 \text{ h/day})(365 \text{ days/year})(80 \text{ years}) = 2.1 \times 10^{11} \text{ cm}^3$$

The work of the left and right ventricles can be determined as follows:

Left ventricle: $W_{LV} = P_{LV} \Delta V_{LV} = (1.33 \times 10^4 \text{ N/m}^2)(2.1 \times 10^5 \text{ m}^3) = 2.8 \times 10^9 \text{ J}$

Right ventricle: $W_{RV} = P_{RV} \Delta V_{RV} = (0.266 \times 10^4 \text{ N/m}^2)(2.1 \times 10^5 \text{ m}^3) = 0.56 \times 10^9 \text{ J}$

The total average work done by the heart, $W_T$, during an average lifetime is

$$W_T = W_{LV} + W_{RV} = 2.8 \times 10^9 \text{ J} + 0.56 \times 10^9 \text{ J} = 3.36 \times 10^9 \text{ J}$$

**11.18** The density of a radiopharmaceutical is 0.75 g/cm³. Determine the mass of 2.0 L of this radiopharmaceutical.

**Solution** The mass of a liquid or fluid is related to the density and volume by

$$\rho = \frac{m}{V}$$

Rearranging and solving for the mass $m$ yields

$$m = \rho V = (0.75 \text{ g/cm}^3)(2000 \text{ cm}^3) = 1500 \text{ g} = 1.5 \text{ kg}$$

**11.19** Blood is pumped from the heart at a rate of 5 L/min into the aorta of radius 2.0 cm. Assuming that the viscosity and density of blood are $4 \times 10^3 \text{ Ns/m}^2$ and $1 \times 10^3 \text{ kg/m}^3$, respectively, determine the velocity of blood through the aorta.

**Solution** The blood flow velocity is related to the volumetric flow rate by

$$\text{Velocity} = \frac{\text{flow rate}}{\text{cross-sectional area}}$$

where

$$\text{Flow rate} = (5 \times 10^{-3} \text{ m}^3\text{/min})(1 \text{ min/60 s}) = 8.33 \times 10^{-5} \text{ m}^3\text{/s}$$

$$\text{Cross-sectional area} = \pi r^2 = (3.14)(0.02 \text{ m})^2 = 1.26 \times 10^{-3} \text{ m}^2$$

Therefore, the blood flow velocity is

$$v = 6.6 \times 10^{-2} \text{ m/s}$$

**11.20** The number of capillary vessels in the human circulation is approximately $1 \times 10^9$ with the diameter and length of each vessel being 8 $\mu$m and 1 nm, respectively. Assuming cardiac output is 5 L/min, determine (a) the average velocity of blood flow through the capillary vessels, (b) the time taken for blood to traverse a single capillary vessel, and (c) the time required for 1 mL of blood to flow through a single capillary vessel at a normal flow rate.

**Solution**

(a) The blood flow velocity through the capillary vessels can easily be determined by

$$v = \frac{Q}{nA} = \frac{5 \times 10^3 \text{ mL/min}}{(1 \times 10^9 \text{ capillary vessels})(3.14)(4 \times 10^{-4} \text{ cm})^2} = 0.17 \text{ cm/s}$$

(b) The time taken for the blood to traverse a single capillary vessel is given by

$$t = \frac{1 \text{ mL}}{(5 \times 10^3/10^9 \text{ capillary vessels}) \text{ mL/min}} = 2 \times 10^5 \text{ min} = 139 \text{ days}$$

(c) The time required for 1 mL of blood to flow through a single capillary vessel at a normal flow rate is

$$t = \frac{1.0 \text{ mm}}{1.66 \text{ mm/s}} = 0.60 \text{ s}$$

**11.21** What is the kinetic energy, KE, per unit volume $V$ of blood that has a speed of 0.5 m/s?

**Solution** The kinetic energy of the blood is given by

$$\text{KE} = \tfrac{1}{2}mv^2 = \tfrac{1}{2}(\rho V)v^2$$

Substituting the values given in the problem yields

$$\frac{\text{KE}}{V} = \tfrac{1}{2}(1000 \text{ kg/m}^3)(0.5 \text{ m/s})^2 = 125 \text{ J/m}^3$$

**11.22** Bernoulli's principle can be applied to a syringe to describe the dynamics of an injection. This problem can be solved by starting with Bernoulli's principle assuming that 1 refers to the position within the body of the syringe and 2 is the position within the throat of the syringe or region prior to entrance into the needle. Derive an expression for the velocity of fluid exiting the throat and entering the needle.

**Solution**

$$P_1 - P_2 = \frac{\rho v_2^2}{2}\left[1 - \left(\frac{A_2}{A_1}\right)^2\right]$$

and solve for $v_2$ to obtain

$$v_2 = \sqrt{\frac{2(P_1 - P_2)}{\rho[1 - (A_2/A_1)^2]}}$$

**11.23** Thirty milliliters of an anesthetic solution drawn into a 5-g syringe is found to have a combined mass of 80 g. Determine the density of the anesthetic solution.

**Solution** The density of the anesthetic solution can be determined from

$$\rho = \frac{m}{V}$$

The mass of the anesthetic solution, $m_a$, is the total mass $m_T$ minus that of the syringe, $m_s$, or

$$m_T = m_a + m_s$$
$$m_a = m_T - m_s = 80 \text{ g} - 5 \text{ g} = 75 \text{ g}$$

The volume $V$ of the anesthetic solution contained in the syringe is

$$V = 30 \text{ mL}$$

Thus, the density of the anesthetic solution can be determined from

$$\rho = 75 \text{ g}/30 \text{ mL} = 2.5 \text{ g/mL}$$

## EXERCISES

**11.1** The surface of the spherical alveolus of radius $r_0$ is covered by a monolayer of a surface-active substance (lipoprotein) which has a much lower surface tension ($\gamma_l = 1.2 \times 10^{-3}$ N/m) than that of water ($\gamma_w = 7.2 \times 10^{-2}$ N/m) in its surroundings. Determine the internal pressure due to the surface tension as a function of the radius of the alveolus.

**11.2** The average velocity of blood flow in the aorta in a resting man is 20 cm/s. The radius of the aorta is 1.2 cm. At rest, the number of opened capillaries is about $n = 12 \times 10^9$. They have on average a length $\ell = 0.75$ mm and a radius $r = 3$ $\mu$m.
**(a)** What is the blood flow velocity in the capillaries?
**(b)** What time does a red blood cell spend in a capillary on average?
**(c)** What is the total area of all the capillary walls? Compare this with the total surface area of all the red blood cells (erythrocytes) in the blood. Consider that, as an approximation, an erythrocyte has a discoidal form with radius $R = 4$ $\mu$m and width $d = 2$ $\mu$m. The concentration of erythrocytes is about $c = 5 \times 10^6/\text{mm}^3$, and the average blood volume $V = 5$ L.

**11.3** Estimate the steady-state mechanical work and power required for the human heart to circulate the blood. Consider the heart with the direction of blood flow as shown in the accompanying figure.

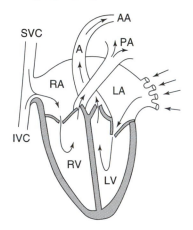

The average volume of blood per stroke is 60 cm$^3$ and the ejection time is 0.2 s. The difference in height from the left ventricle (LV) to the aortic arch (AA) is 15 cm, and the velocity of the blood at the aortic arch is 40 cm/s. The arterial pressure at the aorta (A) ranges from 10,700 N/m² (80 mm Hg) during diastole to 16,000 N/m² (120 mm Hg) during systole. The average pressure in the pulmonary artery (PA) is 2670 N/m² (20 mm Hg) and the difference in height from the right ventricle (RV) to the pulmonary artery is 7 cm.

**11.4** The blood flow in a resting man is 5 L/min. Consider the flow to be steady both in the aorta and in the arterioles and both types of blood vessels to be rigid. The aorta has a radius of 1.2 cm and a length of 60 cm. The number of arterioles is about 60 million. Suppose that each arteriole has a diameter of 20 $\mu$m and a length of 6 mm. The viscosity of the blood is $4 \times 10^{-3}$ Ns/m² in the aorta and $1.5 \times 10^{-3}$ Ns/m² in the arterioles.
**(a)** What is the pressure drop in the aorta and in the arterioles? Calculate the pressure drops also as percentages of the total pressure drop in the systemic circulation (16,000 N/m²).
**(b)** By how many times is the resistance of the arterioles larger than the resistance of the aorta?

**11.5** The radius of the human aorta is $r = 1.2$ cm. The cardiac output in a resting man is 5 L/min; that is, the heart pumps $\Delta V = 5$ L of blood into the aorta in $\Delta t = 1$ min.
**(a)** What is the average flow velocity in the aorta?
**(b)** What is the critical velocity of blood flow in the aorta? The density of blood is $\rho = 1020$ kg/m³, and its viscosity in the aorta is $\eta = 0.004$ Ns/m². The Reynolds number Re = 2320.

**11.6** The diameter of the abdominal aorta of the cat is $d = 3$ mm and the peak velocity of the blood is $v = 0.6$ m/s. If a disturbance is initiated in this vessel (e.g., at the wall), is it likely to propagate or will it die out? The viscosity and the density of the blood are $\eta = 0.003$ Ns/m² and $\rho = 940$ kg/m³, respectively.

**11.7** Assume the maximum force that an eardrum can withstand without breaking is 3.0 N and the area of the eardrum is 1.0 cm².
**(a)** Calculate the maximum tolerable pressure in the middle ear.
**(b)** To what maximum depth could a person dive in fresh water without bursting an eardrum?

**11.8** The left ventricle of the human heart can be approximated by a thick-walled hemispherical shell with inner radius $r_i$ and outer radius $r_o$. The blood pressure inside is $p_i$ and the pressure from the pericardium is $p_o$. On both sides, the pressure is uniform. What is the circumferential wall stress in the heart? Estimate the maximum stress if the systolic pressure of the left ventricle is 20,000 N/m² (150 mm Hg), the average volume is 100 mL, and the volume of the wall is 30 mL.

**11.9** A segment of an artery is narrowed down by an arteriosclerotic plaque to one-fifth of its cross-sectional area. What is the percentage drop in the blood pressure here? Use the typical values $p_1 = 13300$ N/m² (100 mm Hg) for the blood pressure and $v_1 = 0.12$ m/s for the velocity of blood in the healthy artery.

**11.10** Arterial blood pressure is routinely measured by the auscultatory method. An inflatable cuff is wrapped around the arm and attached to an open-tube mercury manometer. The pressure in the cuff above atmospheric level is shown by the difference in the heights of the mercury columns on the two tubes. In practice, the short, wide tube is usually hidden, and only the long, narrow tube can be observed. Considering the geometry given in the figure, what does a 100–mm rise in the mercury column in the narrow tube mean in terms of the pressure of the cuff? The inner radii of the tubes are $R = 7.0$ mm and $r = 1.3$ mm, respectively.

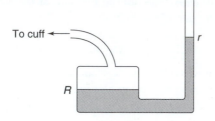

**11.11** The speed of blood in the aorta is 50 cm/s, and this vessel has a radius of 1.0 cm.
(a) What is the rate of blood flow through this aorta?
(b) If the capillaries have a total cross-sectional area of 3000 cm², what is the speed of the blood in them?

**11.12** A hypodermic syringe is filled with a solution whose viscosity is $1.5 \times 10^{-3}$ Ns/m². As the figure shows, the plunger area of the syringe is $8.0 \times 10^{-5}$ m², and the length of the needle is 0.025 m. The internal radius of the needle is $4.0 \times 10^{-4}$ m. The gauge pressure in a vein is 2000 N/m². What force must be applied to the plunger, so that $1.0 \times 10^{-6}$ m³ of solution can be injected in 5.0 s?

Area = $8.0 \times 10^{-5}$ m²

0.025 m

F

**11.13** An alveolus has a radius of $2.5 \times 10^{-4}$ m and experiences a pressure of $1.5 \times 10^{3}$ Pa. Assuming that its shape is almost spherical, calculate the surface tension due to its walls.

**11.14** Calculate the total surface area of the alveoli in the lungs of an adult, given that there are 300 million alveoli and their average radius is $1.0 \times 10^{-4}$ m.

**11.15** Sodium sulfate solution has a surface tension of $7.3 \times 10^{-2}$ N/m and a density of 1500 kg/m³. It is sometimes used as a plant fertilizer. Calculate the maximum diameter of xylem ("plant piping") which would raise the fertilizer to the top of a 0.5-m-high plant. Assume that the contact angle is 0°.

**11.16** The smallest capillaries in trees are 0.01 mm in radius. Given that capillary action can raise sap to a height of only 0.4 m in trees, calculate the radius of a capillary that would raise sap to the top of a 50-m-tall cedar tree.

**11.17** What is the tension in an arteriole wall if the mean blood pressure is 60 mm Hg and the radius is $1.5 \times 10^{-4}$ cm? Compare it to the tension in a capillary wall if the blood pressure is 30 mm Hg and the radius is $5 \times 10^{-6}$ m?

**11.18** Suppose the pressure in a patient's stomach is 18 mm Hg. To what height must the food in a nasogastric feeding arrangement be raised for the pressure due to the weight of the food to be twice the pressure in the stomach? The density of the food ingested is 1250 kg/m$^3$.

**11.19** Calculate the blood pressure in millimeters of mercury in an artery in the brain 0.35 m above the heart. The pressure at the heart is 120 mm Hg and the density of blood is 1050 kg/cm$^3$. Compare this value to the blood pressure in an artery in the foot 1.5 m below the heart.

**11.20** Blood flow is increased during exercise by the dilation of vessel radii together with an increase in pressure. Suppose that the flow rate increases by a factor of 4.5 and blood pressure increases by 40%. Calculate the factor by which the average blood vessel radius must have increased to produce this flow.

**11.21** Calculate the average blood velocity in an artery supplying the brain if its radius is 0.0050 m and the flow rate through it is $4 \times 10^{-6}$ m$^3$/s. What is the average velocity at a constriction in the artery if that constriction reduces the radius by a factor of two? Assume the same flow rate as above.

**11.22** The blood in an artery of radius $5 \times 10^{-3}$ m flows with a speed of 0.15 m/s. This artery subdivides into a large number of capillaries of radius $5 \times 10^{-6}$ m. The flow velocity in the capillaries is $5 \times 10^{-4}$ m/s. Calculate the number of capillaries this artery divides into.

# CHAPTER *12*

# TEMPERATURE AND HEAT

## PHYSICAL BACKGROUND

On the Celsius temperature scale, there are 100 equal divisions between the ice point (0°C) and the steam point (100°C). For most scientific work, the Kelvin temperature scale is the scale of choice. One kelvin is equal in size to one Celsius degree; however, the temperature $T$ on the Kelvin scale differs from the temperature $T_c$ on the Celsius scale by an additive constant of 273.15: $T = T_c + 273.16$. The lower limit of temperature is called absolute zero and is designated as 0 K on the Kelvin scale. The operation of any thermometer is based on the change in some physical property with temperature, for example, the thermal expansion of a substance.

Most substances expand when heated. For linear expansion, an object of length $L_0$ experiences a change in length $\Delta L$ when the temperature changes by $\Delta T$:

$$\Delta L = \alpha L_0 \, \Delta T \tag{12.1}$$

where $\alpha$ is the coefficient of linear expansion. For volume expansion, the change in volume $\Delta V$ of an object of volume $V_0$ is given by

$$\Delta V = \beta V_0 \, \Delta T \tag{12.2}$$

where $\beta$ is the coefficient of volumetric expansion. For an object held rigidly in place, a thermal stress can occur when the object attempts to expand or contract.

The internal energy of a substance is the sum of the internal kinetic, potential, and other kinds of energy that the molecules of the substance have. Heat is energy that flows from a higher temperature object to a lower temperature object. The specific heat capacity $c$ of a substance of mass $m$ determines how much heat $Q$ must be supplied to or removed from the substance to change its temperature by an amount $\Delta T$:

$$Q = cm \, \Delta T \tag{12.3}$$

and $Q < 0$ when heat is transferred from the surroundings to a system while $Q > 0$ when heat is transferred from the system to the surroundings. Heat is measured in units of calories. One calorie is the amount of heat required to raise the temperature of 1 g of water by 1 K. When materials are placed in thermal contact within a perfectly insulated container, the principle of energy conservation requires that heat lost by warmer materials equals heat gained by cooler materials.

Heat of transformation is similar to the specific heat but accounts for changes in the phase of the body. The specific heat of a body assumes that no change in phase occurs during a temperature change. For a substance to change states of matter, heat energy must be added to the substance. The amount of heat required to change the phase of 1 kg of a substance is the heat of transformation $L$. Thus, the total amount of heat $Q$ gained or lost by a substance of mass $m$ during a change between phases is

$$Q = mL \tag{12.4}$$

where $L$ is the heat of transformation (latent heat) characteristic of the substance. The heat of transformation exists in two forms, according to the particular phase transformation: (a) Heat of fusion $L_f$ is the amount of heat energy required to change 1 g of solid matter to liquid and (b) heat of vaporization $L_v$ is the amount of heat energy required to change 1 g of liquid matter to gas.

## TOPIC 12.1 Thermography

Normal body temperature falls into a narrow range, as shown in Figure 12.1. When the temperature of different parts of the body is raised, radiation increases substantially. This is documented using an infrared camera which registers the intensity of infrared radiation (see Chapters 13 and 24) produced at different locations of the human body. A person's overall body temperature can indicate the presence and seriousness of an infection since one of the body's defense mechanisms against disease is to raise its temperature. Skin temperature is lower than core temperature but higher than normal room temperature. It is therefore possible to measure the infrared radiation from a person. Since this radiation's intensity is proportional to absolute temperature to the fourth power (see Chapter 13), the amount of infrared radiation is a sensitive indicator of surface temperature. The technique of measuring infrared radiation and thereby mapping temperature is called thermography. The camera is connected to a color monitor that displays different intensities as different colors. The colored picture is called a thermograph or thermogram.

In medicine thermography is an important diagnostic tool; for example, breast cancer may show up on a thermograph because malignant tissues are often associated with elevated temperatures. Thermography gives an indication of blood supply, since one of the main methods of heat transfer in the body is blood flow. A depressed skin temperature indicates a deficiency in blood flow to a given region. This could be caused by, for example, clotting or stroke. A locally elevated temperature can indicate the presence of a malignant tumor. Such tumors grow very rapidly compared to other tissues, and this requires an increased blood supply.

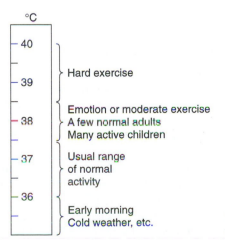

**FIGURE 12.1**  Body temperature during various activities and at particular times.

## TOPIC 12.2 Hypothermia and Low Temperatures in Biology and Medicine

The generally accepted figure for the average normal body temperature (measured in the mouth) is 37°C. When the ambient temperature is lower than body temperature, heat is lost from the body. If the body loses too much heat, circulatory mechanisms are activated which cause a reduction of blood flow to the skin in order to reduce the heat loss. The physiological response to the cold (i.e., shivering) results in an increase in heat production. This form of internal heating is generated from carbohydrate reserves and continues until they are depleted. Once this happens, the body temperature begins to drop again, and if the body temperature drops below 33°C, external heat must be supplied to the person; otherwise severe thermal injuries may arise as a result of hypothermia.

There is an overall slowing down of tissue metabolism when a physiological system is cooled below its normal temperature. There are, in fact, millions of enzyme-assisted reactions which normal metabolic processes depend on in living cells. These reactions are not affected by temperature in the same way; for example, some enzymes work more efficiently at 25°C than at 38°C. Some reactions are inhibited almost completely at the lower temperature. Cooling, therefore, upsets the metabolic balance, causes cumulative biological disturbance and modifies the function of cells; for example, some homeothermic animals have a constant body temperature usually higher than the environment and cannot survive lowering their body temperatures by 10 to 20 K. In some quarters it has been suggested that a small lowering of body temperature might assist patients with circulatory disorders and heart trouble because this would reduce the work load on the heart. However, the risk which arises from disturbing the metabolic balance could make this procedure inadvisable. Surgery using the application of cold is called cryosurgery. If the temperature of the entire body is lowered, the metabolic rate drops and most bodily functions are slowed. This can be advantageous in certain types of surgery. More often, however, cold is used to freeze small regions of the body.

The removal of heat from the body can also be of therapeutic value. The method of removing heat is most often conduction, sometimes convection. Lowered temperature acts as a local anesthetic. Children who are teething are fond of sucking on ice cubes to relieve pain. Swelling can sometimes be reduced by the application of ice packs.

A physiological system such as the organs and tissues of animals or humans, when cooled down below the freezing point of the cellular fluids, which is slightly below 0°C, will be in danger that the water present inside it will freeze. Ice forms first in the external water surrounding the cells. Since water expands in volume on freezing, a certain amount of mechanical deformation of structure will result without inflicting critical damage. Furthermore, inside the cells the water remains in the supercooled liquid state until the temperature has dropped to somewhere between −5 and −15°C. When ice has formed inside the cell, it is nearly always lethal to the cell. The freezing of the water results in dehydration of the cell since the concentration of the various solutes rapidly increases, and such changes in concentration cause violent changes in metabolic rates. It is believed that dehydration is the most important cause of freezing damage. However, cells may be protected from freezing damage by the use of suitable biological "antifreezes." Such substances, known as cryoprotective agents, inhibit the freezing of the cellular water and prevent a lethal increase in the concentration of the solutes. Glycerol is the most important cryoprotective agent, and it has found important use in the long-term storage of blood for transfusion purposes. Blood collected in the normal way and stored at 4°C has a shelf life of only about three weeks. In the freezing technique, the red cells are first separated from the blood plasma and a solution of glycerol is added to the red cells. The resulting mixture is frozen and stored at a very low temperature. At –80°C a storage time of about two years is possible, while at −160°C this rises to five years.

Poikilotherms (whose body temperature is the same as the environment), whether animals, plants, or even microorganisms, are not as seriously affected by a reduced body temperature as the homeotherms. Their metabolic rates, however, suffer a profound alteration. Provided that actual freezing of cellular water does not occur, such changes in poikilotherms are perfectly reversible. This is so even when the temperature is lowered below 0°C. Ice melts at 0°C, but water can exist as a liquid well below this, and the phenomenon is called supercooling. Some insects have recovered after being supercooled to –30°C. If the rate of cooling is too rapid, some types of cells can be killed. This effect is termed thermal shock and is usually attributed to changes in the structure of cell membranes.

Finally, lowered temperatures can serve as a preservative, as in food refrigerators and freezers. Blood, bone marrow, and sperm are among the substances preserved by freezing.

## TOPIC 12.3  Heat Stroke

At the opposite extreme, if the body is subjected to hot temperatures over an extended period, the body becomes overheated and the body temperature begins to rise. Severe and prolonged elevation of body temperature gives rise to the clinical condition known as heat stroke (above about 40°C). The effects of heat stroke are particularly important in the thermal regulation of physiological processes since they are related to metabolic activity. The body metabolism involves the conversion of energy, supplied by food, to heat. When at rest the normal human metabolizes at a rate of approximately 75 W, and under prolonged periods of strenuous exercise this can approach 800 W. If the exercise is outdoors in intense sunlight, the rate increases to about 950 W.

The body maintains a constant temperature due to its ability to dissipate the generated heat efficiently via many mechanisms such as radiation, conduction, convection, evaporation, and respiration (see Chapter 13). The consequences of heat stroke and an elevated body temperature include a cessation of cell growth at $\approx 41°C$ and can result in irreversible chemical damage to vital organs such as the kidneys and the brain at 42°C.

Local and general applications of heat have long been recognized for their therapeutic value. For example, heat applied to a sore muscle can significantly relieve pain. The mechanism of relief by elevated temperature seems to be twofold: relaxation of muscles and increased blood flow. When the temperature of a region is elevated, the body responds by increasing blood circulation to that region to cool it down, that is, to carry away heat by convection. Various forms of electromagnetic radiation are used in heat therapy, including infrared, microwaves, and radio waves. The applications of heat treatments vary since each method of treatment has its own advantages and disadvantages. For example, microwaves are thought to cause cataracts; ultrasound penetrates deeper than microwaves, but it can cause tissue burns near bones; and prolonged application of infrared from a heat lamp can cause swelling.

## SOLVED PROBLEMS

**12.1**  Suppose the air temperature is 34°C, the same as normal skin temperature, and a person rides a bike at a speed of 15 km/h. How many grams of water must this cyclist evaporate each minute to get rid of the body heat produced by this activity? Note that power consumption while cycling is 400 W and the efficiency of cycling is approximately 20%.

**Solution** Since air temperature is the same as skin temperature, conduction and radiation will not transfer any heat. Hence, all of the body heat must be transferred out of the person by the evaporation of perspiration from the skin and water from the lungs. The power consumption while cycling is 400 W ($P_{in} = 400$ W), and the efficiency for cycling is 20%. This means that 80 W is going to useful work, and the remaining 320 W is going to waste heat. Accordingly,

$$P_{out} = eP_{in} = (0.20)(400 \text{ W}) = 80 \text{ W}$$

Moreover, $P_{in} = P_{out} + P_{heat}$, and therefore

$$P_{heat} = P_{in} - P_{out} = 320 \text{ W}$$

By definition, power is energy per unit time. Here energy is in the form of heat:

$$P_{heat} = \frac{Q}{t}$$

so that

$$Q = P_{heat} t = (320 \text{ W})(60 \text{ s}) = 4587 \text{ cal}$$

Now, the energy required to cause a phase change is

$$Q = mL_v$$

Solving for the mass, given that the heat of vaporization of water at body temperature is 580 cal/g, yields

$$m = \frac{Q}{L_v} = \frac{4587 \text{ cal}}{580 \text{ cal/g}} = 7.9 \text{ g}$$

**12.2** How much will the temperature of the cyclist in the previous example increase in 1 h if none of the body heat generated is lost to the surroundings? Assume the cyclists mass is 76 kg and the body's specific heat is 0.83 cal/g °C.

**Solution** The energy required to cause a temperature change is given by

$$Q = cm \, \Delta T$$

Therefore,

$$\Delta T = \frac{Q}{cm} = \frac{(4587 \text{ cal/min})(60 \text{ min})}{(0.83 \text{ cal/g °C})(76 \times 10^3 \text{ g})} = 4.4°C$$

**12.3** Calculate the power in waste heat put into a classroom by 30 students generating 210 W each. How many calories of energy do the students put into the room during a 50-min class?

**Solution** Assuming a value of 210 W generated per person, the power in waste heat is simply

$$P_{heat} = (210 \text{ W/person})(30 \text{ persons}) = 6300 \text{ W}$$

The number of calories can be calculated by multiplying the power by time, or

$$Q = (3.0 \text{ kcal/min/person})(30 \text{ persons})(50 \text{ min}) = 4.5 \times 10^6 \text{ cal}$$

which is enough heat to raise the temperature of 100 kg of water by 4.5°C!

# EXERCISES

**12.1** On a hot day's race, a cyclist consumes 7.0 L of water over the span of 3.5 h. Making the approximation that all of the cyclist's energy goes into evaporating this water as sweat, how much energy in kilocalories did the person use during the race?

**12.2** The latent heat of vaporization of $H_2O$ at body temperature (37.0°C) is $2.42 \times 10^6$ J/kg. To cool the body of a jogger with an average specific heat capacity of 3500 J/kg °C by 1.5°C, how many kilograms of water in the form of sweat must be evaporated?

**12.3** Blood can carry excess energy from the interior to the surface of the body, where the energy is dispersed in a number of ways. While a person is exercising, 0.5 kg of blood flows to the surface of the body and releases 1800 J of energy. The blood arriving at the surface has the temperature of the body interior, 37.0°C. Assuming that blood has the same specific heat capacity as water, determine the temperature of the blood that leaves the surface and returns to the interior.

**12.4** When resting, a person has a metabolic rate of about $4 \times 10^5$ J/h. The person is submerged neck-deep into a tub containing 1000 kg of water at 28°C. If the heat from the person goes only into the water, find the water temperature after half an hour.

**12.5** The density of water at 0°C is 999.8 kg/m$^3$ while that of ice at 0°C is 917.0 kg/m$^3$. What is the pressure generated by expanding ice as it freezes in a biological cell whose major component is water? Assume the cell to be a sphere of radius 5 $\mu$m.

**12.6** A 70-kg man climbs a mountain 1200 m high (measured from the base) in 4 h and uses 9.8 kcal/min.
**(a)** Calculate his power consumption in watts.
**(b)** What is his power output in useful work?
**(c)** What was the efficiency of this man during the climb?

**12.7** **(a)** What percentage of the body's energy is used by the heart and skeletal muscles when sleeping? The power required to maintain the body's metabolism while sleeping is approximately 83 W.
**(b)** How many kilocalories per minute are required for the heart and skeletal muscles?

**12.8** A burn produced by steam at 100°C is more severe than one produced by the same amount of water at 100°C. To verify this, calculate the heat that must be removed from 0.01 kg of water at 100°C to lower its temperature to 40°C. Compare it to the heat removed from the same amount of steam at 100°C.

# CHAPTER *13*

# TRANSFER OF HEAT

## PHYSICAL BACKGROUND

There are three mechanisms of heat transfer, all of which are important in maintaining normal human body temperature.

Convection is the process in which heat is carried by the bulk movement of a fluid. During natural convection, the warmer, less dense part of a fluid is pushed upward by the buoyant force provided by the surrounding cooler and denser part. Forced convection occurs when an external device, such as a fan or pump, causes the fluid to move.

Conduction is the process whereby heat is transferred directly through a material, any bulk motion of the material playing no role in the transfer. The heat $Q$ conducted during a time $t$ through a bar of length $L$ and cross-sectional area $A$ is expressed as

$$Q = \frac{(kA \, \Delta T)t}{L} \tag{13.1}$$

where $\Delta T$ is the difference in temperature between the ends of the bar and $k$ is the thermal conductivity of the material. Materials that have large values of $k$, such as most metals, are known as thermal conductors. Materials that have small values of $k$, such as styrofoam and wood, are referred to as thermal insulators (see Table 13.1). Conduction is the primary source of heat transfer in solids.

Radiation represents the transfer of heat energy by electromagnetic waves which are emitted by rapidly vibrating electrically charged particles. The electromagnetic waves propagate from the heated body or source at the speed of light ($3 \times 10^8$ m/s) (see Chapter 24). Two examples of heat transfer by radiation are (a) the heating of food using a microwave oven and (b) the heat energy received from the sun.

Consider an object kept at a temperature $T$. The rate of emission of heat energy by radiation of the body is

$$Q = Ae\sigma T^4 \tag{13.2}$$

where $A$ is the surface area of the object emitting radiation energy and $e$ is the emissivity of the object surface. Here, $e$ is unitless and varies between 0 (poor absorber of radiation) and 1 (excellent absorber of radiation or blackbody). In equation (13.2), $\sigma$ is the Stefan–Boltzmann constant ($\sigma = 5.67 \times 10^{-8}$ W/m$^2 \cdot$ K$^4$), and $T$ is the absolute temperature, in kelvins. If the object initially held at a temperature $T$ is subjected to an environment at temperature $T_s$, then the net flow of heat from the body to its surroundings is

$$P_{\text{net}} = e\sigma A(T^4 - T_s^4) \tag{13.3}$$

Objects that are good absorbers of radiant energy are also good emitters, and objects that are poor absorbers are also poor emitters. An object that absorbs all the radiation incident upon it is called a perfect blackbody. A perfect blackbody, being a perfect absorber, is also a perfect emitter.

**TABLE 13.1  Thermal Conductivities of Selected Materials (at 20°C Unless Otherwise Noted)**

| Substance | Thermal conductivity, $k$ (J/s · m · °C) |
|---|---|
| Metals | |
| Aluminum | 240 |
| Brass | 110 |
| Copper | 390 |
| Iron | 79 |
| Lead | 35 |
| Silver | 420 |
| Steel (stainless) | 14 |
| Gases | |
| Air | 0.0256 |
| Hydrogen ($H_2$) | 0.180 |
| Nitrogen ($N_2$) | 0.0258 |
| Oxygen ($O_2$) | 0.0265 |
| Other materials | |
| Asbestos | 0.090 |
| Body fat | 0.20 |
| Concrete | 1.1 |
| Diamond | 2450 |
| Glass | 0.80 |
| Goose down | 0.025 |
| Ice (0°C) | 2.2 |
| Styrofoam | 0.010 |
| Water | 0.60 |
| Wood (oak) | 0.15 |
| Wool | 0.040 |

## TOPIC 13.1  Heat Regulation in Animals and Humans

Two main groups of animals may be distinguished with regard to their body temperatures (see also Topics 12.2 and 12.3). The larger of these groups comprises the cold-blooded creatures (poikilotherms) such as lizards and snakes. Within this group the blood temperature follows the environmental temperature. Since animal activity decreases as blood temperature is lowered, poikilotherms become very inactive when placed in cold surroundings. The smaller group comprises the mammals and the birds known as homeotherms. Most mammals have a blood temperature of between 36 and 38°C, while the blood temperature of birds is a little higher, varying between 41 and 43°C. The blood temperature of homeotherms remains constant under widely varying conditions.

Homeothermy involves an automatically regulated balance between heat production in the body of the animal and heat losses from the surface. The regulatory mechanism affects (i) the blood flow to the extremities of the body; (ii) the secretion of sweat from the skin; (iii) the rate of breathing, which affects the rate of water evaporation in the respiratory system; (iv) the bristling of the hair or fluffing of the feathers as the case may be; and (v) the metabolic rate of heat production itself. Processes (i) through (iv) affect the loss of heat while (v) affects heat generation. While small and large mammals maintain the same body temperature, the relative rate of heat production is higher in small animals, since the small animal has a larger body surface compared to its mass. Since heat loss takes place from the surface, an animal must produce heat at a rate equal to the loss to stay warm.

The most obvious means of physically regulating heat losses is to achieve a high degree of thermal insulation between the body and the surrounding air. Fat under the skin serves this purpose since it has a low thermal conductivity. Fur on a mammal and feathers on a bird are good insulators. Furthermore, in extremely cold weather the fur can be bristled and the feathers fluffed to trap air in the layer separating the skin from the outside cold, and air has a very low thermal conductivity. Control of heat loss by increased surface circulation ceases to be effective when the external temperature is equal to the blood temperature. In this case overheating is prevented by an increase in the rate of evaporation of water achieved by increased sweating (as in humans) or in the respiratory tract (as in dogs). In summary, the amount of heat flowing through the body's surface area depends on: (i) the surface area, (ii) the thermal conductivity, (iii) the temperature difference across it, and (iv) its thickness.

Some authors have claimed that a camel is able to convert the fat in its hump into water as well as food. However, if this were so, the animal would have to increase its oxygen intake and would actually lose more water from the surfaces of its lungs than it would gain by converting the fat. To explain this phenomenon, we must take into account three important factors. First, the blood temperature of a camel deprived of water can vary much more than most mammals. This may be 34°C when it is cool up to 40.5°C at the hottest time of day. As a consequence, the difference in temperature between the camel's body and the outside air is much less than it would be for a human. Second is the highly insulating character of the camel's hair. This results in the temperature drop between its skin and the outside air being greater over this relatively small distance. Thus the temperature gradient is lessened by the same factor. A camel will not freely sweat until its blood temperature has reached 40.5°C, that is, only during the very hottest part of the day. However, a man sweats continuously from dusk to dawn. The third factor is that when a camel loses water, only a small proportion comes from the plasma water of the blood. Thus the blood continues to flow freely. A camel can in fact lose 25% of its body weight without doing serious damage to itself. Contrast this again with the situation in humans who will lose a significant amount of plasma water. Furthermore, human blood will become thick, the heart will have to do more work to pump it, and the reduced circulation can no longer dissipate the metabolic heat. This results in physical deterioration when only 5% of body weight is lost, delirium when 10% is lost, and death at 12% loss.

Remarkable adaptation mechanisms have been developed by fish like the mackerel sharks and tunas. Sea creatures obtain their oxygen from the seawater around them because nearly all of them are poikilotherms rather than from the air as mammals do (e.g., whales, porpoises and birds). However, the oxygen in seawater is simply not sufficient to provide the amount of metabolic heat required to maintain a high body temperature. Water has a high thermal capacity compared with air, and the venous blood in the gills loses a lot of heat to the seawater flowing through the gills and cools to the sea temperature. It transpires that this blood loses heat faster to the water than it can absorb oxygen. Thus, even if the oxygen supply could be increased, this still would not increase the body temperature. Tuna and mackerel sharks possess a very efficient modification to their circulatory system which provides a thermal barrier against heat loss. It is called the rete mirabile (wonderful net). This modification is made up of a mass of very fine veins and arteries which acts like a heat exchanger. The warm oxygen-deficient blood in the veins of the rete loses its heat to the cold oxygen-rich blood in the arteries. Thus the arterial blood is warmed long before it gets to the interior of the fish. The venous blood, on the other hand, reaches the sea temperature before it reaches the gills. In the rete veins and arteries are so closely linked that the mutual area of contact is quite large and the heat exchange process very efficient as a consequence. Only heat is exchanged in the rete. No oxygen is exchanged as the walls of the vessels are too thick to allow any diffusion of oxygen. The swimming muscles can operate at a generally increased temperature because they are protected partly by a rete mirabile. Of course, this enormously increases their power output and gives the warm-blooded animal a distinct advantage in the survival stakes.

Bluefin tuna, equally at home in arctic and tropical waters, have been demonstrated to regulate their temperature in essentially the same way as mammals and birds, but the associated mechanism is not so efficient. When the seawater temperature is 20°C, the yellowfin's maximum muscle temperature is about 23°C. When the sea temperature is 31°C, the muscle temperature rises to 34°C, so a constant temperature difference of 3°C between the fish and the environment is preserved. The skipjack tuna maintains a constant difference of 6°C. In the bluefin, on the other hand, muscle temperatures remain between 25 and 32°C almost independently of seawater temperature, but the mechanism whereby this is achieved is not known.

An interesting example of forced convection is blood circulation, which can transfer heat from one part of the body to another. The circulatory system is highly adjustable; blood flow can be increased or decreased to specific areas depending on need. If a person becomes overheated, blood vessels to the surface dilate, carrying more blood to the surface for cooling. Air convection around the body also speeds the rate of heat transfer.

Nature uses air as an insulator by cleverly controlling both conduction and convection. Hair and feathers are designed to trap air in small pockets and greatly slow the rate of heat transfer by convection. The density of uncompressed hair and feathers, which are mostly air, is quite small. Convection is going on in each little chamber, but the distance that the air can move is very small and thus the rate of heat transfer is very small.

## SOLVED PROBLEMS

**13.1** Using dimensional analysis, compare the heat loss per body mass for a small animal like a hummingbird and a large animal like an elephant.

**Solution** The energy production of an animal can be safely assumed to be proportional to the weight of the animal. Equivalently, the heat production is proportional to the volume, or heat production $Q_P$ is proportional to $r^3$, while heat loss $Q_L$ is proportional to $r^2$. Consequently, the ratio of $Q_P/Q_L$ is proportional to $r$. If we consider for simplicity the spherical elephant and the spherical hummingbird, the ratio of heat productions to heat loss of the two animals would be

$$\frac{Q_P/Q_L \text{ (elephant)}}{Q_P/Q_L \text{ (hummingbird)}} = \frac{r \text{ (elephant)}}{r \text{ (hummingbird)}}$$

Clearly, it is much easier for the elephant to keep warm.

**13.2** An effective mechanism for transfer of excessive heat is conduction through body fat. Suppose that heat travels through 0.030 m of fat in reaching the skin, which has a total surface area of 1.7 m² and a temperature of 34.0°C compared to the normal interior temperature of 37°C. Find the amount of heat that reaches the skin in an hour.

**Solution**   The thermal conductivity of body fat is given as $k = 0.20$ J/s · m · °C. According to the formula

$$Q = \frac{(kA\ \Delta T)t}{L}$$

we find

$$Q = \frac{(0.02\ \text{J/s} \cdot \text{m} \cdot °\text{C})(1.7\ \text{m}^2)(37.0°\text{C} - 34.0°\text{C})(3600\ \text{s})}{0.030\ \text{m}} = 12240\ \text{J}$$

**13.3**    The human body of surface area 1.5 m$^2$ with a surface temperature of 23°C radiates heat in the form of energy. If the body is surrounded by an environment of 20°C, determine (a) the total rate of heat energy and (b) the net rate of heat loss from the body. Assume the emissivity $e = 1$.

**Solution**   First note that 23°C = 296 K and 20°C = 293 K

**(a)** The rate of radiation of heat energy is given by

$$Q = Ae\sigma T^4 = (1.5\ \text{m}^2)(5.67 \times 10^{-8}\ \text{W/m}^2 \cdot \text{K}^4)(296\ \text{K})^4 = 652.9\ \text{W}$$

**(b)** The net rate of heat energy loss due to radiation is given by

$$Q = Ae\sigma(T^4 - T_s^4) = (1.5\ \text{m}^2)(5.67 \times 10^{-8}\ \text{W/m}^2 \cdot \text{K}^4)[(296\ \text{K})^4 - (293\ \text{K})^4] = 26.1\ \text{W}$$

**13.4**    What is the total emitted power and the net power loss due to radiation of a human body with a surface temperature of 32°C? Assume that the surface area of the body is 1.5 m$^2$ and the temperature of the surroundings is 20°C.

**Solution**   The total power $P_t$ radiated is represented by the area under the spectral emittance curve of the blackbody (the Stefan–Boltzmann law):

$$P_t = e\sigma A T^4$$

where $\sigma$ is the Stefan–Boltzmann constant and $A = 1.5$ m$^2$ is the surface area of the body. Numerically, for $T = 32°\text{C} = 305$ K, $P_t = 736$ W. The power loss of radiation from the body will be smaller, i.e. 627 W, since the same body will be receiving radiant power from the surroundings according to the same Stefan–Boltzmann law. The net amount of energy radiated per unit time is

$$P_n = e\sigma A(T^4 - T_0^4)$$

where $T_0 = 20°\text{C} = 293$ K is the temperature of the surroundings. The final answer is, $P_n = 736\ \text{W} - 627\ \text{W} = 109\ \text{W}$.

## EXERCISES

**13.1**    Estimate the rate at which the top of a person's head absorbs the sun's energy on a clear day if (a) it is covered with hair ($e = 0.75$), it has a surface area of 250 cm$^2$ (assumed flat), and the person stands straight with the sun making an angle of 45° with the vertical or if (b) it is bald with $e = 0.25$ under the same conditions.

**13.2**    Calculate the distance required for the heat flow by conduction from the blood capillaries beneath the skin to the surface if the temperature difference is 1.5°C? Assume that 250 W must be transferred through the whole body's surface area of 1.5 m$^2$.

**13.3** Suppose the skin temperature of a naked person is 35°C when the person is standing inside a room whose temperature is 22°C. The surface area of the individual is 1.5 m². (a) Assuming the emissivity is 0.75, find the net loss of radiant power from the body. (b) Determine the number of food calories of energy that are lost in 1 h due to the net loss rate obtained in part (a).

**13.4** In the heat conduction equation

$$Q = \frac{(kA\,\Delta T)t}{L}$$

the combination of factors $kA/L$ is called the conductance. The human body has the ability to vary the conductance of the tissue beneath the skin by means of vasoconstriction or vasodilation, in which the flow of blood to the veins and capillaries underlying the skin is decreased or increased, respectively. The conductance can be adjusted over a range such that the tissue beneath the skin is equivalent to a thickness of 0.080 mm of styrofoam or 3.5 mm of air. Estimate the value of the factor by which the body can adjust the conductance.

**13.5** Assuming that the body's core temperature is 37°C and skin surface temperature is 34°C, determine the rate of heat conduction of the human body. The surface area of the body is 1.5 m², and the thickness of human tissue, $h$, averages 0.01 m. Thermal conductivity for tissue is taken to be 0.2 J · s$^{-1}$ · m$^{-1}$ · °C$^{-1}$.

**13.6** Arctic mammals like the polar bear both have a layer of fat plus another insulating layer of air, namely in their pelts. The pelt traps a layer of stationary air. Heat transfer through the pelt is through thermal conduction. (a) Derive a formula for the difference between the inner body and the environmental temperature. Include the thickness of the fat layer and the thickness of the air layer. (b) For simplicity, assume a spherical polar bear with a 0.05-m-thick layer of fat. How thick should the layer of air be for the polar bear to be comfortable at a temperature of −25°C? Estimate the bear's mass to be 1200 kg.

**13.7** In thermography radiant heat is measured from the surface of a body to determine temperature differences. Estimate the percentage difference in the rate of heat transfer by radiation from an area of 1 m² at a temperature of 37°C compared to that at 40°C. Assume the surroundings to be at room temperature.

**13.8** Calculate the rate of heat conduction in watts out of an animal with 0.035 m thick fur. The surface area of the animal is 1.5 m², its skin temperature is 35°C, and the temperature of the surrounding air is 2°C. Heat losses due to convection and conduction can be neglected, and the thermal conductivity of the fur can be assumed to be the same as that for air.

# IDEAL GAS LAW AND KINETIC THEORY

## PHYSICAL BACKGROUND

Each element in the Periodic Table is characterized by an atomic mass. One atomic mass unit (a.m.u.) is exactly one-twelfth the mass of an atom of carbon-12. The molecular mass of a molecule is the sum of the atomic masses of its atoms. One mole of a substance contains Avogadro's number $N_A$ of particles, where $N_A = 6.022 \times 10^{23}$. The mass in grams of one mole of a substance is equal to the atomic or molecular mass of its particles.

The ideal gas law relates the absolute pressure $P$, the volume $V$, the number of moles $n$, and the Kelvin temperature $T$ of an ideal gas according to

$$PV = nRT \tag{14.1}$$

where $R = 8.31$ J/mol $\cdot$ K is the universal gas constant. An alternative form of the ideal gas law is

$$PV = NkT \tag{14.2}$$

where $N$ is the number of particles and $k = R/N_A$ is Boltzmann's constant and is equal to $1.38 \times 10^{-23}$ J/K. A real gas behaves as an ideal gas when the density of the real gas is low enough that its particles do not interact, except via elastic collisions.

According to the kinetic theory of gases, an ideal gas consists of a large number of particles (atoms or molecules) that are in constant random motion. The particles are far apart compared to their dimensions, so they do not interact except when elastic collisions occur. The pressure on the walls of a container is produced by the impact of the particles with the walls. According to the kinetic theory of gases, the absolute temperature $T$ of an ideal gas is a measure of the average translational kinetic energy $\overline{\text{KE}}$ per particle through the relation

$$\overline{\text{KE}} = \tfrac{3}{2}kT \tag{14.3}$$

The internal energy $U$ of $n$ moles of a monatomic ideal gas is

$$U = \tfrac{3}{2}nRT \tag{14.4}$$

In contrast to the mass flow of liquids, diffusion involves the random spontaneous movement of individual molecules (see Figure 14.1). This process can be quantified by a constant known as the diffusion coefficient $D$ of the material, given in general by the Stokes–Einstein equation

$$D = \frac{kT}{f} \tag{14.5}$$

where $k$ is the Boltzmann constant, $T$ is the absolute temperature in kelvin, and $f$ is a frictional coefficient. If the diffusing molecule is spherical, is in low concentration, is larger

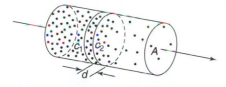

**FIGURE 14.1** Diffusion due to a gradient of concentration.

than the solvent molecules, and does not attract a layer of solvent molecules to itself, then the frictional coefficient is

$$f = 6\pi\eta r \tag{14.6}$$

where $r$ is the radius of the diffusing molecule in meters and $\eta$ is the coefficient of viscosity of the solvent expressed in newton-seconds per square meter. Since $D = kT/f$ and $f = 6\pi\eta r$, then

$$D = \frac{kT}{6\pi\eta r} \quad \text{or} \quad r = \frac{kT}{6\pi\eta D} \tag{14.7}$$

The diffusion constant of a particular molecular species depends on the nature of the molecule and on the solvent. Large molecules have smaller diffusion constants. The diffusion constant $D$ depends also on temperature. In Table 14.1, diffusion constants for several molecules at room temperature in water are listed. Note that the volume of a sphere is $V = \frac{4}{3}\pi r^3$, the mass of a molecule is $m = \rho V$, and the molar mass $M = mN_A$, where $N_A$ is Avogadro's number ($6.02 \times 10^{23}$ molecules/mol). Combining the above equations with equation (14.7), we obtain

$$M = \frac{4}{3}\pi \left(\frac{kT}{6\pi\eta D}\right)^3 N_A \rho \tag{14.8}$$

With all the quantities in this equation in SI units, the molar mass will be in kilograms per mole.

Fick's law states that the rate of diffusion per unit area in a direction perpendicular to the area is proportional to the gradient of concentration of solute in that direction. The concentration is the mass of solute per unit volume, and the gradient of concentration is the change in concentration per unit distance. If the concentration changes from $c_1$ to a lower value $c_2$ over a short length $d$ of the pipe, then the mass $m$ of the solute diffusing down the pipe in time $t$ is given by

$$\frac{m}{t} = DA \frac{c_1 - c_2}{d} \tag{14.9}$$

**TABLE 14.1  Diffusion Constants at Room Temperature in Water**

| Molecule | Diffusion constant $D$ (m²/s) |
|---|---|
| Water | $2 \times 10^{-9}$ |
| Oxygen | $8 \times 10^{-10}$ |
| Glucose | $6 \times 10^{-10}$ |
| Tobacco mosaic virus | $3 \times 10^{-12}$ |
| DNA (molar mass $5 \times 10^6$ g) | $1 \times 10^{-12}$ |
| Protein | $1 \times 10^{-10}$ |
| Hemoglobin | $6.9 \times 10^{-12}$ |

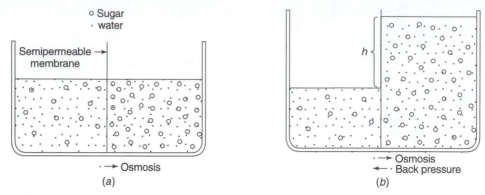

**FIGURE 14.2**  Illustration of osmosis.

This is a simplified version of Fick's law, whose differential form is given by

$$J = -D \frac{dc}{dx} \qquad (14.10)$$

This equation states that $J$, the flux of particles (number of particles passing through an imaginary normal surface of unit area per unit time) is related to the force which is pushing them $(-dc/dx)$.

Osmosis is usually defined as the transport of molecules in a fluid through a semipermeable membrane due to an imbalance in its concentration on either side of the membrane. Osmosis may be by diffusion, but it may also be a bulk flow through pores in the membrane. In either case, water moves from a region of high concentration to a region of low concentration. In Figure 14.2 the pressure on the right is then greater than the pressure on the left by an amount $h\rho g$, where $\rho$ is the density of the liquid on the right and is called the relative osmotic pressure.

The general formula for the osmotic pressure $\Pi$ of a solution containing $n$ moles of solute per volume $V$ of solvent is

$$\Pi = \frac{m}{V} RT = cRT \qquad (14.11)$$

The net osmotic pressure exerted on a semipermeable membrane separating the two compartments is thus the difference between the osmotic pressure of each compartment. This equation is known as van't Hoff's law, and it looks exactly like the ideal gas law, but osmotic pressure refers to the pressure exerted on a semipermeable membrane by an aqueous solution, while the ideal gas law refers to the pressure exerted on the wall of a container by an enclosed gas.

## TOPIC 14.1  Diffusion Through Membranes

Most diffusion processes in biological organisms take place through membranes (see Figure 14.3). All cells and some structures within cells, such as the nucleus, are surrounded by membranes. These membranes are very thin, from $65 \times 10^{-10}$ to $100 \times 10^{-10}$ m across. Most membranes are selectively permeable; that is, they allow only certain substances to cross them because there are pores through which substances diffuse. These pores are so small (from $7 \times 10^{-10}$ to $10 \times 10^{-10}$ m) that only small molecules get through. Other fac-

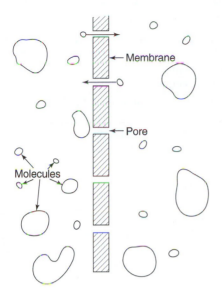

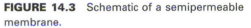

**FIGURE 14.3**   Schematic of a semipermeable membrane.

tors contributing to the semipermeable nature of membranes have to do with the chemistry of the membrane, cohesive and adhesive forces, charges on the ions involved, and existence of carrier molecules.

Diffusion in liquids and through membranes is a slow process. Hemoglobin greatly increases the oxygen capacity of blood, and furthermore excess carbon dioxide causes hemoglobin to release oxygen. Diseases that affect membrane thicknesses (such as fibrosis of the lungs) or the affinity of blood for oxygen (such as carbon monoxide poisoning) reduce the effectiveness of the respiratory system.

We can apply Fick's law in equation (14.10) to the transport of molecules across a membrane which can be a biomembrane, for example, the phospholipid layers surrounding cells, or an artificial membrane such as that used in a dialysis machine. Consider as an example a container of sugar water which is divided by a membrane of thickness $\Delta x$. Assume also that the concentration of glucose on the left side is $c_L$ and that on the right side is $c_R$. The glucose diffusion current $I$ across the membrane, according to Fick's law, is given by

$$I = \frac{m}{t} = \frac{D}{\Delta x} A \, \Delta c \tag{14.12}$$

where $\Delta c = c_R - c_L$, $D$ is the diffusion constant, and $A$ is the cross-sectional area of the membrane.

## TOPIC 14.2   Diffusion in Biology

For biologically important molecules diffusing through water at room temperatures, values of the diffusion constant $D$ range from $1 \times 10^{-11}$ to $100 \times 10^{-11}$ m²/s, the corresponding range of molecular weights being about $10^4$. The diffusion constant has been shown to be related to the temperatures and viscosity of the liquid by

$$D = \frac{kT}{6\pi r\eta} \tag{14.13}$$

where $r$ is the radius of the particle of solute, assumed spherical, $k$ is Boltzmann's constant, and $\eta$ is the medium's viscosity. The radius of a sphere is proportional to the cube root of its mass, and therefore we conclude that $D$ is inversely proportional to the cube root of the mass; see equation (14.8). This explains why for a wide range of molecular weights the values of $D$ are in a comparatively small range. In gases this result does not hold, and $D$ becomes inversely proportional to the square root of the mass of solute particles.

An important characteristic of diffusion processes is the proportionality of the average value of the squared displacement of a diffusing particle to the time elapsed for the diffusion process:

$$\overline{R}^2 = 6Dt \tag{14.14}$$

This equation helps us to determine whether some cellular processes such as DNA transcription and translation are physically feasible. For example, one might ask whether or not rates of diffusion $D = 2 \times 10^{-9}$ m²/s are sufficient to allow 50 amino acids per second to be made into protein at a ribosome. Taking the distance to be the length of a bacterial cell (i.e., 3 $\mu$m) gives

$$t = \frac{\overline{R}^2}{6D} = \frac{(3 \times 10^{-6})^2}{6 \times 2 \times 10^{-9}} = 7.5 \times 10^{-4} \text{ s} \tag{14.15}$$

The process in a real bacterial cell would not be quite this rapid since the cytoplasm will be about five times as viscous as water. This will decrease the diffusion constant to one-fifth of the value used, with the result that the time will be increased by a factor of 5 to $3.8 \times 10^{-3}$ s, which is still very fast. Thus diffusion, while a slow process in a bulk liquid, is a very fast process within the confines of a cell.

## TOPIC 14.3   Osmosis in Biological Organisms

A detailed study of osmosis began in the middle of the nineteenth century with experiments on plant cells. If a plant cell is put into a concentrated solution of sugar, for example, the protoplasts of the cell contract away from the walls. If the cells thus treated are removed and placed in pure water, the protoplasts expand again. This phenomenon is easily observed under a microscope and is known as plasmolysis.

The osmotic pressure $\Pi$ can be found from experiments on weak solutions to be proportional to the concentration of solute as shown in equation (14.11), that is, inversely proportional to the volume of the solution, and also proportional to the absolute temperature. Hence the law of osmosis may be written

$$\Pi V = R'T \tag{14.16}$$

where R' is a constant depending only on the mass of the solute present. It transpires that

$$R' = nR \tag{14.17}$$

where $R$ is the universal gas constant and $n$ is the number of moles of solute present. The solute behaves as if it were a perfect gas, so osmotic pressure arises from the bombardment of the walls of the container by the molecules of sugar in the solution. At higher concentrations equation (14.17) is not valid, and the reasons for this are quite analogous to those that cause the breakdown of the simple gas laws. Reverse osmosis may take place in situations where an external pressure is applied to one compartment such that it exceeds the osmotic pressure (see Figure 14.4).

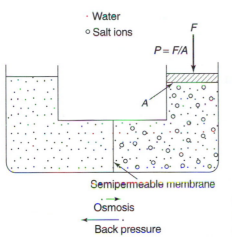

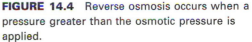

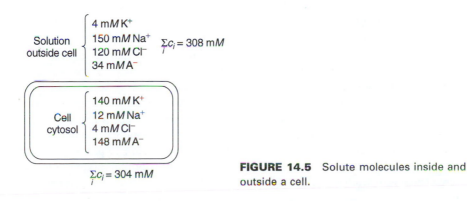

**FIGURE 14.4** Reverse osmosis occurs when a pressure greater than the osmotic pressure is applied.

Osmosis between roots and groundwater is thought to be responsible for the transfer of water into many plants. Groundwater is purer and hence has a higher water concentration than sap, so osmosis moves water into the roots. Water in sap is then transferred by osmosis into cells, causing them to swell with increased pressure. This pressure is called turgor pressure and is partly responsible for the ability of many plants to stand up. Relative osmotic pressure is not large enough to cause sap to rise to the top of a tall tree, however. To do this, the sap would have to have more dissolved materials in it than is found to be the case.

## TOPIC 14.4   Osmotic Pressure of Cells

A variety of solute molecules are contained within cells (see Figure 14.5). The cellular fluid (cytosol) has a chemical composition of 140 m$M$ K$^+$, 12 m$M$ Na$^+$, 4 m$M$ Cl$^-$, and 148 m$M$ A$^-$, where 1 m$M$ stands for a concentration of $10^{-3}$ mol/L. The symbol A stands for protein. Cell walls are semipermeable membranes and permit transport of water but not of solute molecules. We can apply the osmotic pressure concept to cells, but because of the content of the cellular fluid, we need to find the osmotic pressure of a mixture of solute molecules. We use Dalton's law to determine the osmotic pressure inside a cell. A mixture

**FIGURE 14.5** Solute molecules inside and outside a cell.

of chemicals, with concentrations $c_1, c_2, c_3, \ldots$, dissolved in water has the total osmotic pressure equal to the sum of the partial osmotic pressures $\Pi$ of each chemical. Thus

$$\Pi = \Pi_1 + \Pi_2 + \Pi_3 + \cdots = RT(c_1 + c_2 + c_3 + \cdots) \qquad (14.18)$$

The total osmotic pressure inside a cell, $\Pi_{in}$, is therefore

$$\Pi_{in} = RT\, \frac{(140 + 12 + 4 + 148) \times 10^{-3}\ \text{mol}}{1\ \text{L}} \times \frac{1\ \text{L}}{10^{-3}\ \text{m}^3} = 7.8 \times 10^5\ \text{Pa} \quad (14.19)$$

where we used the concentrations given above and a temperature $T = 310$ K since the gas constant $R = 8.31$ J/mol $\cdot$ K. Cell walls would be expected to burst under such large pressures. However, they do not because the exterior fluid also exerts an osmotic pressure. The cell exterior is composed of 4 m$M$ K$^+$, 150 m$M$ Na$^+$, 120 m$M$ Cl$^-$, and 34 m$M$ A$^-$. As a consequence, the total osmotic pressure of the cell exterior, $\Pi_{out}$, is given by

$$\Pi_{out} = RT\, \frac{(4 + 150 + 120 + 34) \times 10^{-3}\ \text{mol}}{1\ \text{L}} \times \frac{1\ \text{L}}{10^{-3}\ \text{m}^3} = 7.9 \times 10^5\ \text{Pa} \quad (14.20)$$

where $\Pi_{out}$ is again a large osmotic pressure but because $\Pi_{in}$ and $\Pi_{out}$ are quite close in values, the osmotic pressure difference between the exterior and interior parts of the cell is very small, as it is the net pressure exerted on the cell wall. For fragile animal cells, it therefore becomes vitally important to keep their interior and exterior osmotic pressures closely matched. The cell therefore has a sophisticated control mechanism to do this.

## TOPIC 14.5    Osmotic Work

If two solutions have the same osmotic pressure, we call them iso-osmotic. However, if the pressures are different, the one at higher pressure is called hypertonic and the one at lower pressure is called hypotonic. When cells are placed in a solution and neither swell or shrink, we call the solution isotonic.

For marine animals the regulation of osmotic pressure differences is of central importance. In the tissues of most marine invertebrates the total osmotic concentration is close to that in sea water; for example, coastal crabs need to support changes in the outside osmotic pressure through various forms of osmoregulation. The salt concentration of seawater is about 500 m$M$. As long as the salt concentration remains near this value, the blood of many crabs is isotonic with that of seawater. When it is outside this range, the system maintains the osmotic pressure difference across its membrane through the activity of proteins in the cell wall—these are called ion pumps and the process is known as osmoregulation. For example, the mitten crab can survive inside concentrations within a range of 100 to 600 m$M$. In the sharks and rays the osmotic pressure of their blood is very close to that of seawater. Strangely, the bony fishes have abandoned this arrangement, and their interior osmotic pressure is about one-third that of seawater. The skin of fishes is semipermeable, and as they constantly lose water, they must compensate for this loss by continually drinking seawater. The gills of fish actively secrete salt ions to avoid the resultant buildup of excess salts. On the other hand, freshwater teleost fishes have an interior osmotic pressure which exceeds that of freshwater. Water continually enters through the skin of these fishes and must be excreted by steady urination. This leads to a loss of salts which must be actively replenished by the active import of salt ions.

The cell composition begins to drift away from its optimal mixture if the ion pumps are chemically destroyed. Across the cell wall the osmotic pressure difference then rises, causing the cell to swell, become turgid, and eventually explode. The cells of bacteria and plants are not osmotically regulated since their cell walls are able to withstand pressures in the range of 1 to 10 atm.

The removal of interior salts by teleost fish and the importation of salt by freshwater fish requires work to be done. The minimum work performed when $n$ moles of solute are transferred from one solution with a concentration $c_1$ to a solution with concentration $c_2$ is given by

$$W = nRT \ln \frac{c_1}{c_2} \qquad (14.21)$$

In the above equation $c_1$ might be the salt concentration in the tissue of a fish and $c_2$ the salt concentration in seawater. In this case $c_2 > c_1$, so the osmotic work done by the seawater should be negative—the physical reason for this being that energy is required to move salt molecules from a solution of low concentration to one of high concentration. The actual work done by the fish in excreting salt will be the reverse of that in equation (14.21). Thus, by measuring the number of moles of salt secreted per second by the gills or in the urine of sea animals, and by knowing the salt concentrations inside and outside the fish, one can compute the osmotic work done by sea animals using equation (14.21).

Osmotic pressure is also used by plants and trees. For example, in the spring, excess sugar is secreted by the roots of trees. This turns the water in its capillaries into a sugary syrup. Tree roots have a high osmotic pressure inside them which leads to absorption of water from the soil. A key role is also played, it is believed, by osmotic pressure in the growth of plants. The openings on the surfaces of cell leaves, called stomata, are bordered by guard cells which can regulate their internal pressure by controlling the potassium concentration. Water absorption causes these cells to swell under osmotic pressure and the stomata are closed.

In the human body osmoregulation is extremely important because unlike plant cells, eukaryotic cells in higher organisms cannot withstand large osmotic pressure differences. The blood has a high concentration of proteins compared with the surrounding tissues. Furthermore, blood capillaries are permeable to small molecules such as water, oxygen, carbon dioxide, and salt but not to proteins. The capillary–tissue osmotic pressure difference sucking water out of tissue is approximately 23 mm Hg. The capillaries are pressurized by the blood pressure with respect to the tissue, and the average blood pressure is close to 23 mm Hg. It turns out that the osmotic pressure difference is neatly offset by the hydrostatic pressure difference, but this delicate balance may be easily upset by such situations as high blood pressure or starvation. High blood pressure can lead to the accumulation of fluids in tissues, a condition known as edema. When we drink water or consume salt, we may be in danger of upsetting our delicate osmotic balance, which can be prevented through a regulatory mechanism provided by ADH (antidiuretic hormone). This hormone rapidly alters the water permeability of tubes connected to the urinary duct in response to changes in salt concentration.

## TOPIC 14.6   Regulation of Fluid between Cells (interstitial fluid)

In humans and other animals the interstitial fluid is regulated by exchange of substances with blood in capillaries. Many substances are moved across capillary walls, but the transport of water (i.e., osmosis) is of immediate interest. Figure 14.6 shows the process under average conditions. Blood contains water, glucose, electrolytes (dissolved salts), gases, proteins, red and white blood cells, waste materials such as urea, and so on. Blood enters

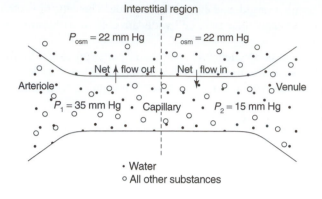

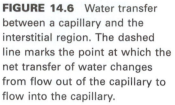

**FIGURE 14.6**   Water transfer between a capillary and the interstitial region. The dashed line marks the point at which the net transfer of water changes from flow out of the capillary to flow into the capillary.

from the left in the figure, and because the capillary is small in diameter, there is a significant pressure drop along it. Pressures are typically 35 mm Hg at the entrance and 15 mm Hg at the exit. Water is more concentrated in the interstitial fluid than in the blood, but the typical osmotic pressure difference between blood and interstitial fluid is 22 mm Hg. Reverse osmosis therefore occurs near the entrance of the capillary. Blood pressure near the capillary's exit is less than the relative osmotic pressure, and osmosis carries water from the interstitial region into the capillary. The overall result is an exchange of water while the total amount of water in the interstitial region remains constant.

In the human body waste products of metabolism are removed by osmosis in the kidneys. This process is called dialysis. Dialysis is the diffusion of substances other than water through semipermeable membranes and occurs in kidneys and many other places in biological organisms. Small molecules are usually involved in dialysis since membranes tend to be impermeable to large molecules. Fick's law applies to the direction and rate of dialysis, Reverse dialysis, also called filtration, occurs when pressure on the high-concentration side is large enough to reverse the normal direction of dialysis. Kidney function and the effect of diuretics are two examples of dialysis in humans. When the kidneys receive blood, it flows into them via a cluster of capillaries called the glomerulus, which is enclosed in a double membrane (see Figure 14.7). From the glomerulus there is a convoluted narrow tube which is in close contact with other capillaries along its length. The tubule (the narrow tube), glomerulus, and membrane are known as a nephron. In each human kidney there are about one million nephrons. In a healthy person the glomerulus is both a semipermeable membrane and a filtering membrane which prevents the passage of blood colloids. A hydrostatic pressure is required to force the solvent through the membrane in the process of filtration. This is opposed to the process of osmosis, in which the separation of solvent and solute generates an osmotic pressure in the opposite direction. To cause a flow through the tubules, a pressure of about 40 mm Hg is required. The osmotic pressure of the blood colloids is about 30 mm Hg. Hence a pressure of at least 70 mm Hg is required for the kidneys to function properly. It is obvious that a prolonged fall in blood pressure could be dangerous. Liquid flowing into the tubules contains both valuable salts and waste materials. The valuable materials and about 99% of the fluid, in its convoluted passage, are returned to the bloodstream, but the waste is prevented from doing so and eventually is excreted as urine. If one kidney fails and the other is still healthy, it is so efficient that it can cope with the job of cleansing the whole blood supply. A condition called uremia results if both fail and death becomes a possibility. A human being cannot survive more than three weeks with uncleansed blood.

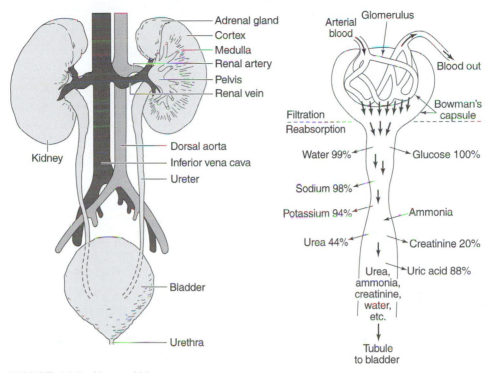

**FIGURE 14.7**   Human kidney system.

## TOPIC 14.7   Gas Exchange in Animals: Breathing and Diffusion

The process of gas exchange in land animals is similar to that in plants, and the gas exchange organs are situated inside the body cavity. This not only protects the organs from mechanical damage but serves the essential purpose of preventing excessive evaporation of moisture. Frequently these organs are well supplied with glands whose job it is to replenish any water that does get lost. There are two distinct ways in which these general requirements are fulfilled. The anthropods (e.g., insects, spiders, and centipedes) exchange gases between tissues and air in an elaborate network of tubes called trachea. In the vertebrates lungs are the gas exchange organ.

Figure 14.8 schematically represents gas exchange with the circulatory system. The partial pressure of oxygen in the lungs is only 105 mm Hg, less than that in the atmosphere, because air in the lungs is only partially replaced with each breath. Similarly, the partial pressure of carbon dioxide in the lungs is greater than in the atmosphere. Gases move in the directions of the arrows, in each case from high to low concentration (partial pressure). Oxygenated blood travels through the arteries to the capillaries, where it is transferred to cells having lower oxygen concentration. Oxygen-poor blood travels back through the veins to the lungs, where the process begins again. A similar cycle for the removal of carbon dioxide is also shown.

We demonstrate the general principles of gas exchange by reference to humans, and in Figure 14.9 we show, in diagrammatic form, the main features of the human gas exchange mechanism. The air is inhaled through the nose, passes down the trachea, and then branches off into the lungs via the bronchi. Each bronchus proliferates into smaller

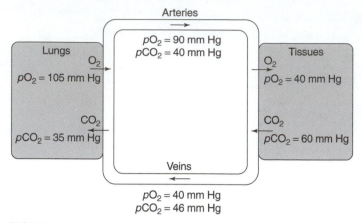

**FIGURE 14.8** Schematic of the interchange of gases between the lungs and blood and between the blood and tissues in the body.

bronchia, and these lead in turn to the tiny bronchioles. This whole structure very much resembles a tree with the trachea as the trunk, bronchi as the branches, and bronchioles as the twigs. The bronchioles terminate in a network of many millions of tiny sacs known as alveoli which have very thin elastic walls (see Figure 14.10). If these sacs were flattened out, their total surface area would be about 60 $m^2$. The vital process of the diffusion of oxygen and carbon dioxide takes place over this enormous area.

Gas transfer into and out of the blood is a primary function of the respiratory system. Diffusion of gases between air in the lungs and blood proceeds in the direction from high to low concentration, and the rate of diffusion is greatest when the difference in concentration is greatest. Diffusion obeys Fick's law here, but the actual rate of exchange is greatly affected by the presence of hemoglobin in the blood. Hemoglobin has a chemical

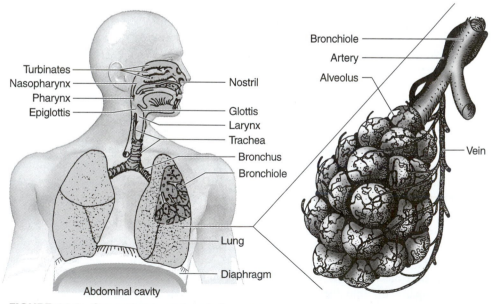

**FIGURE 14.9** Gas exchange system in humans.

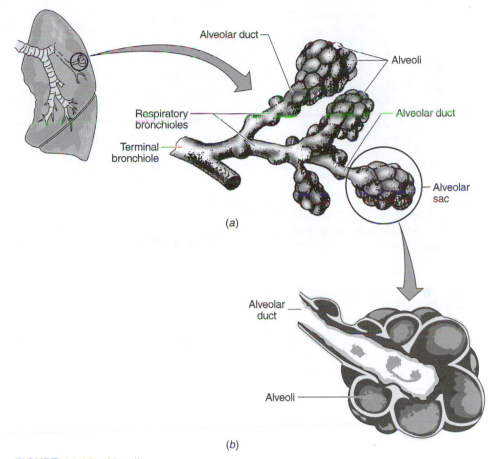

(a)

(b)

**FIGURE 14.10** Alveoli.

affinity for oxygen, and its presence allows the blood to carry far more oxygen than it otherwise could. The rate of exchange also depends on blood acidity, the presence of carbon dioxide, and temperature which is cleverly designed to enhance oxygen transfer when it is most needed. The rate at which diffusion takes place across the walls of the alveoli is proportional to the pressure difference. This explains why it is difficult to breathe at high altitudes.

The large area of the alveoli can be reduced by disease; for example, in pneumonia, lymph and mucus in the bronchioles and the alveoli reduce the gas exchange area and the patient suffers from oxygen starvation. Oxygen may be used to allow him or her to breathe by bringing about sufficient diffusion in the restricted area of the alveoli. There is a breakdown in the alveolar walls when a patient has the disease emphysema, which also reduces the gas exchange area. This breakdown also occurs gradually during the lifetime of a patient and is severely aggravated by cigarette smoking and industrial air pollution.

Blood is carried into the lung by the pulmonary artery, which terminates in a maze of tiny capillaries which are in very close contact with the alveoli. As the capillaries and alveoli are so close, molecules of oxygen pass through the walls of the alveoli and the walls of the capillaries and attach to the hemoglobin in the blood. The metabolic combustion of oxygen in cells generates carbon dioxide, which moves in the reverse direction

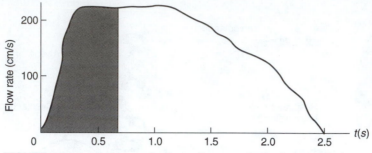

**FIGURE 14.11**   Variation of volume flow rate in the lungs with time.

and is finally exhaled. An improvement to lung ventilation is facilitated and improved by the presence of the diaphragm. This is a muscular partition separating the abdominal and thoracic cavities and acts essentially as a bellows to draw the air into the lungs.

## TOPIC 14.8   Lung Functioning

In this topic we discuss the functioning of the lung using the concepts of gas dynamics. In Figure 14.11 we display an example of the volume flow rate versus time through the lungs, and it can be seen to rise to a fairly steady value almost immediately and then fall off toward the end of the expiration period. The expiration time is 2.5 s, and the total area under the flow rate–time curve gives the total volume of gas expired. This is called the tidal volume, and here it is about $450 \times 10^{-6}$ m$^3$. Simultaneously the nitrogen concentration can be measured as a function of time, as shown in Figure 14.12. The concentration of nitrogen is zero at the beginning of the expiration and then rises to a final steady value of about 72%. We can understand these observations using a very simple model of the lung. We assume that the dead space is made up of a piece of glass tubing and the volume of the alveoli is a partially inflated balloon attached to it. Typically a healthy person will have a dead space volume of $150 \times 10^{-6}$ m$^3$ while the alveolar volume at the beginning of inspiration is about $2550 \times 10^{-6}$ m$^3$. When we breathe normally, both the dead space and the alveoli are full of air at its normal concentration of 80% nitrogen. Therefore, at the beginning of inspiration the volume of nitrogen in the lungs is $0.8(2550 + 150) \times 10^{-6}$ m$^3 = 2160 \times 10^{-6}$ m$^3$. When the subject breathes in $450 \times 10^{-6}$ m$^3$ of pure

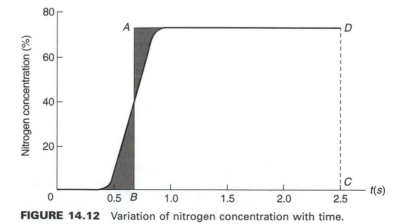

**FIGURE 14.12**   Variation of nitrogen concentration with time.

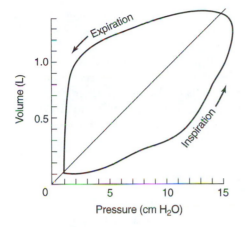

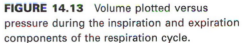

**FIGURE 14.13** Volume plotted versus pressure during the inspiration and expiration components of the respiration cycle.

oxygen, it flushes air down from the dead space into the alveoli until, at the end of inspiration, the alveoli contain $2160 \times 10^{-6}$ m³ of nitrogen in a total expanded volume of $(2550 + 450) \times 10^{-6}$ m³ $= 3000 \times 10^{-6}$ m³. Thus the new concentration in the alveoli is $(2160/3000) \times 100 = 72\%$ and the dead space is completely filled with $150 \times 10^{-6}$ m³ of pure oxygen. Initially, only pure oxygen is breathed out and the concentration of nitrogen is zero. Then, when the dead space has been emptied, the gas which emerges is from the alveoli and its concentration is 72% nitrogen.

If, at the air boundary between the dead space and the alveolar volume, no mixing took place, then, as soon as the gas in the dead space had been expired, the nitrogen concentration would instantly rise to the final alveoli concentration of 72% (shown by the vertical line in Figure 14.12). In this ideal case the dead space would simply be the volume corresponding to the shaded area under the flow rate–time curve. However, mixing does occur, and instead of an instantaneous rise, there is a fairly rapid but gradual rise.

To determine the amount of work done by breathing, in clinical practice pressure–volume curves for inspiration and expiration are drawn such as those in Figure 14.13.

## TOPIC 14.9  Gas Exchange in Terrestrial Organisms

### Nitrogen-Fixing Bacteria

Nitrogen is an essential component of proteins, and some bacteria in the soil and in the nodules of certain plant roots are capable of extracting and "fixing" nitrogen from the atmosphere into compounds which can be used in protein synthesis. They are called nitrogen-fixing bacteria. Some blue-green algae can also do this. Nitrogen can also be generated by lightning since oxides of nitrogen are produced which are then washed down into the soil. Here they form nitrates, which plants can manufacture into proteins. Green plant cells need carbon dioxide to undergo photosynthesis, and land animals need oxygen to breathe. Mechanisms of gas exchange operate in both these processes and depend on physical concepts. Furthermore, in both cases water vapor in the air plays a vital role.

### Gas Exchange in Plant Leaves

The energy in sunlight is utilized by green plants to transform carbon dioxide—an inorganic compound of low energy content—into compounds with a high energy content. This process, called photosynthesis, globally produces, it is estimated, at least $1.4 \times 10^{14}$ kg of organic matter in a year. About 10% of this is produced by green plants on land and the rest by microorganisms (e.g., bacteria, algae, diatoms, and dinoflagellates) in the sea. In Figure 14.14 we show a cutaway section of a typical leaf.

In the process of photosynthesis the leaf absorbs carbon dioxide through the stomata and oxygen is given back through the stomata. When the process is taking place, the $CO_2$ content of the air spaces between the cells of the spongy layer (see Figure 14.14) falls below the 0.03% present in the atmosphere, causing more $CO_2$ to diffuse through the stomata. The moisture film around each cell dissolves the $CO_2$, which then passes into the cell itself, where the process of photosynthesis converts it to glucose and oxygen according to the equation

$$6CO_2 + 12H_2O \rightarrow C_6H_{12}O_6 + 6O_2 + 6H_2O \tag{14.22}$$

In fact, the oxygen is produced from the water and not directly from the $CO_2$, and also reaction (14.22) makes it clear that water must be present for the process to take place. The air that diffuses into the spongy layer tends to dry out the moisture film, and the leaf loses substantial amounts of water to the atmosphere. The rate of loss of this latter process depends on the relative humidity of the air outside. Diffusion of water through the stomata is referred to as transpiration. When the relative humidity is low enough (e.g., on a very hot summer day), the rate of loss of water by transpiration may be greater than the rate at which replacement water can be supplied from the roots of the plant. A very cleverly designed self-protection mechanism takes over in such circumstances when the loss of water causes the guard cells to wilt so that they close the stomata. This characteristi-

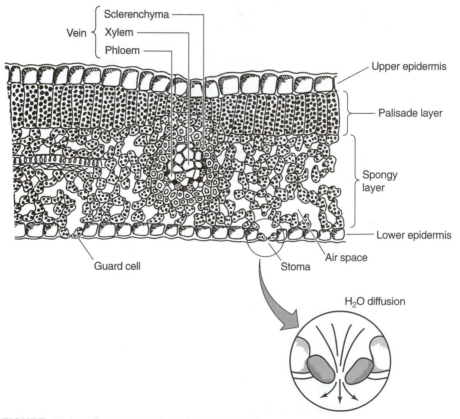

**FIGURE 14.14** Cutaway section of a typical leaf.

cally reduces the rate of loss by transpiration. This is an example of what is called botanical feedback.

## TOPIC 14.10   Oxygen Consumption by Aerobic Bacteria

Oxygen is supplied by diffusion for the metabolism of aerobic bacteria. Consider a single bacterium in an aqueous solution which contains a certain amount of dissolved oxygen. Let us suppose that the concentration of oxygen far from the bacterium has its equilibrium value, $c_0$, which for oxygen dissolved in water is about 0.2 mol/m$^3$. Suppose also that the bacterium is spherical and has a radius $R$ and is constantly consuming oxygen and that at the surface of the bacterium oxygen molecules are adsorbed very efficiently. As a consequence, the oxygen concentration just outside the bacterium must be close to zero. The concentration difference between the surface of the bacterium and infinity will set up a diffusion current $I$ of oxygen from infinity to the bacterium. The oxygen concentration $c(r)$, where $r = 0$ is the origin of the coordinate system, must depend only on the distance $r$ from the origin. At the surface of the bacterium, where $r = R$, the concentration must vanish, so $c(R) = 0$. The oxygen molecule diffusion current density $J(r)$ also can only depend on $r$. To find out what the functions $c(r)$ and $J(r)$ are, we note that the number of oxygen molecules is constant provided the product of $J(r)$ and the area $A = 4\pi r^2$ of a spherical surface at $r$ is constant, which is just the diffusion current $I$ since $J = I/A$. Therefore

$$J(r) = -\frac{I}{4\pi r^2} \tag{14.23}$$

the minus sign indicating that current is flowing inward from large to small $r$. The concentration $c(r)$ may be also written as

$$J(r) = -D\frac{dc}{dr} \tag{14.24}$$

Equating equations (14.23) and (14.24), we find that

$$\frac{dc(r)}{dr} = \frac{I/D}{4\pi r^2} \tag{14.25}$$

Equation (14.25) may be solved by separation of variables to give the concentration $c(r)$ as

$$c(r) = -\frac{I/D}{4\pi r} + \text{const} \tag{14.26}$$

We find the constant in equation (14.26) by recalling that far from the bacterium (i.e., $r \rightarrow \infty$) $c(r) = c_0$. Hence

$$c(r) = -\frac{I/D}{4\pi r} + c_0 \tag{14.27}$$

Since the oxygen concentration at the surface of the bacterium (at $r = R$) must be zero, so $c(R) = 0$. Thus in equation (14.27), putting $c(R) = 0$ gives

$$I = 4\pi RDc_0 \tag{14.28}$$

Equation (14.28) determines the maximum number of oxygen molecules which can be taken in by the bacterium per second. The number of oxygen molecules required per unit volume by a bacterium is called the metabolic rate $M$ and for bacteria is about 20 mol/m$^3$ · s. The incoming diffusion current $I$ must equal or exceed $M$ times the volume, $V = \frac{4}{3}\pi R^3$, of the bacterium for it to function normally. Hence

$$I > \frac{4}{3}\pi R^3 M \tag{14.29}$$

Equation (14.28) combined with equation (14.29) gives an inequality for the size of $R$. That is,

$$R < \sqrt{\frac{3Dc_0}{M}} \tag{14.30}$$

which provides an upper bound for the size of the bacterium. Putting in values for the parameters, one finds that $R < 10$ $\mu$m. Bacteria large compared to this size will simply not get enough oxygen by pure diffusion. This limit is, in fact, about the size of a typical spherical bacterium. Larger cells can be produced by making them very long, keeping the diameter small (i.e., cylindrical). This greatly increases the diffusion current, and diffusion imposes no size limitation on the length of the cylindrical bacteria. Large bacteria are indeed cylindrical rather than spherical.

## TOPIC 14.11 Active Transport

The discussion of molecular phenomena and biological processes so far has included the general categories of diffusion, osmosis, reverse osmosis, dialysis, and reverse dialysis (filtration). All these processes for transporting substances on the cellular level are passive in nature. The driving energy for passive transport comes from molecular kinetic energy or pressure. There is another class of transport phenomena, called active transport, in which the living membrane itself supplies energy to cause the transport of substances.

Biological organisms sometimes need to transport substances from regions of low concentration to regions of high concentration—the direction opposite to that in osmosis or dialysis. Of course, sufficiently large back pressure causes reverse osmosis or reverse dialysis, but there are known instances in which substances move in the direction that reverse osmosis or reverse dialysis would take them even where existing pressures are insufficient to cause reverse osmosis or reverse dialysis. In these instances, active transport must be taking place, which means that living membranes expend their own energy to transport substances. Active transport can also aid ordinary osmosis or dialysis and explains why some transport proceeds faster than expected from osmosis or dialysis alone.

Active transport is extremely important in nerve cells. Changes in the concentration of electrolytes across nerve cell walls are responsible for nerve impulses. After repeated nerve impulses, significant migration has occurred and active transport "pumps" the electrolytes back to their original positions. For a detailed discussion of these processes we refer the reader to Chapter 20.

## TOPIC 14.12 Atmospheric Pressure Variations and Physiology

Living on the earth's surface means that a constant pressure of the column of air above us is exerted on all objects. At sea level this amounts to $P = 1$ atm $= 760$ mm Hg $= 1.01 \times 10^5$ N/m$^2$. However, pressure changes with height above sea level. For the first few thousand meters above sea level the pressure will drop about 25 mm Hg per 300 m of height. This can be obtained from the formula

$$P(h) = P_0 + \rho g h \tag{14.31}$$

where $h$ is the height above sea level and $\rho$ is the density of air. However, with an increase in height the density of air markedly changes. As a result, a more appropriate but also more complicated formula states that

$$\ln P(h) = \ln P_0 - 10^{-3} \frac{Mgh}{RT} \tag{14.32}$$

where $M$ is the molecular mass of the gas, $R = 8.31$ J/mol $\cdot$ K, and $T$ is the absolute temperature in kelvins. There is an extra complication which has not been accounted for and it is the temperature variation with height. For example, Mexico City lies at 1500 m above sea level. The corresponding change in the air density is that from $\rho = 1.2$ kg/m$^3$ at sea level to $\rho = 1.0$ kg/m$^3$ in Mexico City. With this in mind and using equation (14.31), one obtains the value of atmospheric pressure there as $P(h) = 0.85 \times 10^5$ N/m$^2$. Going to the extreme height on the earth, that is, climbing the tallest mountain, Mount Everest, at $h = 8880$ m above sea level, and using equation (14.32) for calculations gives a result that says $P/P_0 = 0.331$. Therefore, the atmospheric pressure there drops below a third of its normal value, posing a serious health hazard to the climber. Almost all living organisms on earth above sea level exist at pressures between 350 and 760 mm Hg. At any altitude where the pressure is less than 520 mm Hg, which starts at approximately 3000 m above sea level, exertion is likely to result in distress. Breathing pure oxygen will help, but this, too, has a limit of applicability. At 14,000 m above sea level atmospheric pressure is reduced to only 130 mm Hg; even with 100% oxygen content there would barely be enough oxygen crossing the molecular walls in the lungs to maintain consciousness. At $h = 21,000$ m the atmospheric pressure equals that of water vapor on the body, leading to an astonishing effect of boiling body fluids and resulting in rapid death.

## TOPIC 14.13 Drag Forces on Swimming Organisms

Between 1910 and 1920 an English biologist, J. S. Haldane, was interested in the effects of size on animals. When an insect falls down from a great height, it drifts slowly with a constant speed while a small mammal travels much faster and, although dazed on impact, will survive the fall. A human being would obviously be killed and an elephant would be splattered. Clearly, the larger the animal, the more important gravity becomes. The impact speed would be the same in a vacuum; thus frictional forces become important and are progressively more so the smaller the animal.

Let us now consider the motion of a bacterium moving through water assuming that its initial velocity at time $t = 0$ is zero. An object suspended in a fluid weighs less than the same object in air because of buoyancy from the liquid (see Chapter 11). The apparent weight of an object of mass $m$ is given by

$$W = mg\left(1 - \frac{\rho_{fluid}}{\rho_{object}}\right) \tag{14.33}$$

where $\rho_{fluid}$ is the density of the fluid and $\rho_{object}$ the density of the object. Often, $\rho_{fluid} < \rho_{object}$, so that $W$ is positive and the object sinks. For a bacterium, $\rho_{object}$ is about 1200 kg/m$^3$, which exceeds the density of water (here $\rho_{fluid} \cong \rho_{object}$ is about 1000 kg/m³) so the bacterium will sink. Let us adopt a coordinate system with the positive $y$ axis pointing upward and denote the vertical position of the bacterium by $y(t)$. When an object moves through a fluid, it experiences a drag force which for a small sphere is given by Stokes's law in the form

$$F_{drag} = 6\pi\eta Rv \tag{14.34}$$

where the radius of the sphere is $R$ and $\eta$ is the viscosity (for seawater this is approximately $1.07 \times 10^{-3}$ in SI units). Thus, applying Newton's second law, we have

$$m\frac{dv}{dt} = W - 6\pi\eta Rv \tag{14.35}$$

This may be modified by changing to a different dependent coordinate, $x$, given by

$$x = 6\pi\eta Rv - W \tag{14.36}$$

and equation (14.35) becomes

$$\frac{dx}{dt} = -x\frac{6\pi\eta R}{m} \tag{14.37}$$

Equation (14.37) is easy to integrate by the separation-of-variables method and gives, on changing back to the velocity variable $v$ rather than $x$,

$$v = v_{term}(1 - e^{-t/\tau}) \tag{14.38}$$

The velocity $v$ is downward and the component $v_{term}$ is the terminal velocity. The quantity $\tau$ is called the drift time and is given by

$$\tau = \frac{m}{6\pi\eta R} \tag{14.39}$$

We can estimate this time by using the fact that the mass of a spherical bacterium is

$$m = \tfrac{4}{3}\pi R^3 \rho_{bacterium} \tag{14.40}$$

the density of the bacterium being estimated as $\rho_{bacterium} = 1200$ kg/m³, so for $R = 10^{-6}$ m, $m = 5 \times 10^{-15}$ kg. The drift time is therefore

$$\tau = \frac{m}{6\pi\eta R} \cong \frac{5 \times 10^{-15}\text{ kg}}{6 \times \pi \times (1.07 \times 10^{-3}\text{ kg/ms}) \times (10^{-6}\text{ m})} = 2.5 \times 10^{-7}\text{ s} \tag{14.41}$$

so after less than a microsecond, the bacterium reaches its terminal velocity. One can easily check that the velocity of the bacterium becomes $v_{term}$ for very long times and corresponds to zero acceleration. Thus, after a very short time, the bacterium no longer accelerates or the term $F = ma = m\,dv/dt$ in Newton's second law becomes effectively very small compared to the effects of drag. In other words, at a microbiological level the viscous force is very strong, and the physical effects of the inertial mass $m$ are very weak.

This principle holds good also for macromolecules such as DNA and proteins. For practical purposes, inertial effects play almost no role for cells and macromolecules, so that, under a gravitational or electrical force, macromolecules and cells move effectively at the terminal velocity.

For a spherical bacterium, the terminal velocity is

$$v_{\text{term}} = \frac{mg(1 - \rho_{\text{fluid}}/\rho_{\text{object}})}{6\pi\eta R} \tag{14.42}$$

as can be seen directly from equation (14.35) by simply neglecting $m\, dv/dt$. We find, with the above data, that for a bacterium this is roughly $10^{-7}$ m/s. Thus, to get to the bottom of a pond which was 1 m deep would require about $10^7$ s (115 days!). The *Escherichia coli* bacterium can actually swim by itself at about $10^{-3}$ m/s, which is 10,000 times faster than its terminal velocity. In the case of microscopic swimmers such as mobile bacteria, algae and spermatozoa, the Reynolds number is a very small quantity, typically $10^{-3}$ or less. Clearly this is a regime that is dominated by viscous forces and inertial effects become negligible. The ability of the organism to glide is lost completely. Thus microscopic cells are usually driven by slender fibers called flagella. These flagella are flexible and often beat in a helical wave which is asymmetrical.

Let us now consider particles in air (e.g., pollen or bacterial spores). The viscosity of air is about $1.8 \times 10^{-5}$ SI units (i.e., much smaller than water), so as a consequence the terminal velocity of a pollen grain of radius $10^{-5}$ m in air is higher than a bacterium in water (about 0.01 to 0.1 m/s). Airborne pollen needs wind to move; otherwise it would settle down.

The expenditure of energy as a fish swims can depend on a power of velocity greater than 1. To minimize this serious expenditure of energy, fish have low drag coefficients; for example, mackerel has a drag coefficient of only 0.004, as does the sea lion and the seal. When human beings swim, their drag coefficient is much higher and about 0.03. In Table 14.2 we present a table of drag coefficients for animals that move through liquid media.

If we follow a streamline along the surface of a fish, we see that the fish squeezes flow lines running along the side of the fish. This increases the flow velocity and either reduces or increases the local pressure by an amount proportional to $\rho v^2$. We can plot the excess pressure $P(r)$ along the flow line (compared with pressure far from the fish) and $\rho v^2$ and obtain a coefficient $C_p(r)$, where

$$C_p(r) = \frac{P(r)}{(1/2)\rho v^2} \tag{14.43}$$

This is called the pressure coefficient. This coefficient does not depend on the swimming velocity $v$ since the positive pressure at the front of the fish and the negative pressure along the side of the fish are both proportional to $\rho v^2$. In Figure 14.15 we have plotted this coefficient as a function of distance along the side of the fish. At the front of the fish $C_p = 1$. The excess pressure along the side of the fish is negative, so $C_p$ is negative along the side.

A parameter which best describes the swimming process is the Reynolds number, which is proportional to the inertial force and inversely proportional to the viscous force:

$$\text{Re} = \frac{\rho v D}{\eta} \tag{14.44}$$

The Reynolds number describes the hydrodynamics of a swimmer, telling us whether the forces which the swimmer produces go into overcoming inertial or viscous forces in the fluid environment. The effect of increasing both $D$ and $v$ is to dramatically increase the Reynolds number and hence the fraction of the total exerted force that goes into inertial effects. The result is to give a big boost through the fluid. For large organisms such as dolphins, evaluation of the Reynolds number yields values of $10^4$ or higher. This is considerably above the so-called turbulence limit. The dolphin has, however, a very smooth

**TABLE 14.2** Reynold's Number (Re) and Drag Coefficient (*f*) for Animals That Move through Liquid Water

| Animal | Re | *f* |
|---|---|---|
| Fleas, *Ctenophthalamus* | 65–205 | 0.96 |
| Fruit fly, *Drosophilia virilis* | 300 | 1.16 |
| Locust, *Schistocerca gregaria* | 8,000 | 1.47 |
| Marine isopods | | |
|   *Idotea wosnesenskii* | 2,700 | 0.08 |
|   *I. resecata* | 5,500 | 0.055 |
| Dytiscid beetles | | |
|   *Acilius sulcatus* | 8,600 | 0.28 |
|   *Dytiscus marginalis* | 15,000 | 0.33 |
| Tadpole, *Rana catesbiana* | 1,000–2,500 | 0.36–0.74 |
| Frogs | | |
|   *Hymenochirus boettgeri* | 1,500–8,000 | 0.11–0.24 |
|   *Rana pipiens* | 17,000–40,000 | 0.05–0.06 |
| Crabs, *Callinectes sapidus* | 10,000 | 0.3 |
| Ducks, various underwater | 420,000 | 0.028 |
| Cephalopod, *Nautilus* | 100,000 | 0.48 |
| Falcon, *Falco peregrinus* | 380,000 | 0.24 |
| Fish | | |
|   Trout, *Salmo gairdneri* | 50,000–200,000 | 0.015 |
|   Mackerel, *Scomber* | 100,000 | 0.0043 |
| Saithe, *Pollachius virens* | 500,000 | 0.005 |
| Penguin, *Pygoscelis papua* | $1 \times 10^6$ | 0.0044 |
| Marine mammals | | |
|   Sea lion, *Zalophus californianus* | $2 \times 10^6$ | 0.0041 |
|   Seal, *Phoca vitulina* | $1.6 \times 10^6$ | 0.004 |
| Human, swimming | $1.6 \times 10^6$ | 0.035 |

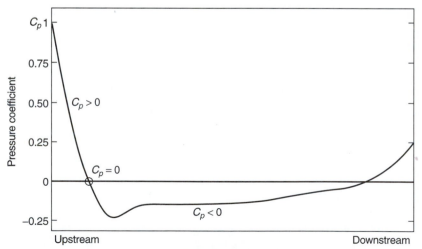

**FIGURE 14.15** Pressure coefficient versus distance along the surface.

delicate skin to minimize turbulence so that under most swimming conditions the flow remains close to streamline. The Reynolds number for very small fish such as minnows is about 100. While they clearly are in the streamline limit, viscous effects are relatively much more important and limit the ability of small fish to glide extensively.

## SOLVED PROBLEMS

**14.1** How much oxygen is taken up into the lung from air at a temperature of 20°C within 1 min in a relaxed state of respiration? The partial pressure of the oxygen in the air is 21 kPa. The volume and frequency of inspiration are 0.5 L and 15 $\text{min}^{-1}$, respectively.

**Solution** Within 1 min, the number of inspirations is 15; thus, the total volume of inspired air amounts to $V = 15 \times 0.5$ L $= 7.5$ L. The oxygen in the air occupies this volume with a partial pressure $p = 21$ kPa. Consider the oxygen to be an ideal gas. The amount of oxygen can be calculated from the ideal gas law as

$$n = \frac{pV}{RT} = \frac{(21 \text{ kPa})(7.5 \times 10^{-3} \text{ m}^3)}{(8.31 \text{ J/mol} \cdot \text{K})(293 \text{ K})} = 0.065 \text{ mol}$$

As the molecular mass of oxygen $M = 32$ g/mol, the mass of oxygen inspired during 1 min under normal conditions is $m = nM = 2.07$ g.

**14.2** The diffusion coefficient for potassium ions crossing a biological membrane 10 nm thick is $1.0 \times 10^{-16}$ $\text{m}^2$/s. What flow rate of potassium ions would move across an area $100 \times 100$ nm if the concentration difference across the membrane is 0.50 $\text{mol/dm}^3$?

**Solution** We use Fick's law:

$$\frac{\Delta n}{\Delta t} = -\frac{DA}{\Delta x}(C_2 - C_1) = -\frac{(1.0 \times 10^{-16})(10^2)^2(10^{-9})^2(-0.5 \times 10^3)}{10 \times 10^{-9}}$$

$$= 5 \times 10^{-20} \text{ mol/s} = 3 \times 10^4 \text{ ions/s}$$

**14.3** What would be the time to travel the 3-$\mu$m length of a bacterial cell for a carbon monoxide hemoglobin?

**Solution** Carbon monoxide hemoglobin (molar mass 68,000) has a diffusion coefficient of $6.2 \times 10^{-11}$ $\text{m}^2$/s in water. In more viscous cytoplasm, $D$ would be $(6.0 \times 10^{-11})/5$ $\text{m}^2$/s, and the time to travel the 3-$\mu$m length of a bacterial cell would be

$$t = \frac{\overline{R}^2}{6D} = \frac{5(3 \times 10^{-6})^2}{6 \times 6.2 \times 10^{-11}} = 0.12 \text{ s}$$

**14.4** Calculate the expected osmotic pressure for a 5 $M$ solution of sucrose in water at 30°C and the height of an equivalent water column.

**Solution** Sucrose ($C_{12}H_{22}O_{11}$) has $M = 342$ g/mol. A 5 $M$ sucrose solution has 5 mol in 1 kg of water. The solution can be considered as 5 $\text{mol/dm}^3 = 5000 \times 342$ $\text{g/m}^3$. Then

$$\Pi = \frac{RTC}{M} = \frac{8.3 \text{ J/mol} \cdot \text{K} \times 303 \text{ K} \times 5000 \text{ mol/m}^3 \times 342 \text{ g/mol}}{342 \text{ g/mol}} = 125 \text{ atm}$$

The height of a corresponding column of water would be

$$h = \frac{125 \times 10^6 \text{ N/m}^2}{9.8 \text{ m/s}^2 \times 1000 \text{ kg/m}^3} = 1276 \text{ m}$$

**14.5** A mass of 5 g of a substance of molar mass 250 g is dissolved in 600 cm³ of water at 27°C. Calculate the osmotic pressure of the solution.

**Solution**

$$\Pi = \frac{(5/250) \text{ mol} \times 8.3 \text{ J/mol} \cdot \text{K} \times 300 \text{ K}}{6 \times 10^{-4} \text{ m}^3} = 8.3 \text{ N/m}^2$$

**14.6** In the lungs, the respiratory membrane separates tiny sacs of air called alveoli (absolute pressure $= 1 \times 10^5$ N/m²) from the blood in the capillaries. The average radius of the alveoli is 0.125 mm, and the air inside contains 14% oxygen. Assuming that the air acts as an ideal gas at body temperature (310 K), find the number of oxygen molecules in one of the sacs.

**Solution** The volume of a sac is $V = \frac{4}{3}\pi r^3$. The ideal gas law explicitly contains the total number $N$ of molecules:

$$N = \frac{PV}{kT} = \frac{(1.00 \times 10^5 \text{ N/m}^2)[\frac{4}{3}\pi(0.125 \times 10^{-3})^3]}{(1.38 \times 10^{-23} \text{ J/K})(310 \text{ K})} = 1.9 \times 10^{14}$$

The number of oxygen molecules is 14% of this value, or $2.7 \times 10^{13}$.

**14.7** Large amounts of water can be given off by plants. It has been estimated, for instance, that a single sunflower can lose up to a half-liter of water a day during the growing season. Inside the leaf, water passes from the liquid phase to the vapor phase at the walls of the mesophyll cells, as shown in the figure. The water vapor then diffuses through the intercellular air spaces and eventually exits the leaf through small openings, called stomatal pores. The diffusion constant for water vapor in air is $D = 2.4 \times 10^{-5}$ m²/s. A stomatal pore has a cross-sectional area $A = 8.0 \times 10^{-11}$ m² and a length $L = 2.5 \times 10^{-5}$ m. The concentration of water vapor on the interior side of a pore is $C_2 = 0.022$ kg/m³, while that on the outside is $C_1 = 0.011$ kg/m³. Determine the mass of water vapor that passes through a stomatal pore in 1 h.

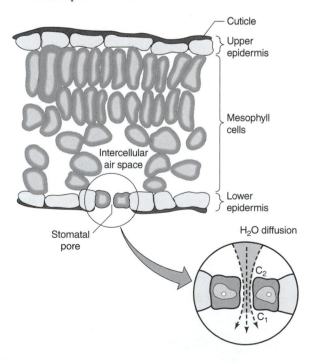

**Solution**  Fick's law of diffusion shows that

$$m = \frac{(DA\,\Delta C)t}{L} = \frac{(2.4 \times 10^{-5}\ \mathrm{m^2/s})(8.0 \times 10^{-11}\ \mathrm{m^2})(0.022 - 0.011)\ \mathrm{kg/m^3}(3600\ \mathrm{s})}{2.5 \times 10^{-5}\ \mathrm{m}}$$

$$= 3.0 \times 10^{-9}\ \mathrm{kg}$$

**14.8**  The carbon monoxide–hemoglobin complex has a diffusion coefficient $D_w = 6.2 \times 10^{-11}$ $\mathrm{m^2/s}$ in water. How long will it take to diffuse a distance $d_w = 1$ cm through water? What time is required for it to diffuse a distance $d_c = 1\ \mu\mathrm{m}$ in an *E. coli* cell, where the viscosity is 5 times as great as the viscosity of water?

**Solution**  The time needed to cover a distance $d_w$ through diffusion can be expressed as

$$t = \frac{d_w^2}{4D}$$

Numerically, $t_w = 4 \times 10^5\ \mathrm{s} = 4.7$ days. We now evaluate it for the small *E. coli* bacterial cell, where $d_c = 1\ \mu\mathrm{m}$ and whose cytoplasm is 5 times as viscous as water. This will decrease the diffusion coefficient to one-fifth of the value for water: $D_c = 5D_w = 3.1 \times 10^{-10}\ \mathrm{m^2/s}$. It will take a CO–hemoglobin molecule $t_c = 8.1 \times 10^{-4}\ \mathrm{s} = 0.8$ ms to diffuse the length of a bacterial cell.

**14.9**  The typical distance between the neighboring cells of a leaf is $x = 50\ \mu\mathrm{m}$. The ATP produced by respiration in the mitochondrium of a cell has to be transported to the neighboring cell. Should the mitochondrium itself or the separated ATP molecule be transported to the neighboring cell? The diffusion coefficient of ATP is $D_{ATP} = 3 \times 10^{-10}\ \mathrm{m^2/s}$ and the mitochondrium is considered to be a sphere of radius $r_m = 0.3\ \mu\mathrm{m}$ and density $\rho_m = 1100\ \mathrm{kg/m^3}$.

**Solution**  The molar mass of mitochondria is calculated as:

$$M_m = N_A \rho_m V_m = \tfrac{4}{3}\pi N_A \rho_m r_m^3 = 8 \times 10^{10}\ \mathrm{g/mol}$$

Since the ATP molecule consists of 10 carbon atoms, 13 oxygen atoms, 3 phosphorus atoms, 5 nitrogen atoms, and 12 hydrogen atoms, its molar mass $M_{ATP} = 503$ g/mol. The diffusion coefficients of molecules of similar shapes are related inversely to the third power of their molar masses, so:

$$D_m = \sqrt[3]{\frac{M_{ATP}}{M_m}}\,D_{ATP}$$

The diffusion coefficient of the mitochondrium calculated above is $D_m = 5.5 \times 10^{-13}\ \mathrm{m^2/s}$. The square root of the mean-square displacement of a diffusing particle in one dimension during time interval $t$ is calculated as:

$$x = \sqrt{4Dt}$$

For both particles, the numerical values of $x$ and $D$ are known, and so the time needed to diffuse to the neighboring cell can be calculated. The data are listed in the table below. Since ATP needs a 540 times shorter time than the mitochondrium to cover the same (50-$\mu$m) distance via diffusion, it is more reasonable for the leaf to detach the ATP molecule from the mitochondrium and let the ATP diffuse to the neighbor.

| Quantities | ATP | Mitochondrium |
|---|---|---|
| $M$ (g/mol) | 503 | $8 \times 10^{10}$ |
| $D$ (m²/s) | $3 \times 10^{-10}$ | $5.5 \times 10^{-13}$ |
| $t$ (s) | 2.1 | 1140 |

**14.10** What is the temperature dependence of the diffusion coefficient of a spherical biomolecule? How many times is the diffusion coefficient increased when the temperature is elevated by $\Delta T = 10$ K? (This ratio is often called $Q_{10}$.)

**Solution** The diffusion coefficient of a macromolecule in a solution may be expressed by the Einstein relationship

$$D = \frac{kT}{6\pi\eta r}$$

The temperature dependence arises not only from the numerator but also from the denominator, as the viscosity is strongly temperature dependent. Simultaneously with the diffusion of the macromolecule, the solvent is also displaced. Therefore, the diffusion modifies the solvent–solvent interaction, which requires activation energy $A$. Such an activation-type process is described by the form of the viscosity:

$$\eta = \eta_\infty e^{A/RT}$$

Here $\eta_\infty$ is the viscosity at very high temperature. If the expression for $\eta$ is substituted into the equation for $D$ above, then the explicit form of the temperature dependence of the diffusion constant will be obtained:

$$D = \frac{k}{6\pi\eta_\infty r} Te^{-A/RT}$$

According to the definition, the expression of $Q_{10}$ can be given as

$$Q_{10} = \frac{D(T + 10\ K)}{D(T)} = \frac{T + 10\ K}{T} \exp\left(\frac{10\ K}{T + 10\ K}\frac{A}{RT}\right)$$

and the actual value of $Q_{10}$ depends on $T$ and the activation energy $A$.

## EXERCISES

**14.1** A cell biologist is investigating the one-dimensional diffusion of motor proteins across a thin microtubule (a protein filament). By analyzing diffusion over a time of 25 s, the researcher obtained the following table of displacements:

| Number of particles | Approximate displacement ($\mu$m) |
|---|---|
| 12 | −52 |
| 35 | −25 |
| 50 | 0 |
| 35 | 25 |
| 12 | 52 |

**(a)** What is the mean displacement of the particles?

(b) What is the mean-square displacement of the particles?

(c) What is the diffusion coefficient of the particles?

**14.2** Suppose a spherical virus has a radius of 50 nm. What is the expected diffusion coefficient of such a particle in water at 25°C?

**14.3** Two spherical enzymes, A and B, have respective diffusion coefficients of $7.5 \times 10^{-11}$ and $4.5 \times 10^{-11}$ m²/s. Their densities are $\rho_A = 1.25 \times 10^3$ kg/m³ and $\rho_B = 0.85 \times 10^3$ kg/m³. What is the ratio of their molar masses?

**14.4** The diffusion coefficient for sodium ions crossing a biological membrane 12 nm thick is $1.5 \times 10^{-18}$ m²/s. What flow rate of sodium ions would move across an area 12 nm $\times$ 12 nm if the concentration difference across the membrane is 0.60 mol/dm³?

**14.5** The diffusion coefficient of potassium across the same membrane is 100 times greater than for sodium. If the concentration on one side is 100 mmol/dm³ and on the other is 10 mmol/dm³, what would be the potassium flow rate across 10.0 nm² of the membrane?

**14.6** Two kinds of spherical protein molecules are being compared. One has twice the surface area of the other. What is the ratio of the diffusion coefficient of the smaller to that of the larger? What would be the ratio of their diffusion coefficients if they differed in mass by a factor of 2 but had the same surface area?

**14.7** (a) How long does it take on average for a molecule of glucose to move 0.01 m in water via a diffusion process? (b) Calculate the same diffusion time for an oxygen molecule moving through a 0.005 m thick tear layer in the veinless cornea of the eye.

**14.8** The air at sea level contains 20.9% of oxygen and has a total pressure of $1.01 \times 10^5$ N/m². A deep-sea diver breathes gas mixture at a pressure of $2 \times 10^6$ N/m². Calculate the required percentage of oxygen in the mixture in order to maintain the same oxygen partial pressure as at sea level.

**14.9** Assuming that the lungs hold 2 L of air at the physiological temperature of 37°C, calculate the number of moles of air in the lungs.

**14.10** The diffusion coefficient of a large protein molecule is found to be $2.5 \times 10^{-13}$ m²/s in water at 20°C. Assuming it is spherical with a density of 1500 kg/m³, compute its molar mass.

**14.11** If the diffusion coefficient for CO–hemoglobin in water is $6.5 \times 10^{-11}$ m²/s and for CO–hemoglobin in an unknown liquid is $16.5 \times 10^{-11}$ m²/s, what is the viscosity of the liquid?

**14.12** If the diameter of a cell were increased five times, what would be the change in the diffusion time for a protein in the cell?

**14.13** The concentration of glucose (a sugar) outside a bacterial membrane of thickness 10 nm is found to be 0.60 mol/dm³. Measurement of the amount of glucose crossing the bacterial membrane yields a flux of 1.00 mol/m² · s. Assuming that this glucose flux is solely due to diffusion, determine the diffusion coefficient of glucose. Assume that 6% of the bacterial membrane area is pores.

**14.14** Once inside a cell glycerol molecules have a diffusion coefficient of about $2.1 \times 10^{-10}$ m²/s. Approximately how long would it take molecules to diffuse from the membrane to the center of the cell, which is a biconcave disc about 10 $\mu$m in diameter and about 1 $\mu$m thick in the center.

(a) if they entered at the periphery of the biconcave disc?

(b) if they entered at the center of the biconcave disc?

**14.15** A red blood cell of surface area 150 $\mu m^2$ is placed in a solution of 0.5 M glycerol. If the diffusion coefficient across the membrane is $1.2 \times 10^{-12}$ m²/s, what is the initial number of molecules crossing the cell wall per unit time? Assume the membrane is 12 nm thick.

**14.16** The osmotic pressure of a normal red blood cell is 8 atm. A solution of salt water is being prepared for injection into a patient who has suffered blood loss. The salt that is used, NaCl, dissociates into $Na^+$ and $Cl^-$ in water. The solution must be isotonic with normal red blood cells at the body temperature of 37°C.
(a) How many moles, $n$, of NaCl should be put into 1 L of distilled water to prepare the solution?
(b) If, mistakenly, only $\frac{1}{4}n$ moles of salt are put into the water, what would be the net osmotic pressure exerted on the wall of a red blood cell placed in the solution?

**14.17** Typically the size of animal cells is about 10 $\mu$m. These cell surfaces withstand osmotic pressure differences up to about $10^4$ Pa. What can be concluded from this about the maximum elastic tension of animal cell walls?

**14.18** Many freshwater animals have an interior salt concentration higher than that of their surroundings. Water constantly moves in through the skin of these animals. They must excrete the excess water by urinating, which requires osmotic work. Below is a table of the salt concentrations in the blood and urine of some freshwater animals, as well as the amount of urine produced per hour. Assume that the animals live in freshwater with a salt concentration of 6 m$M$. Calculate the osmotic work done by these animals per hour and compare it with the metabolic rates given in the table.

| Animal | Salt concentration in blood (mmol/L) | Salt concentration in urine (mmol/L) | Production rate (mL/h) | Metabolic rate of urine (cal/h) |
|---|---|---|---|---|
| Mitten crab | 320 | 320 | 0.1 | 14 |
| Crayfish | 420 | 124 | 0.1 | 10 |
| Clam | 42 | 24 | 0.5 | 1.2 |

**14.19** ATP Synthase
The enzyme ATP synthase is located in the walls of mitochondria and transforms a low-energy chemical compound ADP into a high-energy compound ATP which takes up about 7.3 kcal/mol of energy per ADP-to-ATP transformation. There is an $H^+$ concentration difference across the mitochondrial wall. Assume that the concentration ratio between the two sides of the wall is 1000. During ATP synthase activity, protons move through the enzyme from the high-concentration side to the low-concentration side. Evaluate the maximum osmotic work done by the solution during the movement of one proton. Find the minimum number of protons required per ADP-to-ATP transformation.

**14.20** A solution containing 25.0 mg/cm³ of a pure protein at $T = 300$ K has an osmotic pressure of 150 Pa. Assuming ideal behavior, what would be the molar mass of the protein?

**14.21** Oxygen for hospital patients is kept in special tanks, where the oxygen has a pressure of 65.0 atm and a temperature of 288 K. The tanks are stored in a separate room, and the oxygen is pumped to the patient's room, where it is administered at a pressure of 1.00 atm and a temperature of 297 K. What volume does 1.00 m³ of oxygen in the tanks occupy at the conditions in the patient's room?

**14.22** Compare the number of molecules in the standard dosage of pain killers Tylenol and Advil. Tylenol uses 325 mg of acetaminophen ($C_8H_9NO_2$) as the standard dose, while Advil uses 200 mg of ibuprofen ($C_{13}H_{18}O_2$).

# CHAPTER **15**

# *THERMODYNAMICS*

## PHYSICAL BACKGROUND

A thermodynamic system is the collection of objects on which attention is being focused, and the surroundings are everything else. The state of a system is the physical condition of the system, as described by values for physical parameters, often pressure, volume, and temperature.

The total energy of a system—the sum of the potential and kinetic energies—is represented by its internal energy $E$. The internal energy of a system can be altered either by performing work on it or by adding heat to it. Thus the change in internal energy is related to the heat energy transferred to the system $Q$ and the work done by the system $W$ according to the first law of thermodynamics and is given by

$$\Delta E = Q - W \tag{15.1}$$

The first law of thermodynamics is the conservation-of-energy principle applied to heat, work, and change in the internal energy. The internal energy is called a function of state, because it depends only on the state of the system and not on the method by which the system came to be in a given state. Heat and work are not functions of state, because they depend on how the system is changed from one state to another.

Thermal processes are quasi-static when they occur slowly enough that a uniform pressure and temperature exist throughout the system. An isobaric process is one that occurs at constant pressure. The work $W$ done when a system changes at a constant pressure $P$ from an initial volume $V_i$ to a final volume $V_f$ is given as

$$W = P(V_f - V_i) \tag{15.2}$$

An isochoric process is one that takes place at a constant volume, and no work is done in such a process. An isothermal process is one that occurs at a constant temperature. An adiabatic process is one that takes place without the transfer of heat. The work done in a quasi-static thermal process is given by the area under the pressure–volume graph for the process.

When $n$ moles of an ideal gas change quasi-statically from an initial volume $V_i$ to a final volume $V_f$ at a constant Kelvin temperature $T$, the work done is expressed by

$$W = nRT \ln\left(\frac{V_f}{V_i}\right) \tag{15.3}$$

When $n$ moles of a monatomic ideal gas change quasi-statically and adiabatically from an initial temperature $T_i$ to a final temperature $T_f$, the work done is given by

$$W = \tfrac{3}{2}nR(T_i - T_f) \tag{15.4}$$

Along with the ideal gas law (discussed in Chapter 14), an ideal gas also obeys the relation

$$P_i V_i^{\gamma} = P_f V_f^{\gamma} \tag{15.5}$$

in an adiabatic process, where $\gamma = C_p/C_v$ is the ratio of the specific heat capacities at constant pressure and constant volume, respectively.

The molar specific heat capacity $C_x$ of a substance under the condition of constant variable $x$ (e.g., $P$, $V$) determines how much heat $Q$ is added or removed when the temperature of $n$ moles of the substance changes by an amount $\Delta T$:

$$Q = C_x n\, \Delta T \tag{15.6}$$

For a monatomic ideal gas, the molar specific heat capacities at constant pressure and constant volume are, respectively,

$$C_p = \tfrac{5}{2}R \quad \text{and} \quad C_v = \tfrac{3}{2}R \tag{15.7}$$

where $R$ is the ideal gas constant. For any type of ideal gas $C_p - C_v = R$.

A reversible process is one in which both the system and its environment can be returned to exactly the initial states they were in before the process occurred. All spontaneous processes (such as the conduction of heat) and any process involving friction are irreversible.

In terms of heat flow, the second law of thermodynamics declares that heat flows spontaneously from a substance at a higher temperature to a substance at a lower temperature and does not flow spontaneously in the reverse direction. In terms of entropy, the second law of thermodynamics states that the total entropy of a closed system does not change when a reversible process occurs and increases when an irreversible process occurs.

The change in entropy $\Delta S$ for a process in which heat $Q$ enters or leaves a system reversibly at a constant Kelvin temperature $T$ is

$$\Delta S = \left(\frac{Q}{T}\right)_R \tag{15.8}$$

where the subscript $R$ stands for "reversible." Irreversible processes cause energy to be degraded in the sense that part of the energy becomes unavailable for the performance of work. The energy that is unavailable for doing work because of an irreversible process is

$$W_{\text{unavailable}} = T_0\, \Delta S_{\text{universe}} \tag{15.9}$$

where $\Delta S_{\text{universe}}$ is the total entropy change of the universe and $T_0$ is the Kelvin temperature of the coldest reservoir. Table 15.1 gives relationships between certain characteristic energies.

**TABLE 15.1  Equivalences Between Various Energy Units**

| |
|---|
| 1 erg = $10^{-7}$ J |
| 1 eV = $1.6 \times 10^{-19}$ J |
| 1 kcal/mol = $6.9 \times 10^{-21}$ J |
| 1 $k_B T$ (at room temperature) = $4.14 \times 10^{-21}$ J |
| 1 $k_B T$ (at room temperature) = 0.6 kcal/mol |
| 1 J = 0.239 cal |

## TOPIC 15.1   Biochemical Energy Generation

In order to properly discuss the generation of biochemical energy, some thermodynamic concepts need to be introduced. In particular, the Gibbs free energy is defined as

$$G = E + pV - TS \qquad (15.10)$$

where $E$ is the internal energy, $p$ pressure, $V$ volume, $T$ absolute temperature (in kelvin), and $S$ entropy. One of the most important biochemical reactions in the context of energy generation in animals is the oxidation of glucose:

$$C_6H_{12}O_6 + 6O_2 \rightarrow 6CO_2 + 6H_2O \qquad (15.11)$$

in which 1 mol of glucose produces $\Delta G = +686$ kcal of the Gibbs free energy, which is the maximum work obtainable from this reaction. In total, 180 g of glucose reacts with 134 L of $O_2$ to produce 686 kcal of energy; that is, 5.1 kcal is produced per liter of $O_2$. Work can be obtained indirectly either by (a) burning the glucose and using the heat released to run a heat engine or (b) using glucose as a step in a series of complex reactions releasing work at the end of the series. This has been utilized in animal cells via the so-called Krebs cycle. In it, for every mole of glucose metabolized, 38 mol of ATP is formed from ADP in the reaction

$$ADP + phosphate \rightarrow ATP \qquad (15.12)$$

The overall reaction can be written as

$$Glucose + 6O_2 + 38\ ADP + 38\ phosphate \rightarrow 38\ ATP + 6CO_2 + 6H_2O \quad (15.13)$$

and requires an input of 382 kcal of energy. However, at each ATP hydrolysis reaction, which is

$$ATP \rightarrow ADP + phosphate \qquad (15.14)$$

that is, inverse to reaction (15.12), 8 kcal of energy is made available. Consequently, with 38 ATP molecules one stores 304 kcal for available work, of which only about 50% is converted into useful work (as in muscle contraction), the rest being lost to heat production. Thus, the overall efficiency of biochemically based molecular "engines" in living cells is on the order of 20%.

Next we examine fat breakdown. We suppose 302 g of fat reacts with 414 L of $O_2$ according to

$$C_3H_5O_3(OC_4H_7)_3 + 18.5O_2 \rightarrow 15CO_2 + 13H_2O \qquad (15.15)$$

when 1941 kcal of energy is produced. This then corresponds to 4.7 kcal/L $O_2$.

In these two examples the number of kilocalories per liter of $O_2$ is fairly close. For protein breakdown a similar picture emerges. On average 4.9 kcal is produced per liter of $O_2$.

## TOPIC 15.2   First Law of Thermodynamics and Living Organisms

In this section we apply the first law of thermodynamics to the human body. The system includes all the food stored in the body. Complex chemical compounds (e.g., sugars) are continually being broken down by a living organism. The human body performs work on its environment and also produces heat. The first law of thermodynamics may be expressed as

$$\frac{\Delta E}{\Delta t} = \frac{\Delta Q}{\Delta t} - \frac{\Delta W}{\Delta t} \qquad (15.16)$$

In equation (15.16), $\Delta t$ is a short time interval and we interpret the other terms as follows. The rate at which the body does work on its environment is denoted by $\Delta W/\Delta t$. We assume that $\Delta W/\Delta t$ is positive. The rate at which the body absorbs heat from its environment is described by the term $\Delta Q/\Delta t$. Knowing the heat capacity of air enables us to find $\Delta Q/\Delta t$. The rate at which chemical energy is lost by the system, because of the breakdown of food and the production of waste products such as $CO_2$ and urea, is represented by $\Delta E/\Delta t$. This is a negative quantity called the catabolic rate, which is contributed to, for example, by the breakdown of glucose and fat.

The catabolic rate in kilocalories per mole per second can now be related to the oxygen consumption rate of the person tested:

$$\frac{\Delta E}{\Delta t} = 4.9 \frac{\Delta O_2 \, (L)}{\Delta t} \qquad (15.17)$$

In equation (15.17), $\Delta O_2 \, (L)/\Delta t$ is the volume of oxygen (in liters) consumed per second by the person. Using a simple breathing apparatus, it is easy to measure this oxygen consumption rate, so we can determine the catabolic rate using equation (15.17).

The efficiency $e$ of molecular activity is defined as the work done per second divided by the catabolic rate:

$$e = \frac{\Delta W/\Delta t}{|\Delta E/\Delta t|} \qquad (15.18)$$

Perfectly efficient use of energy means that $e = 1$, while a complete waste of catabolic energy implies that $e = 0$. Both the numerator and denominator in equation (15.18) can be measured. Table 15.2 has a particular activity listed—only cycling and shoveling—with the power output, the catabolic rate, and the efficiency given for each activity. The efficiency rate depends on how many muscles are used cooperatively to produce work and on the energy used by other organs. The more muscles are used to produce useful work compared with the activity of the other organs, the more efficient the body is (see Chapter 6 for details). A single muscle is only about 25% efficient in converting food energy to work; the rest goes to thermal energy. The body can thus never be more than 25% efficient, since other organs consume energy and produce no work. Leg muscles are the largest in the body; hence cycling and climbing stairs are relatively efficient processes (20%). Shoveling uses mostly arm and shoulder muscles and has an efficiency of only about 3%.

The body needs special cooling mechanisms such as sweating and vasodilation (a drug, agent, or nerve that can cause dilation of the walls of blood vessels) to get rid of excess heat during exercise. Table 15.3 presents catabolic rates for a 65-kg male and the associated consumption of oxygen. The catabolic rate during sleep is due to the activity of the internal organs and is called the basal, or resting, catabolic rate. The rate of con-

**TABLE 15.2 Power Output, Catabolic Rate, and Efficiency for Cycling and Shoveling**

| Task | Power output, $\Delta W/\Delta t$ (W) | Catabolic rate, $\Delta E/\Delta t$ (W) | Efficiency $e$ |
|------|------|------|------|
| Cycling | 112 | 505 | 0.19 |
| Shoveling | 17.5 | 570 | 0.03 |

**TABLE 15.3  Catabolic Rate and Oxygen Consumption During a Variety of Activities**

| Activity | Oxygen consumption (kcal/min) | Power (W) | Rate of energy consumption (L $O_2$/min) |
|---|---|---|---|
| Sleeping | 1.2 | 83 | 0.24 |
| Sitting at rest | 1.7 | 120 | 0.34 |
| Standing relaxed | 1.8 | 125 | 0.36 |
| Walking slowly (4–5 km/h) | 3.8 | 265 | 0.76 |
| Cycling (13–18 km/h) | 5.7 | 400 | 1.14 |
| Shivering | 6.1 | 425 | 1.21 |
| Playing tennis | 6.3 | 440 | 1.26 |
| Swimming breaststroke | 6.8 | 475 | 1.36 |
| Ice skating (14–15 km/h) | 7.8 | 545 | 1.56 |
| Climbing stairs (116 $\text{min}^{-1}$) | 9.8 | 685 | 1.96 |
| Cycling (21 km/h) | 10.0 | 700 | 2.00 |
| Playing basketball | 11.4 | 800 | 2.28 |
| Cycling, professional racer | 26.5 | 1855 | 5.30 |

version of food energy to some other form is called the metabolic rate. The total energy conversion rate of a person at rest is called the basal metabolic rate (BMR) and is divided among various systems in the body, as shown in Table 15.4. Table 15.4 also lists oxygen consumption rates in milliliters per minute. Energy consumption is directly proportional to oxygen consumption, since the digestive process is basically one of oxidizing food. Approximately 4.9 kcal of energy is produced for each liter of oxygen consumed, independent of the type of food. Because of this, some physiological measurements use the oxygen consumption rate as a measure of an individual's energy production rate. The dependence of BMR on mass applies both to humans and to other warm-blooded animals, as Figure 15.1 illustrates.

The digestive process is quite effective in metabolizing food; only about 5% of the caloric value of foods is excreted in the feces and urine without being utilized by the body. The body stores excess food energy by producing fatty tissue. Food is but one type of chemically stored energy. The energy content of foods is given in units called kilocalories (also called dietary calories); 1 kcal = 4186 J. Table 15.5 lists the number of kilocalories per gram for some common foods and fuels.

**TABLE 15.4  Basal Metabolic Rate and Oxygen Consumption Rates**

| Organ of BMR | Power consumed at rest | | Oxygen consumption | |
|---|---|---|---|---|
| | kcal/min | W | mL $O_2$/min | Percent |
| Liver and Spleen | 0.33 | 23 | 67 | 27 |
| Brain | 0.23 | 16 | 47 | 19 |
| Skeletal muscle | 0.22 | 15 | 45 | 18 |
| Kidney | 0.13 | 9 | 26 | 10 |
| Heart | 0.08 | 6 | 17 | 7 |
| Other | 0.23 | 16 | 48 | 19 |

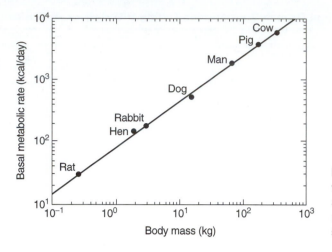

**FIGURE 15.1** Plot of BMR versus mass for various animals shows a straight-line relationship on a logarithmic scale indicating BMR $\alpha$ $m^{3/4}$.

## TOPIC 15.3 Physics of Animal Thermoregulation

In this section we discuss, from the point of view of thermodynamics, how the body adapts to thermal stress (for additional information we refer to Topics 12.2, 12.3, and 13.1). The brain, for example, can be irreversibly damaged when the temperature exceeds 42°C for a prolonged period. Our body temperature starts to drop following submersion in ice water and may lead to hypothermia and death. Animals have a variety of mechanisms to avoid such stress, for example, sweating, shivering, cooling by flapping of wings, fluffing of feathers to improve thermal insulation, vasodilation, and vasoconstriction (a drug agent or nerve that causes narrowing of the walls of blood vessels). Feverish patients may go through cycles of shivering and sweating due to constant switching between different

**TABLE 15.5 Kilograms per gram of Common Foods and Fuels**

| Food | Energy (kcal/g) | Food | Energy (kcal/g) |
|---|---|---|---|
| Average carbohydrate | 4.1 | Eggs | 1.63 |
| Average protein | 4.1 | Grapes | 0.69 |
| Average fat | 9.3 | Hamburger, lean | 1.63 |
| | | Ice cream, chocolate | 2.22 |
| Apples | 0.58 | Lard (fat) | 9.30 |
| Beans, Kidney | 1.18 | Milk, whole | 0.64 |
| Beer | 0.42 | Milk, low-fat | 0.42 |
| Big Mac | 2.89 | Oranges | 0.49 |
| Butter | 7.20 | Peanuts, roasted | 5.73 |
| Carrots | 0.42 | Peas | 0.71 |
| Celery | 0.15 | Potato, baked | 0.93 |
| Cheese, cheddar | 4.00 | Sirloin, lean | 1.66 |
| Cheese, cottage | 1.06 | Sugar | 4.00 |
| Chocolate | 5.28 | Tomato | 0.22 |
| Coffee, black | 0.008 | Tuna, in oil | 1.97 |
| Cola, carbonated | 0.36 | Wine | 0.85 |
| Corn flakes | 3.93 | | |

thermal control strategies. Body temperatures are maintained by mammals to be, as far as possible, independent of the temperature of the environment (the ambient temperature). Reptiles usually have a body temperature close to the ambient temperature, but under extreme conditions they too may be forced to adopt thermal control strategies.

The first law of thermodynamics relates the amount of heat lost by a system, $\Delta Q$, to a decrease in internal energy, $\Delta E$, of the system and to the work done by the system, $\Delta W$. Thus

$$\Delta Q = \Delta E + \Delta W \tag{15.19}$$

For a mammal, the amount of metabolic energy released by the burning of, for example, carbohydrates is $\Delta E$. The amount of work done by the muscles and other organs is described by $\Delta W$.

The heat current $I$ out of an organism is defined to be the heat $\Delta Q$ lost by the mammal divided by the time it took to lose the heat. Thus

$$I = \frac{\Delta E}{\Delta t} \tag{15.20}$$

follows from the expression of the first law of thermodynamics in equation (15.19) provided we assume that no work is being done by the mammal. The quantity on the right-hand side of equation (15.20) is the BMR. This quantity depends on the animal's body mass $M$ according to an empirical formula called Kleiber's law, where

$$\left( \frac{\Delta E}{\Delta t} \right)_{\text{rest}} = mM^{0.75} \tag{15.21}$$

The prefactor $m$ for mammals is about 3.4 in SI units. Using Newton's law of cooling for the heat current $I$ in equation (15.20), we find that the temperature of a resting organism can be expressed as

$$T(\text{organism}) = T(\text{environment}) + \frac{1}{Ah_c} mM^{0.75} \tag{15.22}$$

where $A$ is the body surface area and $h_c$ is the coefficient of heat transfer. The term $T(\text{organism})$ is the temperature of the inner organs and not the skin. The temperature of the environment, $T(\text{environment})$, is the temperature outside the skin or pelt of the animal. Here we are considering the transport of heat from the inner organs across any layer of fat and across the pelt to the outer environment. We use equation (15.22) to discuss the thermal control strategies in various animals.

Let us first consider seals or whales in the Arctic. The ambient temperature is that of seawater and is close to freezing. If the target temperature of the organism is 37°C, then the second term in equation (15.22) must make this up. Aquatic mammals in the Arctic rely on a layer of fat to insulate them thermally from the near-freezing environment. Heat transport through this layer of fat is by thermal conduction. For the sake of simplicity, consider a walrus to be a sphere with a radius of 1 m. The mass $M$ is about $10^3$ kg, the surface area being $4\pi$ square meters. We suppose the outer part of the sphere is a layer of fat of thickness $L$. From equation (15.22) for heat transfer we have

$$T(\text{organism}) = T(\text{environment}) + \frac{L}{A\kappa_{\text{fat}}} mM^{0.75} \tag{15.23}$$

Solving for $L$, one finds $L = 8$ cm, quite close to the observed thickness of the layer of fat for a walrus.

Now suppose the mammal has a certain body mass $M$ and it would like its layer of fat to be as thin as possible. We now find the best value for the body surface area $A$. To reduce $L$ without altering the body temperature, $A$ must be reduced in equation (15.23). For a given volume, the smallest surface area is a sphere. Aquatic mammals in Arctic seas, from the point of view of heat transfer, should therefore have spherical shapes, and while seals, sea lions, and walruses are not exactly spherical, they are much more compact than land animals like cats and dogs. We now inquire if there is an advantage for aquatic mammals to be big. We recall that body area scales with animal size $L$ as $L^2$ while the body mass scales as $L^3$. Thus the ratio $M^{0.75}/A$ depends on body size $L$ as

$$\frac{M^{0.75}}{A} = \frac{(L^3)^{0.75}}{L^2} = L^{1/4} \tag{15.24}$$

This expression increases slowly with $L$, and thus larger aquatic mammals do not need as thick a layer of blubber as smaller aquatic animals. An aquatic mammal could increase its metabolic rate above the resting rate by exercising and thus increase body temperature.

Now let us consider mammals in the tropics where the environment can be very hot. If the environment's temperature approaches the target body temperature (37°C), the second term of equation (15.23) must be reduced to prevent overheating. Let us suppose that the tropical mammal has no layer of fat and many blood vessels in the skin to assist cooling so that the skin temperature is equal to the temperature of the organism. To reduce the second term of equation (15.23), the surface area $A$ must be made as large as possible. Mammals living in the tropics have surprisingly large surface areas—the large ears of the elephant and the long neck of the giraffe being but two examples.

Let us take the case of an animal resting in perfectly stationary air with no convection. The animal is treated, once again, as a sphere of radius $R$, so the coefficient of heat transfer $h_c = \kappa(\text{air})/R$. Inserting this last expression into equation (15.23) gives

$$T(\text{organism}) - T(\text{environment}) = \frac{mM^{0.75}}{4\pi R\kappa(\text{air})} \tag{15.25}$$

Assuming a density of 1000 kg/m³ for the animal, we obtain

$$T(\text{organism}) - T(\text{environment}) = 5637R^{5/4} \tag{15.26}$$

in SI units. Again the temperature difference between the environment and the organism grows with size. For a sphere of diameter 0.01 m in air, one obtains a temperature difference of a few degrees. For a 0.10-m-diameter sphere it is of order 100 degrees and for a sphere with a diameter of 1 m the difference is 1000 degrees. However, we have neglected convection as a heat transfer mechanism. The corresponding heat transfer coefficients in air for a sphere of about 1 m in diameter is found to be about 1 for free convection and 10 for forced convection (in SI units). Our estimate earlier had been 0.025 using $h_c = \kappa(\text{air})/R$. For forced convection this is about a factor of 1000 too small. Discounting the factor 1000, we find that the temperature difference becomes a few degrees, which is much more reasonable. Table 15.6 gives $m$ coefficients in Kleiber's law, which plays an important role in heat regulation.

Finally, let us look at thermoregulation of the human body. Heat in this case is carried off by a combination of thermal conduction through tissue and convection through the bloodstream. In the case of humans the BMR is about 100 W. For intense activity $\Delta E/\Delta t$ can rise to about 500 W. If we do not sweat, convection and radiation are the dominant heat transfer mechanisms. Again, if we do not sweat, the experimentally measured heat transfer coefficient of the human body is about 15 W/m² · K. If we apply equation (15.23) using a surface area of about 1.7 m², an environment temperature of 20°C, and a

**TABLE 15.6 Prefactor *m*
Coefficients in Kleiber's Law**

| Organisms | *m* (SI units) |
|---|---|
| Mammals | 3.4 |
| Marsupials | 2.3 |
| Lizards | 0.38 |
| Salamanders | 0.038 |

$\Delta E/\Delta t$ of 100 W, we obtain a body temperature of about 24°C. Thus, to reach a body temperature of 37°C, we must increase our metabolic rate $\Delta E/\Delta t$ to about 425 W. At lower ambient temperatures $\Delta E/\Delta t$ would have to be correspondingly higher to keep the temperature at 37°C. We do tend to eat more in colder climates. Our heat production can range between 500 and 1000 W if we are intensely active. In this situation, for an ambient temperature of 20°C we would then have a body temperature above 37°C. It is therefore unavoidable in this situation for overheating to occur when we prolong our intense activity. This is where sweating comes in, which increases the transfer coefficient from 15 to about 300. The excess temperature can be reduced by as much as 20 degrees by this mechanism.

Work performed during muscle contraction is accompanied by a temperature increase of the muscle fiber; that is, the energy delivered by ATP is converted partly to thermal energy. A small amount of this heat is conducted away and some is radiated. Breathing also cools the body because of the water vapor exhaled as a by-product, convection being a very effective cooling process. When these processes are saturated, sweating is triggered to provide additional cooling. In fact, the body maintains a nearly constant temperature by eliminating excess thermal energy as rapidly as possible. Blood flow increases when blood vessels dilate, pores open so that sweating increases, and respiration increases. In hot weather these mechanisms may not be sufficient for extreme aerobic activities (e.g., marathon running), so athletes may pour water over themselves to increase cooling, although it is not unusual for marathon runners to experience a rise in temperature of about 4°C during a race. One of the most useful protections against losing heat is a warm hat, as up to 40% of the thermal energy lost by the body exits through the head.

## TOPIC 15.4   Entropic Elasticity of DNA

The length of a strand of DNA, $L_0$, depends on the number of its base pairs. Human DNA strands have a length $L_0$ of about 1 m, but experiments are performed on shorter strands of bacterial or viral DNA (typically $L_0$ here is of the order 50 to 100 $\mu$m). Between base pairs the spacing is approximately 3.4 Å (1 Å = $10^{-10}$ m) so the number of base pairs is of order $2 \times 10^5$. A technique called the optical trap is used to keep two micrometer-sized silicon beads in place at the end of a DNA strand and allows careful measurement of both the force $F$ exerted on the trapped beads by the DNA strand and the spacing $L$ between the beads. A force–extension plot between $F$ and $L$ can therefore be determined, and in Figure 15.2 a plot for a 50-$\mu$m strand is shown. An approximate relation between $F$ and $L$ is given as

$$F(L) \cong kL \quad \text{for } L << L_0 \tag{15.27}$$

where $k$ is a constant to be determined. The above relationship is in the form of Hooke's law as discussed at length in Chapter 10. However, there is an important difference be-

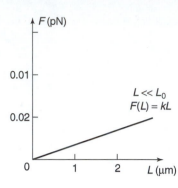

**FIGURE 15.2** Force–extension plot for a DNA strand.

tween human DNA and a stretched solid in that the applied force is not zero even when $L$ is much less than $L_0$. A strand of DNA attempts to make the spacing between its end-points as short as possible by curling up since the coil agitates constantly due to thermal fluctuations resulting from collisions with water molecules. The thermal energy carried by each molecule is $k_B T$; that is, proportional to the ambient temperature $T$ and at room temperature ($T = 300$ K), the thermal energy is about $4 \times 10^{-21}$ J. The effect of thermally induced curling is called entropic elasticity. The constant $k$ in equation (15.27) is the effective spring constant of the DNA strand given by

$$k = \frac{3k_B T}{4\xi_p L_0} \tag{15.28}$$

where the quantity $\xi_p$ is called the persistence length and is roughly equal to the typical size of a coil, which is of order 500 Å. For larger applied forces, L begins to approach the strand length and equation (15.27) is no longer valid, and $F$ increases rapidly once $L$ exceeds the strand length of $50 \times 10^{-6}$ m. In this regime forces are typically of order $10^{-11}$ N.

## SOLVED PROBLEMS

**15.1** How many grams of fat will an idle person gain in a day by consuming 2500 kcal of food? An idle person burns approximately 1.22 kcal/min, and digesting fat produces 9.3 kcal/g.

**Solution** The energy requirement in kilocalories can be calculated as

$$(1.22 \text{ kcal/min})(60 \text{ min/h})(24 \text{ h/day}) = 1757 \text{ kcal/day}$$

so the number of excess kilocalories is

$$2500 \text{ kcal} - 1757 \text{ kcal} = 743 \text{ kcal}$$

Assuming that the same amount of energy is used to store a gram of fat as is obtained from digesting a gram of fat, we see that

$$\text{Mass gain} = 743 \text{ kcal} \times \frac{1.0 \text{ g fat}}{9.3 \text{ kcal}} = 80 \text{ g fat}$$

**15.2** Compare the amount of oxygen consumed in 1 h by a cyclist traveling at 21 km/h with that used by a person sleeping. What would you expect the ratio of the respiratory rates to be in these two activities?

**Solution** Table 15.3 gives the rate of oxygen consumption when cycling. Thus

$$\ell_{O_2} = (2.00 \text{ L O}_2/\text{min})(60 \text{ min}) = 120 \text{ L O}_2$$

For sleeping,

$$\ell_{O_2} = (0.24 \text{ L O}_2/\text{min})(60 \text{ min}) = 14.4 \text{ L O}_2$$

The ratio is therefore

$$\frac{120 \text{ L O}_2}{14.4 \text{ L O}_2} = 8.33$$

**15.3** In a normal, adult human, the tubules of the kidneys reabsorb 800 mmol of glucose per 24 h. In this glucose transport from the tubular fluid toward the blood, the glucose is concentrated 100-fold.
   **(a)** What is the work done by the tubular cells while concentrating glucose?
   **(b)** What is their average power?

**Solution**

**(a)** The concentration work in an isothermal, reversible process is

$$W = nRT \ln \frac{c_2}{c_1}$$

where $n = 800$ mmol, $R$ is the universal gas constant, $T = 310$ K is the body temperature, and $c_2/c_1 = 100$. On substituting these values into the equation, we obtain $W = 9.5$ kJ.
   **(b)** The average power $P = W/t = 0.1$ W.

**15.4** The dead space in the human respiratory system is about 150 mL. At rest, the respiratory minute volume is 6 L. What is the amount of air that reaches the alveoli:
   **(a)** If a human breathes shallowly, but frequently, e.g., 25 times per minute?
   **(b)** If a human breathes deeply, but less frequently, e.g., 10 times per minute?
   Suppose that the respiratory minute volume is the same in both cases.

**Solution**

**(a)** In one breath, 6000 mL/25 = 240 mL is inspired. The amount of air that reaches the alveoli in 1 min is

$$25(240 - 150) \text{ mL} = 2.250 \text{ L}$$

**(b)** In one breath, 6000 mL/10 = 600 mL is inspired. The amount of air that reaches the alveoli in 1 min is then

$$10(600 - 150) \text{ mL} = 4.500 \text{ L}$$

**15.5** While doing physical exercise, a male athlete produces $P = 300$ W heat power. Assuming that this is consumed by the evaporation of sweat from his skin, how much water would be evaporated per hour? The latent heat of vaporization of water at $T_2 = 100°C$ is $L_w = 2.26 \times 10^6$ J/kg, the temperature of the skin is $T_1 = 20°C$, and the specific heat of water is $c_w = 4186$ J/kg · °C.

**Solution** In an hour, the physical exercise generates $E = P\Delta t = 300\text{W} \times 3600 \text{ s} = 1.0 \times 10^6$ J heat energy, which will be used to heat water of mass $m$ from 20 to 100°C and to convert water to steam at the boiling point of water:

$$E = mc_w(T_2 - T_1) + mL_w$$

From this, the mass of water evaporated from the skin in 1 h can be found as

$$m = \frac{E}{c_w(T_2 - T_1) + L_w} = 0.385 \text{ kg}$$

Only $mc_w(T_2 - T_1) = 0.13 \times 10^6$ J of energy is needed per hour to increase the temperature from 20 to 100°C, but a significantly larger amount, $mL_w = 0.87 \times 10^6$ J/h, is needed for the evaporation of water at 100°C.

**15.6**   What is the work of breathing during normal inspiration? Use the relaxation pressure curve of the total respiratory system shown. The difference between the measured pressure values and the outer pressure (called the transmural pressure or the intrapulmonary pressure), $P_i$, is plotted against the inhaled volume (called the tidal volume, $V_t$).

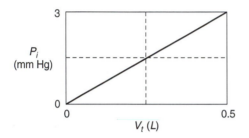

**Solution**   The two major components of the work of breathing are the elastic work required to inflate the respiratory system ($W_e$) and the work required to maintain the air flow against the frictional resistance ($W_f$). The former can be obtained from the relaxation pressure curve as the area under the curve approximated as a straight line:

$$W_e = \tfrac{1}{2} P_i V_t = 0.1 \text{ J}$$

where $P_i = 3$ mm Hg $= 407$ Pa is the intrapulmonary pressure and $V_t = 0.5$ L $= 5 \times 10^{-4}$ m$^3$ is the tidal volume. The second type of work is calculated with the approximation of laminar flow as

$$W_f = \Delta \overline{P} \, V_t = 0.05 \text{ J}$$

where $\Delta \overline{P}$ is the average pressure drop along the airway during quiet inspiration. The total work is then 0.15 J.

# EXERCISES

**15.1**   Suppose an exercising weightlifter loses 0.250 kg of water through evaporation, the heat required to evaporate the water coming from the weightlifter's body. The work done in lifting weights is $2.50 \times 10^5$ J.
(a) Assuming that the latent heat of vaporization of perspiration is $2.42 \times 10^6$ J/kg, find the change in the internal energy of the weight lifter.
(b) Determine the minimum number of food calories that must be consumed to replace the loss of internal energy.

**15.2**   An alcohol rub can rapidly reduce an elevated body temperature in a patient. The heat energy lost by the person is due to the evaporation of alcohol. Find the number of grams of alcohol that must be evaporated from the surface of a 75-kg person to reduce body temperature by 2.5°C. The heat of vaporization for alcohol is 204 cal/g. The specific heat capacity for the human body is 0.83 cal/g · °C.

**15.3** The temperature of a scalpel can reach 150°C in sterilization procedures. Given a 100-g steel scalpel, determine the amount of heat that must be removed to reduce its temperature from 150 to 30°C given that $c_{steel} = 0.11$ cal/g · °C.

**15.4** For how long must a 70-kg male perform very heavy activity to use up the energy provided by the consumption of 1.5 kg of fat?

**15.5** Assume you know only that the effective spring constant $k$ of DNA depends on the thermal energy $k_B T = 4 \times 10^{-21}$ J, the strand length $L_0$, and the persistence length $\xi_p$. Use dimensional analysis to derive

$$k = \frac{3k_B T}{4\xi_p L_0}$$

where $\xi_p$ is the persistence length, approximately equal to the size of the coil and $L_0$ is the strand length. You may assume that $k$ is inversely proportional to $L_0$.

**15.6** When dietary intake is insufficient, body fat is metabolized at a rate of 9.3 kcal/g. If brisk walking consumes energy at a rate of 5 kcal/min, how many minutes would you have to walk for to lose 2 kg of body fat? If you plan on doing this at a rate of 1 h of walking a day, how long will it take to accomplish your goal?

**15.7** A 70-kg patient is to be cooled to 28°C for surgery by being placed in ice water. The power output of this patient is 60 W. It takes 30 min to bring her temperature down and the surgery lasts 3 h. How many kilograms of ice must melt to do this, assuming all other forms of heat transfer are negligible and that the ice water stays at 0°C?

**15.8** During heavy exercise 2 L of blood are pumped to the surface of a person per minute to carry away core heat. If the blood is cooled by 3.0°C at the surface, what is the rate of heat transfer due to blood flow? You may assume that the specific heat of blood is the same as that for water and that the density of blood is 1050 kg/m³.

**15.9** Suppose a man is losing heat to the environment at the rate of 350 W. His body temperature is 2.5°C below normal, and he begins to shiver. If his mass is 75 kg, how long will it take for his temperature to rise to normal?

**15.10** Following strenuous exercise a person has a temperature of 39°C and is giving off heat at the rate of 60 cal/s. How long will it take for this person's temperature to return to 37°C if her mass is 60 kg?

**15.11** (a) Calculate the energy used by a person who does 100 chin-ups. This person's mass is 70 kg and he raises his center of gravity 0.60 m during each chin-up. (b) What is the total power consumed if the chin-ups are done in 5.0 min? Assume the body's efficiency of 20% in this activity.

# WAVES AND SOUND

## PHYSICAL BACKGROUND

Waves are physical oscillations that occur as a direct result of a disturbance in an elastic medium and can be represented by a sinusoidal function (see Figure 16.1). The physical characteristics of a wave include the following:

Amplitude $A$ of a wave represents the maximum displacement of particles within the elastic medium from its point of equilibrium. Amplitude is a measure of displacement and is expressed in units of length.

Frequency $f$ of a wave represents the number of complete cycles that pass a defined point per second. Frequency is expressed in units of hertz (Hz), or cycles per second.

Period $T$ is the time required to observe one complete wave cycle. The period is related to the frequency by

$$T = \frac{1}{f} \tag{16.1}$$

Period is expressed in seconds.

Wavelength $\lambda$ of a wave is the distance between adjacent peaks. As the wavelength decreases, more complete cycles are observed per unit time, and therefore the frequency of the wave increases. Wavelength is expressed in units of length, or $L$. The relationship between wavelength and frequency is given by

$$\lambda = \frac{v}{f} \tag{16.2}$$

Speed $v$ is the speed of the propagating wave and is obtained from equation (16.2) as

$$v = \lambda f \tag{16.3}$$

Phase $\phi$ represents the angular difference in position for two points on a vibrating wave.

Two main types of waves are transverse and longitudinal waves. In a transverse wave, the disturbance is perpendicular to the direction of travel of the wave. In a longitudinal wave, the disturbance is parallel to the line along which the wave travels.

The speed of a wave depends on the properties of the medium in which the wave travels. For a transverse wave on a string that has a tension $F$ and a mass per unit length $m/L$, the wave speed is

$$v = \sqrt{\frac{F}{m/L}} \tag{16.4}$$

**190**

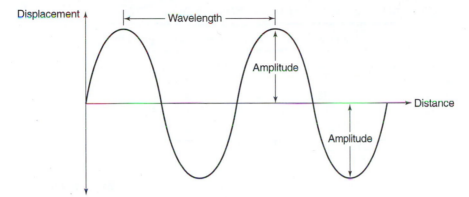

**FIGURE 16.1**   Graphical representation of a wave in terms of its displacement as a function of distance from the source.

Sound is a longitudinal wave that consists of alternating regions of greater than normal pressure (condensation) and less than normal pressure (rarefaction). Each cycle of a sound wave includes one condensation and one rarefaction. A sound wave with a large frequency is interpreted by the brain as a high-pitched sound, while one with a small frequency is interpreted as a low-pitched sound. The pressure amplitude of a sound wave is the magnitude of the maximum change in pressure measured relative to the undisturbed pressure. The larger the pressure amplitude, the louder the sound.

The speed of sound $v$ depends on the properties of the medium. In an ideal gas, the speed of propagating sound is

$$v = \sqrt{\frac{\gamma kT}{m}} \qquad (16.5)$$

where $\gamma = c_p/c_v$ is the ratio of the specific heat capacities at constant pressure and constant volume, $k$ is Boltzmann's constant, $T$ is the absolute temperature, and $m$ is the mass of a molecule of the gas. In a liquid, the speed of sound is given as

$$v = \sqrt{\frac{B_{ad}}{\rho}} \qquad (16.6)$$

where $B_{ad}$ is the adiabatic bulk modulus and $\rho$ is the mass density. For a solid in the shape of a long slender bar, the expression for the speed of sound is

$$v = \sqrt{\frac{Y}{\rho}} \qquad (16.7)$$

where $Y$ is Young's modulus.

The intensity of a sound wave is the power $P$ that passes perpendicularly through a surface of area $A$: $I = P/A$. The SI unit of intensity is watts per square meter (W/m$^2$). The smallest sound intensity that humans can detect is known as the threshold of hearing and is about $I_0 = 1 \times 10^{-12}$ W/m$^2$ for a 1-kHz sound. When a source emits sound uniformly in all directions and no reflections are present, the intensity is inversely proportional to the square of the distance from the source.

The intensity level $\beta$ (in decibels) is used to compare a sound intensity $I$ to the sound intensity $I_0$ of a reference level:

$$\beta = 10 \log\left(\frac{I}{I_0}\right) \qquad (16.8)$$

TABLE 16.1 **Attenuation of Sound Waves in Various Media**

| Material | Frequency (Hz) | Attenuation length $x$ (m) |
|---|---|---|
| Water | 20 | $10^5$ |
|  | $10^6$ | 20 |
| Muscle | $10^6$ | $4 \times 10^{-2}$ |
| Bone | $10^6$ | $4 \times 10^{-3}$ |
|  | $3.5 \times 10^6$ | $6 \times 10^{-4}$ |

Sound waves produce molecular motion in the material through which they propagate. Friction leads to loss of intensity: Energy is being dissipated. If a sound wave with an intensity $I_0$ hits the surface of a medium, then the sound intensity is reduced as it enters deeper and deeper into the medium because of attenuation. At a distance $x$ inside the medium, the sound intensity $I(x)$ is given by

$$I(x) = I_0 e^{-x/x_0} \tag{16.9}$$

The constant $x_0$ is called the attenuation length and represents the measure of depth a sound wave will travel into a material. It depends very strongly on both the frequency and the type of medium involved. Table 16.1 illustrates these properties.

The Doppler effect is the change in frequency detected by an observer because the sound source and the observer have different velocities with respect to the medium of sound propagation. If the observer and source move with speeds $v_o$ and $v_s$, respectively, and if the medium is stationary, the frequency $f'$ detected by the observer is

$$f' = f\left(\frac{1 \pm v_o/v}{1 \mp v_s/v}\right) \tag{16.10}$$

where $f$ is the frequency of the sound emitted by the source and $v$ is the speed of sound. In the numerator, the plus sign applies when the observer moves toward the source and the minus sign applies when the observer moves away from the source. In the denominator, the minus sign is used when the source moves toward the observer and the plus sign is used when the source moves away from the observer.

## TOPIC 16.1  The Physics of Hearing

The purpose of hearing in humans is to record sound and transform this into an electrical signal which is sent by the nerves to the brain. Our ears act as a sound recording with an enormous range of sound intensities. Their ability to distinguish different sound frequencies is astonishing. We first briefly describe the structure of the ear, which, in many ways, is quite bizarre.

The ear consists of three parts: the outer or external ear, the middle ear, and the inner ear. The familiar shell-like structure is called the pinna. The pinna is connected to the external auditory canal, which is about 2.5 cm long and filled with air. This canal ends at the tympanic membrane (eardrum), which is stretched across the canal. Thus the outer ear comprises the pinna and the auditory canal.

The middle ear, or tympanic cavity, is an air-filled cavity inside the temporal bone of the head. It is closed off on one side by the tympanic membrane and the other side by two openings called the oval window and the round window. The middle ear is spanned by three small bones: the malleus, or hammer; the incus, or anvil; and the stapes, or stirrup. The first of these bones, the malleus, is in contact with the tympanum and the last, the incus, is in contact with the oval window. Thus the middle ear comprises the eardrum, or tympanum, malleus, incus, stapes, and Eustachian tube. These bones transmit force exerted on the eardrum to the inner ear through the oval window. Because they form a lever system with a mechanical advantage of about 2, the force delivered to the oval window is multiplied by 2. Furthermore, the oval window has an area about 1/20th that of the eardrum; thus the pressure created in the fluid-filled inner ear is about 40 times that exerted by sound on the eardrum. This system enables the ear to detect very low intensity sounds.

The middle ear also offers some protection against damage from very intense sounds. Muscles supporting and connecting the three small bones contract when stimulated by very intense sounds and reduce the force transmitted to the oval window by a factor of about 30. The reaction time for this defense mechanism is at least 15 ms, so it cannot protect against sudden increases in sound intensity, such as from gunfire.

In Figure 16.2 we present a simplified picture of the anatomy of the ear. The inner ear, or labyrinth, is further inside the temporal bone and is filled with a fluid called endolymph which is similar to the fluid inside cells. The cochlea is the key structure of the inner ear. It has the form of a snail (*cochlea* means "snail" in Latin) and is a long spiral tube with about 2.5 turns. It starts from the oval window and ends at the round window. In the cochlea is the basilar membrane, which is narrow and thick near the oval window and wide and thin near the round window. The organ of Corti transduces the sound signal into an electrical signal and rests on the basilar membrane. It consists of about 15,000 cochlea hair cells, and each of these has about 50 to 100 hairs. The inner ear contains three semicircular canals, the vestibule, and the Eustachian tube. Sensory impulses from the inner ear pass to the brain via the vestibulocochlear nerve.

The sound wave which impinges on the ear is a pressure wave, and the key purpose of the outer and middle ears is to guide the sound signal from the exterior to the oval win-

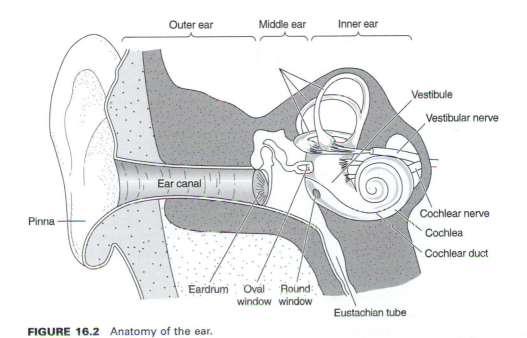

**FIGURE 16.2**  Anatomy of the ear.

dow. The sound wave sets the oval window in motion, and this in turn causes motion in the endolymph fluid. There is an impedance mismatch at the oval window between the fluid-filled inner ear and the air-filled middle ear cavity. If the sound impedances of the two media are very different, then a large amount of sound intensity may be lost. Only 2% of the signal intensity passes the oval window if we use this criterion alone. However, the outer and middle ear compensate for this loss of signal strength by preamplifying the sound signal as much as possible, thus generating a high-intensity sound signal at the oval window. The amplification begins by focusing the sound wave. The power $P$ delivered by the sound wave is the total energy passing a unit cross section per second. Therefore, $P = IA$, where $A$ is the cross-sectional area of the tube carrying the sound and $I$ the intensity. Assuming the sound wave conserves energy, the power is fixed along the channel. The sound intensity $I(A)$ then varies inversely with the cross-sectional area $A$. Thus

$$I(A) = \frac{P}{A} \tag{16.11}$$

Hence, by narrowing the channel, the sound intensity can be enhanced. The pinna collects sound like a horn and guides it to the auditory canal, which amplifies the sound intensity. In fact, this has a radius of only 3 mm. Furthermore, the auditory canal acts like a resonant cavity for sound signals, and this assists the transmission of sound.

The next part of the amplification works differently in that the sound wave applies a pressure oscillation to the tympanic membrane, $\Delta P(\text{tymp})$. The maximum force $F$ on this membrane is thus equal to $F = \Delta P(\text{tymp})A(\text{tymp})$, where $A(\text{tymp})$ is the surface area of the tympanic membrane, that is, about 55 mm². Recall that the ear is connected mechanically to the oval window by three small bones. Hence the oval window also feels an oscillating force $F$. The area of the oval window is $A(\text{oval}) = 3.2$ mm². The force on the oval window is $F = \Delta P(\text{oval})A(\text{oval})$. The ratio of pressure amplitudes is therefore

$$\frac{\Delta P(\text{oval})}{\Delta P(\text{tymp})} = \frac{A(\text{tymp})}{A(\text{oval})} = 17 \tag{16.12}$$

The sound intensity increases as the square of the pressure amplitude so the middle ear magnifies the intensity by a factor of $(17)^2 = 289$, so this more than makes up for the signal loss due to impedance mismatch. One might ask why have three small bones and why do we not possess a single rod structure? A rodlike bone would deliver destructively large forces for intense sound. It transpires that the mechanical structure of the ossicles is designed such that high force levels on the oval window produced by loud sounds are damped out considerably.

## TOPIC 16.2  Sound Perception

One meaning of the word *perception* is awareness through the senses. Some 30,000 nerve endings participate in sending sound information to the brain from the cochlea. The human ear has a remarkable sensitivity and range. It can detect sounds varying in intensity by a factor of $10^{12}$—from threshold to those causing damage—and in frequency from 20 to 20,000 Hz. Yet the perception of sound differs in important ways from the actual physical properties of sound.

Suppose we listen to a perfect sinusoidal sound generated by a tuning fork. We perceive it as a single sustained tone. Tones are characterized by two quantities: loudness and pitch. Mathematically a sinusoidal sound wave is characterized by just two quantities, namely the amplitude $\Delta P$ of the pressure wave and the frequency $f$ of the wave. Tests on subjects have shown that the amplitude $\Delta P$ of the pressure wave corresponds to loudness. This is not surprising since $\Delta P$ is a measure of the force with which the eardrum is pushed back and forth. High-pressure amplitudes correspond to a loud sound. The frequency $f$ corresponds to the pitch of the tone.

When a singer produces a pure tone (e.g., middle C), this is not a sinusoidal wave. The waveform is much more complicated. If you sound the same note on the piano, on a trumpet, or on a guitar, then you perceive different sounds, even if they have the same pitch and loudness. In each case the repeat pattern of the wave has a different detailed shape. We refer to this as the sound quality. The perceived sound quality of a tone is in fact determined by the particular values of the so-called Fourier amplitudes so that changes in the Fourier amplitudes of a tone are perceived as changes in sound quality. See Chapter 17 for more information on the superposition of waves.

The decibel scale corresponds fairly well to the human perception of loudness. Decibels are physically measurable and are fairly representative of the perceived loudness of sounds. The smallest difference in intensity an average person can sense is about 1 dB, and an intensity difference of 3 dB is easily discernable. Loudness depends strongly on frequency as well as intensity. Two sounds of different frequencies but equal intensities rarely sound equally loud. This is because the ear is more sensitive to some frequencies than others. Very large intensities are needed for sound to be audible near the extremes of the normal range of hearing—approximately 100 dB at 20 or 20,000 Hz, for example.

In Figure 16.3 we display Fletcher–Munson curves, which are plots of sound intensities in decibels versus frequency in hertz. The sound intensity is that needed at each frequency to make the sound appear to have the same loudness. The lowest curve shows intensity levels at which sound frequencies just become audible. The labels on the curves indicate intensity levels at 1000 Hz. Concentrating on the threshold of the hearing curve, we see that the intensity level of a 100-Hz sound must be about 37 dB greater than the intensity level at 1000 Hz. Thus the ear is less sensitive to a 100-Hz sound than a 1000-Hz intensity level. The graphs indicate that the ear is most sensitive in the range from about 1 to 5 kHz. At higher and lower frequencies it becomes progressively less sensitive. Constant loudness curves become more horizontal as the loudness increases; that is, when you play a stereo at loud levels, you hear low, middle, and high frequencies about equally. Turning down the volume control so the sound becomes quieter, the high and low frequencies seem to disappear. At the threshold of hearing at $f_0 = 4$ kHz, the sound intensity is about $I_0 = 10^{-12}$ W/m$^2$.

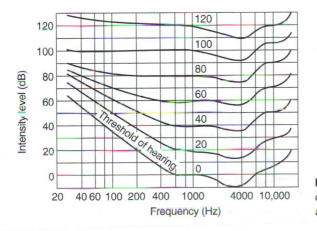

**FIGURE 16.3** Fletcher–Munson curves of intensity versus frequency at the same loudness (in dB).

The displacement amplitude $y_0$ of the air molecules gives us a lower limit to the displacement of the eardrum and is found from the formula (with $I = I_0$ and $f = f_0 = 4000$ Hz)

$$y_0 = \sqrt{\frac{I}{2\pi^2 \rho f^2 v}} \quad \text{as} \quad y_0 = 2.7 \times 10^{-3} \text{ nm} \tag{16.13}$$

which is astoundingly small. Pain occurs when sound intensities reach about 120 dB, a value which corresponds to about 1.0 W/m². At this intensity, the sound amplitude is

$$y_0 = \sqrt{\frac{1.0}{2\pi^2 (1.3)(4000)^2(340)}} = 2.7 \times 10^{-6} \text{ m} \tag{16.14}$$

which is a remarkably small distance. It is on the same order as the shortest length which can be seen with the best light microscope.

The combined effect of the eardrum and the ossicles, the small bones of the middle ear, is to convert the sound energy into mechanical energy. The lever action of these ossicles leads to more forceful displacement of the oval window, although a reduction in amplitude simultaneously takes place. The process is analogous to the operation of a pressure intensifier, as shown in Figure 16.4, where the oil at pressure $P_1$ on the area $A_1$ of the piston exerts a force

$$F_1 = P_1 A_1 \tag{16.15}$$

This force is transmitted through the piston of area $A_2$ to oil at pressure $P_2$. Since

$$A_1 P_1 = A_2 P_2 \tag{16.16}$$

then

$$P_2 = \frac{A_1}{A_2} P_1 \tag{16.17}$$

and the pressure is amplified by the factor $A_1/A_2$.

The main amplification of sound takes place in the middle ear and is due to the fact that the oval window is 20 to 30 times smaller than the eardrum. The energy of the vibration in the air outside the eardrum is communicated to the perilymph in the vestibular canal by the ossicles. The force moving the eardrum is multiplied by the mechanical advantage of the ossicles into a force two or three times larger, which the stirrup exerts on the oval window, and consequently the pressure behind the oval window is between 40 and 90 times larger than the pressure at the eardrum. With the twofold amplification produced by the ear canal, this means that sounds in the frequency range 3 to 4 kHz can be amplified 180 times in favorable cases. This is the pressure amplification. The intensity amplification is the square of this, or approximately 32,000 times.

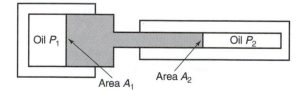

**FIGURE 16.4** Pressure intensifier.

# TOPIC 16.3   Medical Applications of Ultrasound

Ultrasound refers to sound above the human audible limit of about 20 kHz, and it is used in medicine for three purposes: therapeutic, destructive, and diagnostic. In therapy, ultrasound is used to produce heating in the tissues, and intensities of the order of 10 to 100 kW/m² are used. In surgery much higher powers, of 1 to 40 MW/m², are sufficient to destroy unwanted tissue. Ultrasonic techniques are routinely used in neurology, cardiology, and obstetrics and are becoming more and more important in several other fields.

In diagnostic applications, a transmitter produces high-frequency sound pulses which are directed into the body. When a pulse encounters a boundary between two tissues that have different densities, reflections occur. Thus a picture or sonogram of the inner structure may be obtained by scanning ultrasonic waves across the body and detecting echoes generated from various locations. This technology is extensively employed in obstetrics to examine a developing fetus which is surrounded by the amniotic sac of fluid and can be distinguished from other anatomical features. Thus the position of the fetus, its size, and possible abnormalities may be detected.

Ultrasound can also be used to investigate malignancies in the brain, kidney, liver, and pancreas. For example, neurosurgeons use a cavitron ultrasonic surgical aspirator (CUSA) to remove brain tumors once thought to be inoperable. The slender tip of the CUSA probe is caused to vibrate by ultrasonic waves at approximately 23 kHz. The probe shatters any section of the tumor which it touches and saline solution is used to flush out the fragments. The usefulness of the CUSA probe is that a surgeon can selectively remove small pieces of malignant tissue without damaging surrounding healthy tissue.

It is even possible, in the case of internal hemorrhaging, to identify the bleeding area and obtain a rough estimate of blood loss. Monitoring the real-time movement of pulsating structures like heart valves may also be undertaken using ultrasound.

When ultrasound is used to determine anatomical features or foreign objects, the wavelength used must be of the same size or smaller than the object to be detected. Typically high frequencies range from 1 to 15 MHz, so for a 5-MHz pulse the wavelength is

$$\lambda = \frac{v}{f} = \frac{1540 \text{ m/s}}{5 \times 10^6 \text{ Hz}} = 3.08 \times 10^{-4} \text{ m} \tag{16.18}$$

or 0.3 mm where we have taken the velocity of sound through tissue to be $v = 1540$ m/s.

The intensity of ultrasound used for medical diagnostics is kept low to avoid tissue damage. Intensities of about $10^{-2}$ W/m² are used and cause no ill effects. Most of the energy carried in by the ultrasonic wave is converted to thermal energy, but $10^{-2}$ W/m² causes negligible heating of tissues, unlike X-rays, which always do some tissue damage. Ultrasound is therefore often used in obstetrical applications.

Ultrasound of considerably higher intensity is used for therapeutic purposes. Ultrasonic diathermy is deep heating using ultrasound of intensities of 1 to 10 W/m². Care must be taken in ultrasonic diathermy to avoid hot spots, which destroy tissue rather than just warm it a few degrees above body temperature, although ultrasound of intensity $10^3$ W/m² is used in some medical procedures to destroy cancerous tissues or shatter gallstones.

In the next topic we discuss the applications of ultrasound to diagnostics in the case of high risk of stroke.

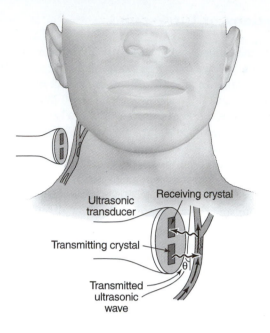

Ultrasonic transducer

Receiving crystal

Transmitting crystal

$\theta$

Transmitted ultrasonic wave

**FIGURE 16.5**   Ultrasonic transducer to measure blood flow in the carotid artery in the neck.

## TOPIC 16.4   Assessment of Stroke Risk and Ultrasound

Stroke is the clinical term which describes the severe reduction or even cessation of blood flow to or in the brain. Calcified deposits may accrue on the inner wall of vessels which directly supply blood to the brain, and one of the major causes of stroke is the obstruction of a carotid artery in the neck. The presence and severity of abnormal obstructions which could lead to a stroke may be detected through a knowledge of blood flow through pertinent vessels. Ultrasound can be used directly by making use of the Doppler effect. Sound waves which range from 20 kHz to 20 MHz in frequency are utilized. An ultrasound transducer is placed directly over the region of interest, that is, the carotid arteries, which generates sound waves that propagate through the skin of the patient. Ultrasound of frequency $f_0$ is directed toward the surface of a blood vessel, interacts with the red cells in the blood, and is reflected back toward the transducer at a Doppler-shifted frequency $f_D$, given by

$$f_D = \frac{f_0 v}{(v \pm v_B)(2 \cos \theta)} \tag{16.19}$$

where $v$ is the speed of sound in blood (1540 m/s), $\theta$ is the angle between the axis of sound propagation and the flowing blood, the factor 2 is to correct for the transit time both to and from the source, and $v_B$ is the blood flow velocity. The plus sign corresponds to the situation where $v_B \cos \theta$ is the component of velocity of the blood away from the transducer and the minus sign when $v_B \cos \theta$ is toward the transducer. In Figure 16.5 we show the ultrasonic transducer in use.

## TOPIC 16.5   Doppler Flowmeter

Another application of ultrasound is in the measurement of the speed of blood. The Doppler flowmeter measures the speed of blood flow and emits a continuous sound whose frequency is typically about 5 MHz. This sound is reflected by the red blood cells and its frequency changed because the cells are moving. The part of the meter which detects the reflected sound measures its frequency, which is Doppler shifted relative to the transmitted signal's frequency. The blood flow speed can be determined from this frequency shift. For flow speeds of 0.1 m/s the frequency shift can be approximately 600 Hz. The flowmeter can be used to detect blood flow in vessels which have narrowed or possibly died due to disease since the blood flow speeds are increased in such regions. The blood flow speed in a fetal heart may be detected as early as 8 to 10 weeks after conception.

## TOPIC 16.6   Complexity of Structure of Ears in Nature

Mammals have evolved from aquatic ancestors, and the complexity of mammalian ears is a result of the difficult transition from sea to land. In contrast, insects have always been land creatures, and their "ears" have a wonderful simplicity by comparison with mammals. In Figure 16.6 we show a section of the ear of a moth, although such ears in insects may even be located on a leg. The eardrum is exposed directly to the air. It forms the outer wall of an air-filled cavity. Through the cavity run the auditory nerves, which go directly from the eardrum to the brain without the complications of the middle and inner ear in humans. This insect ear is very efficient and, despite its simplicity, responds to ultrasonic frequencies. In the grasshopper the ears are on its legs. This allows the grasshopper to vary the distance between the ears providing a better location ability.

## TOPIC 16.7   Echolocation: Imaging by Sound

Ultrasound is generated using an oscillating quartz crystal, the oscillations being driven by an electrical current. Crystals like quartz which are deformed electrically are called piezoelectric, and devices to generate sound in this way are called sound transducers. The transducer is arranged to be at some distance $\ell$ from the object we wish to image, which must be made of a material in which the velocity of sound is different from the surrounding medium. The sound wave generated by the transducer is partially reflected from the object and an echo is produced. Assuming that the original pulse of sound is emitted from the transducer at $t = 0$, the reflected wave will return to the generator after a time $\Delta t = 2\ell/v_s$ seconds, where $v_s$ is the velocity of sound. Thus $\Delta t$ is a convenient measure of

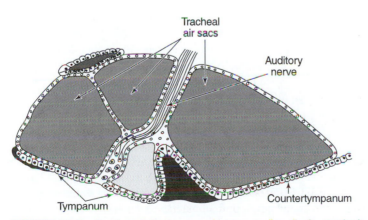

**FIGURE 16.6**   Section of the tympanal organ, or "ear," of a moth situated on the metathorax.

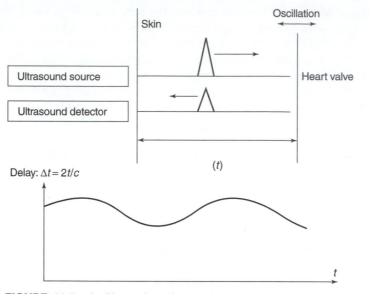

**FIGURE 16.7** An illustration of how ultrasound can give information, via the delay time $\Delta t$ of the working of a heart valve.

the distance to the object from the transducer. Figure 16.7 illustrates how the time delay can give real-time information of the working of a heart valve.

The oscillation of the time delay as a function of time is due to the variation in time of the distance $\ell(t)$ between the sound source and the heart valve as the valve moves to and fro. We call this mode of using an ultrasound probe type A. That is, although it is used to measure distances, no attempt is made to construct an image. It is of interest to note that in the muddy waters of the Yellow River in China lives a freshwater dolphin which is practically blind and relies entirely on echolocation for navigation. In the present era river traffic has produced so much sound pollution that their echolocation system no longer works and they are close to extinction.

The sound produced by the transducer may encounter more than one interface, and as a consequence more than one reflected pulse will be detected. The principle of ultrasound imaging uses this idea. Sound pulses are reflected from the boundary between the organ and surrounding tissue. As an example, consider reflection from the heart. The sound wave is reflected twice, that is, once when it enters the heart and once when it leaves. Thus two echo pulses are produced. From the corresponding time delays one can find the spacing between the two surfaces. Figure 16.8 illustrates the principle of this method using the eye as an example. The position of the sound probe and the incident angle of the sound are constantly shifted when ultrasound is used in three dimensions to image an organ. The distance $\ell$ between reflecting surfaces for each position, angle, and reflected signal is fed into a computer which then reconstructs the shape of the reflecting object using this information.

The length through which a sound wave is attenuated, called the attenuation length, is longest for low-frequency sound waves. In water 20-Hz sound waves can have an attenuation length which is longer than the size of an ocean; for example, whales can communicate across an ocean by this means. This is often referred to as the singing of whales. On the other hand, high-frequency sound attenuates much faster; for example, a 1-MHz

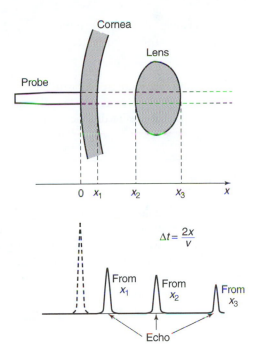

$$\Delta t = \frac{2x}{v}$$

**FIGURE 16.8**  Multiple reflections in the eye.

sound wave is attenuated in 20 m in water but only 4 mm in bone. A 3.5-MHz sound wave has an even smaller attenuation length of 0.5 mm in bone. Thus, if an ultrasound pulse has to cross bone material, its frequency must not exceed 1 MHz, and hence ultrasound imaging of the brain is difficult.

## TOPIC 16.8  Echolocation of Bats

Bats have an extreme adaptation, that is, an inbuilt radar system. They generate ultrasonic waves having an average frequency of about 50 kHz. They are emitted from their snouts in short pulses which are not only of high intensity but are also highly directional. The objects which the bats want to locate are very small, and a high intensity is needed because only a small proportion of energy is reflected back to the bat's ears.

Bats emit sound pulses at a rate of several times per second in order to locate obstacles and their prey, that is, insects. The intensity of these pulses is approximately equivalent to shotgun blasts. Largely due to the efficiency of echolocation, the bat can catch several hundred flying insects per hour. The pulse is both amplitude and frequency modulated and thus has some similarities with radio communication. The frequency modulation ranges between 30 kHz and about 70 kHz, corresponding to a wavelength of the carrier in the range of about 5 to 11 mm since the speed of sound is 340 m/s. Wavelengths $\lambda$ which are equal to or less than the size of the obstacle are scattered more effectively, and hence for small objects such as a mosquito the reflected pulse will have a relatively large short-wavelength component. The pulses have a typical duration of 60 ms for a resting bat and thus correspond to the envelope's wavelength $\Lambda = vT = (340 \text{ m/s})(60 \times 10^{-3} \text{ s}) = 20.4$ m. When freely flying and searching for insects, the bat sends out pulses at a rate of about four per second and the pulse duration is about 1.5 ms. The firing rate determines the maximum range of the echolocating system, which is half of the distance traveled by the pulse and the echo in the period between two successive pulses. Therefore,

$$d = \tfrac{1}{2}vt = \tfrac{1}{2}(340 \text{ m/s})(0.25 \text{ s}) = 43 \text{ m} \qquad (16.20)$$

If the bat detects an echo from an insect, the firing rate increases to as high as 200 pulses per second and the duration of each pulse decreases to less than $10^{-3}$ s. Based on equation (16.25), we can conclude that at this rate the bat–insect distance must be less than 0.85 m.

## TOPIC 16.9   Echolocation of Dolphins

Like bats, dolphins developed an efficient method of locating objects using ultrasound. Dolphins emit high-frequency, ultrasonic waves at about $10^6$ Hz in the form of whistles or clicks, sending possibly as many as 2000 clicks per second. The sounds originate in air sacs in the nasal cavity and propagate forward through the dolphin's forehead. Echoes of the sound from the object or prey bounce back to the dolphin and travel through the lower jaw to the inner ear. Echolocation, which is the name given to this phenomenon, is particularly efficient for dolphins because sound travels almost five times as fast in water than in air.

## TOPIC 16.10   Noise Reduction and Traffic

There are three aspects to noise reduction: distance from a source, attenuation by absorption, and reduction of sound output from the source. The last of these is the most effective. Since sounds come from a vibrating object, it follows that the smaller the area of the object, the less air it can interact with and the lower the sound level it will create. The more rigid the materials of which the vibrating object is made, the smaller amplitude its surface will have, producing smaller pressure waves in the air. The vibrating objects should be cushioned from contact with other objects they might cause to resonate. Absorption is used in earmuffs and acoustical ceiling tile. Absorptive materials usually are soft, with holes or crevices, so that the sound must make multiple reflections to bounce off them or get through them. Distance usually helps, unless the source broadcasts mostly in one direction.

Two main processes contribute to the decrease of the sound intensity as one moves away from the source of sound. The first process involves spherically symmetric propagation of sound waves from its source leading to the attenuation of its intensity $I$ according to

$$I = \frac{I_0}{r^2} \qquad (16.21)$$

However, in actual material media, sound creates random movement of the molecules encountered, which gradually decreases the energy contained in the wave. This process is called dissipation and can be enhanced by more strongly absorbent materials such as clothing, rugs, curtains, and molecules of air with different results depending on the humidity. For example, dry air dissipates far less than damp foggy air. Sound is more strongly dissipated in a forest than over a still lake. The effect of dissipation is rather accurately described by the following exponential relationship:

$$I = I_0 e^{-ar} \qquad (16.22)$$

where $a$ is a characteristic coefficient for a given medium. The larger the value of $a$, the stronger the dissipation. Combining these two effects gives

$$I = \frac{I_0 e^{-ar}}{r^2} \tag{16.23}$$

A special case has to be made for linear (as opposed to point) sources of sound. A common example is a steady roar from a continuous flow of traffic on a highway. In such cases

$$I = \frac{I_0 e^{-ar}}{r} \tag{16.24}$$

leading to a longer range effect of noise. Traffic experts obtained an empirical formula for road noise (in decibels) $\beta$ depending on the following factors: (a) $v$, the speed of traffic in kilometers per hour, and (b) $N$, the number of cars passing a road segment per day (24 h) and at a reference distance of 30 m to the road. For an arbitrary distance $D$ in meters, the formula yields

$$\beta = 25 \log_{10} v + 10 \log_{10} N - 10 \log_{10}\left(\frac{D}{30}\right) - 26 \tag{16.25}$$

For example, for a busy highway with $N = 100,000$, $v = 90$ km/h, one obtains approximately 60 dB at $D \cong 700$ m from the highway.

## SOLVED PROBLEMS

**16.1** What is the maximum distance that a bat can detect an object of cross-sectional area of $10^{-4}$ m² if the intensity of the pulse 0.01 m from the bat's mouth is 0.01 W/m² and the threshold of hearing of the bat is about 0.01 W/m²? See the accompanying diagram.

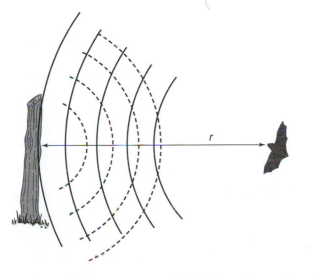

**Solution**  The power output of the bat can be evaluated from

$$P = 4\pi r^2 I = [4\pi (0.01)^2](0.01) = 4\pi \times 10^{-6} \text{ W}$$

If this output power is assumed to be radiated in all directions as in the figure above, then the power striking the object is given by

$$P_1 = \frac{a}{A} P$$

where $a$ is the area of the object and $A$ represents the area of the sphere outlined in the figure whose radius $r$ is equal to the distance between the bat and the object. Thus

$$P_1 = \frac{10^{-4} \times \left(4\pi \times 10^{-6}\right)}{4\pi r^2} = \frac{10^{-10}}{r^2}$$

Subsequently, the object now becomes the source of the echo. If we assume that all of the power striking the source is reflected, then the intensity of the echo at a distance $r$ from the object is given by

$$I = \frac{P_1}{4\pi r^2}$$

For the bat to hear, the echo $I_1$ must be at least $10^{-12}$ W/m². Therefore,

$$10^{-12} = \frac{10^{-10}}{4\pi r^4}$$

Hence,

$$r^4 = \frac{100}{4\pi}$$

or $r = 1.7$ m.

16.2   Ultrasound has a speed of 1500 m/s in tissue.
(a) Calculate the smallest detail visible with a 2.0-MHz ultrasound.
(b) To what depth can the sound probe effectively?
(c) How long does it take the echo to return to the probe from a depth of 0.10 m?

**Solution**
(a) The smallest detail visible has a size about equal to one wavelength of the sound. The relationship between propagation speed, frequency, and wavelength is

$$v_w = f\lambda$$

Solving for $\lambda$ gives

$$\lambda = \frac{v_w}{f} = \frac{1500 \text{ m/s}}{2.0 \times 10^6 \text{ Hz}} = 0.75 \times 10^{-3} \text{ m}$$

Thus details less than a millimeter should be discernible.
(b) The effective depth is 200 wavelengths. Therefore,

$$200\lambda = 200(0.75 \times 10^{-3} \text{ m}) = 0.15 \text{ m}$$

(c) The time for an echo to return is the time for its round trip. Since

$$s = vt$$

we obtain

$$t = \frac{s}{v} = \frac{0.20 \text{ m}}{1500 \text{ m/s}} = 1.3 \times 10^{-4} \text{ s}$$

*original*

**16.3** Ultrasound is commonly used to assess the development of a fetus within the womb of an expectant mother. In such an examination, an 8-MHz ultrasonic beam is focused on the abdomen, emitting ultrasonic waves that propagate through the abdomen and reflect from the wall of the fetal heart, which is moving toward the ultrasonic receiver as the heart beats. The reflected sound wave is then mixed with the transmitted sound wave, resulting in a Doppler-shifted frequency by 10 kHz. The speed of sound in body tissue is 1500 m/s. Determine the speed of the fetal heart. *result: smaller than original*

**Solution** We use Doppler's equations for a fixed source and an observer moving toward the source, or

$$f_o = f_s \left( \frac{v + v_0}{v} \right)$$

where $f_s$ = 8 MHz, $f_o$ = 8 MHz + 10 kHz = 8.01 MHz, and $v$ = 1500 m/s. Solving the above equation for $v_0$ and substituting the numbers, we have

$$v_0 = \frac{f_o}{f_s} v - v = \left( \frac{8.01 \text{ MHz}}{8.00 \text{ MHz}} \right) 1500 \text{ m/s} - 1500 \text{ m/s} = 1.875 \text{ m/s}$$

**16.4** A person has a 40-dB loss of hearing at a given frequency. What is the sound intensity experienced if the normal auditory threshold $I_0 = 5 \times 10^{-12}$ W/m$^2$ at the given frequency?

**Solution** If we take $I_0$ as the reference intensity, then the intensity $I$ required can be characterized by an intensity level $L$ = 40 dB, for which

$$L = 10 \log \frac{I}{I_0}$$

From the equation, $I$ can be obtained as

$$I = I_0 \times 10^{L/10} = 5 \times 10^{-8} \text{ W/m}^2$$

**16.5** Calculate the displacement of the human eardrum by sound at the threshold of hearing at $f$ = 4 kHz ($I_h = 1 \times 10^{-12}$ W/m$^2$) and the threshold of pain ($I_p$ = 1 W/m$^2$). The density of the air is $\rho$ = 1.3 kg/m$^3$ and the velocity of sound in the air is $v$ = 340 m/s.

**Solution** The energy of the sound wave is contained in the harmonic oscillation motions of the particles of the air and is the sum of the total energies of the air particles. The kinetic energy of a particle of mass $m$ oscillating with angular frequency $\omega = 2\pi f$ and amplitude $A$ in the plane wave of the sound is $\frac{1}{2}m\omega^2 A^2 \cos^2 \omega t$ and the potential energy is $\frac{1}{2}m\omega^2 A^2 \sin^2 \omega t$. The total energy $W = \frac{1}{2}m\omega^2 A^2$. In a volume element $\Delta V$ of mass $\Delta m$ of the medium, the time-averaged energy is $\Delta W = \frac{1}{2}\Delta m\, \omega^2 A^2$, and hence the time-averaged energy density $\varepsilon = \Delta W/\Delta V$ is given by

$$\varepsilon = 2\pi^2 \rho f^2 A^2$$

The intensity of the sound $I$ is defined by the amount of energy striking a unit area per second: $I = \varepsilon v$ or

$$I = 2\pi^2 \rho f^2 A^2 v$$

The displacement amplitude $A$ can be calculated if all the other quantities are known in the above equation. At the threshold of hearing at $f$ = 5 kHz, the sound intensity

$I_h = 1 \times 10^{-12}$ W/m², which gives $A_h = 2.7 \times 10^{-3}$ nm for the lower limit of displacement of the eardrum. At the threshold of pain, the sound intensity reaches $I_p = 1$ W/m² and the displacement amplitude of the eardrum amounts to $A_p = 2.7$ $\mu$m.

**16.6** The mean velocity of erythrocytes in blood vessels is determined by means of the Doppler shift of ultrasound. A small segment of the vessel is irradiated with an ultrasound source of frequency $f_s$ at an angle $\alpha$ relative to the direction of blood flow, and the ultrasound scattered by the red blood cells is observed by a detector at frequency $f_D$ from direction $\beta$. What is the mean velocity of the red blood cells in the observed portion of the vessel?

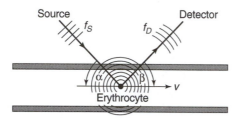

**Solution** The ultrasound source is at rest while the blood cells move along the vessel with velocity $v$. The component in the direction of the connecting line is $v \cos \alpha$, which gives the speed of displacement of the erythrocytes from the source at the point of observation. The cell detects the ultrasound at frequency $f'$, different from that emitted by the source $f_s$, according to

$$f' = f_s\left(1 - \frac{v \cos \alpha}{c}\right)$$

Here, $c$ is the speed of the ultrasound. The red blood cells scatter the ultrasound in all directions and hence serve as a source of radiation from the point of view of the detector outside. The situation is characterized by a moving source and a resting detector. The source (red blood cells) approaches the detector at a speed $v \cos \beta$, and thus the detected frequency is

$$f_D = \frac{f'}{1 - (v \cos \beta)/c}$$

On substituting $f'$ into this equation, we obtain

$$f_D = f_s\frac{1 - (v \cos \alpha)/c}{1 - (v \cos \beta)/c}$$

In practice, $\beta = 90°$ since the detector is placed directly above the investigated section of the blood vessel, giving

$$f_D = f_s\left(1 - \frac{v \cos \alpha}{c}\right)$$

The Doppler shift is defined as the difference in the frequencies of the emitted and detected ultrasound:

$$\Delta f = f_s - f_D = f_s\frac{v \cos \alpha}{c}$$

Hence, the mean velocity $v$ of the erythrocyte can easily be expressed via

$$v = \left(1 - \frac{f_D}{f_s}\right)\frac{c}{\cos \alpha}$$

Therefore, the average velocity of the erythrocytes in the blood vessel is proportional to the Doppler shift.

**16.7**  The alternating pressure in a therapeutic application of ultrasound is superimposed on the static pressure, so that the resulting pressure changes periodically around the static pressure value. Suppose the frequency of the ultrasound in a treatment is $f = 830$ kHz, the irradiated power output is $P = 15$ W, and the irradiating area of the ultrasound-emitting head is $A = 6$ cm$^2$.

(a) Calculate the intensity of the ultrasound in watts per square meter.

(b) What is the effective value of the alternating sound pressure in muscle? The density of muscle is $\rho = 1040$ kg/m$^3$ and the sound velocity in it is $c = 1570$ m/s.

(c) What is the distance between the sites of highest and lowest pressure in muscle?

(d) What is the pressure amplitude and the highest value of pressure difference in muscle?

**Solution**

(a) The intensity is

$$I = \frac{P}{A} = 2.5 \times 10^4 \text{ W/m}^2$$

(b) The effective value of the alternating sound pressure $P_{\text{eff}}$ in the case of sound waves can be defined as

$$P_{\text{eff}} = \frac{P_{\text{max}}}{\sqrt{2}}$$

where $P_{\text{max}}$, the pressure amplitude, relates to the intensity

$$I = \frac{P_{\text{max}}^2}{2\rho c}$$

From the two equations above, we obtain

$$P_{\text{eff}} = \sqrt{I\rho c} = 2 \times 10^5 \text{ N/m}^2$$

(c) The distance between the sites of highest and lowest pressure is half of the wavelength:

$$\frac{\lambda}{2} = \frac{c}{2f} = 0.95 \text{ mm}$$

(d) The pressure amplitude is

$$P_{\text{max}} = \sqrt{2}P_{\text{eff}} = 2.8 \times 10^5 \text{ N/m}^2$$

and the highest value of pressure difference is twice the pressure amplitude.

# EXERCISES

**16.1**  Suppose a bat tries to locate a moth by emitting sound pulses and detecting the echo. Assume that the ultrasound frequency is 10 MHz.
**(a)** If the echo returns after 0.25 s, how far is the moth from the bat?
**(b)** The frequency of the returning echo pulse is 10.25 MHz. What is the velocity of the moth? Is it moving toward or away from the bat?
**(c)** The bat moves toward the moth with a velocity of 5 m/s and keeps emitting echo pulses. What is the frequency of the sound pulses as detected by a stationary observer? What is the frequency of the sound pulses as detected by the moth?

**16.2**  A bat emits an echolocating pulse of $\Delta t = 2$ ms duration with intensity $I_1 = 1 \times 10^{-2}$ W/m$^2$ measured at $r_1 = 1$ cm from its mouth. The threshold of hearing of the bat at a frequency of 60 kHz is close to that for humans at 4 kHz, that is, $I_h = 1 \times 10^{-12}$ W/m$^2$. What are the acoustical power and energy of the pulse? What is the maximum distance at which the bat can detect a bug with effective area $A = 10$ mm$^2$? What is the upper limit of the repetition rate of the echolocating system?

**16.3**  Assume that the human ear is most sensitive to sound of frequency 3 kHz. The threshold of hearing at this frequency is usually measured as $I_0 = 3 \times 10^{-13}$ W/m$^2$. What is the energy that a sound of threshold intensity transports to the eardrum with a surface of about $A = 50$ mm$^2$ in 200 ms?

**16.4**  A person is speaking at an acoustical power of moderate conversation of $P_0 = 10$ $\mu$W in free space. At what distance from the person would the intensity of the sound be at the threshold level ($I_h = 1 \times 10^{-12}$ W/m$^2$)?

**16.5**  A musician with a keen sense of absolute pitch is able to recognize a musical sound of 1 kHz after a demonstration time of only 4 ms. How many periods of the sound wave are needed for the recognition of the pitch?

**16.6**  What is the minimum safe distance from a 5-W speaker? Assume the eardrum's area is $0.9 \times 10^{-4}$ m$^2$ and the maximum power the ear can tolerate without damage is $10^{-6}$ W.

**16.7**  The ear can differentiate between sounds that arrive as little as 1 ms apart. What does it correspond to in terms of the distance between two objects generating sounds?

**16.8**  Ultrasound can be applied to the deep heating of tissue. Assuming the ultrasound intensity to be 5000 W/m$^2$ and the surface area of a transducer to be 10 cm$^2$, estimate the time needed to emit $10^4$ J of energy to the tissue.

**16.9**  A beat frequency of 120 Hz is produced by an ultrasonic echo mixed with its original frequency of 2.6 MHz following a reflection from the heart wall of a patient. Determine the velocity of the heart wall.

**16.10**  The middle ear has a lever system that transmits sound to the inner ear and operates at a mechanical advantage of 2. The area the increased force is exerted on is 0.06 that of the eardrum. With this information calculate the factor by which the pressure created in the cochlear fluid is greater than that of the sound in the eardrum.

**16.11**  An 80-dB sound is absorbed by an eardrum 0.007 m in diameter for 1 h. How much energy does the eardrum absorb in that time?

**16.12**  Ultrasonic scanners determine distances to objects in a patient by measuring the times for echoes to return. What is the difference in time for echoes from tissue layers in a patient that are 0.020 and 0.022 m beneath the surface, respectively. The speed of sound in tissue is 1500 m/s.

**16.13** (a) What frequency ultrasound should be used to see details as small as 0.0015 m in tissue? The speed of sound in tissue is 1500 m/s. (b) To what depth is this sound effective as a diagnostic probe?

**16.14** The speed of sound in tissue is 1500 m/s. An ultrasonic wave sent into blood is partly reflected back toward the source by blood cells. If the returning echo has a frequency 500 Hz higher than the original 2.1-MHz frequency, what is the velocity of the blood?

**16.15** (a) Calculate the maximum net force on an eardrum due to a sound wave having a maximum gauge pressure of $2 \times 10^{-3}$ N/m$^2$ if the diameter of the eardrum is 0.0085 m. (b) Assuming the mechanical advantage of the hammer, anvil, and stirrup is 2, calculate the pressure created in the cochlea. Recall that the stirrup exerts its force on the oval window, which has an area 0.05 times that of the eardrum.

**16.16** A person has an overall hearing loss of 42 dB. By what factor must sound be amplified for it to seem normal in loudness to this person?

# CHAPTER 17

## PRINCIPLE OF LINEAR SUPERPOSITION AND INTERFERENCE PHENOMENON

### PHYSICAL BACKGROUND

The principle of linear superposition states that when two or more waves are present simultaneously at the same place, the resultant wave is the sum of the individual waves. Constructive interference occurs at a point when two waves meet there crest to crest and trough to trough, thus reinforcing each other. Destructive interference occurs when the waves meet crest to trough and cancel each other. When the waves meet crest to crest and trough to trough, they are exactly in phase. When they meet crest to trough, they are exactly out of phase.

Diffraction is the bending of a wave around an obstacle or the edges of an opening. The angle through which the wave bends depends on the ratio of the wavelength $\lambda$ to the width $D$ of the opening; the greater the ratio $\lambda/D$, the greater the angle.

Beats are the periodic variations in amplitude that arise from the linear superposition of two waves that have slightly different frequencies. When the two waves are sound waves, the variations in amplitude cause the loudness to vary at the beat frequency, which is the difference between the frequencies of the waves.

A transverse, or longitudinal, standing wave is the pattern of disturbance that results when oppositely traveling waves of the same frequency and amplitude pass through each other. A standing wave has places of minimum and maximum vibration called, respectively, nodes and antinodes. Under resonant conditions, standing waves can be established only at certain frequencies $f_n$, known as the natural frequencies. For a string that is fixed at both ends and has a length $L$, the natural frequencies are

$$f_n = n\left(\frac{v}{2L}\right) \tag{17.1}$$

where $v$ is the speed of the wave on the string and $n$ is any positive integer 1, 2, 3, . . . . For a gas in a cylindrical tube open at both ends, the natural frequencies are given by the same expression, where $v$ is the speed of sound in the gas and $n = 1, 2, 3, . . . .$ However, if the cylindrical tube is open at only one end, the natural frequencies are

$$f_n = n\left(\frac{v}{4L}\right) \tag{17.2}$$

where $n = 1, 3, 5, . . . .$

A complex sound wave consists of a mixture of a fundamental frequency and overtone frequencies. According to the Fourier theorem, a sufficiently smooth function can be represented by a superposition of periodic functions (harmonics) possessing frequencies related by an integral proportionality constant

$$f(t) = \sum_{n=1} \left[ A_n \cos\left(\frac{2\pi nt}{T}\right) + B_n \sin\left(\frac{2\pi nt}{T}\right) \right] \tag{17.3}$$

where $A_n$ and $B_n$ are called Fourier amplitudes and $\omega_n = 2\pi n/T$ are the corresponding frequencies (harmonics).

## TOPIC 17.1  Generation of Human Voice

Human voice is an example of a sound wave. It is produced by the vocal cords, as described below. There is a narrow V-shaped opening in the larynx whose width can be controlled by the cricothyroid muscles. The sides of the opening are often referred to as the "vocal cords," which are slackened by the contraction of the thyroarytenoid muscles (see Figure 17.1). When a person is not speaking, the vocal cords are relaxed and air freely passes through the gap during breathing. When speaking or singing, the vocal cords are under tension and the V-shaped gap is narrowed to a slit. Initially, let us suppose the slit is closed. When air is exhaled, the air pressure behind the slit increases, the slit opens up, and air rushes through. This means that the air pressure behind the slit then drops and the slit closes again under the elastic tension of the cords. The air pressure then builds up again, and the pressure inside and outside them goes through the same sequence again. This phenomenon is oscillatory or periodic if it is made up of an endless repeat of a basic pattern of events with time.

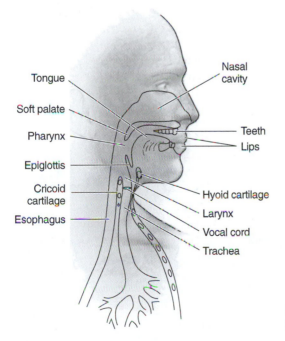

**FIGURE 17.1**  Schematic structure of the human voice-generating organs.

## TOPIC 17.2 The Perception of Sound

Sound is generated by the vocal cords, as explained above. Its perception by the ears is, however, quite subjective. Pitch is the perception of frequency. Most people have good relative pitch; that is, they can tell that one sound has a higher or lower frequency than another. Frequencies usually must differ by 0.3% or more to be told apart. For example, 1000 and 1003 Hz are noticeably different in pitch. A few people have what is called perfect pitch; that is, they can identify musical notes in addition to telling which note has the higher frequency. This ability is rare even among musicians. A person with a poor sense of pitch is called tone deaf. Pitch perception does not depend on sound intensity; high- and low-intensity sounds of the same frequency are sensed to have the same pitch.

Multiple-frequency sounds are often perceived subjectively (see Figure 17.2). A number of terms are used to describe multiple-frequency sounds, such as *noise*, *music*, *rich*, *shrill*, and *mellow*. Humans are able to recognize individual frequencies played simultaneously even though the combined sounds may be complicated in appearance. In addition to recognizing a musical instrument or voice, most people find it easy to tell that several keys are being played in a piano chord, for example. This ability to unravel a sound depends to some extent on the person, whether nearly equal frequencies are involved, and the relative intensity of each frequency. The ability is almost certainly due to the mechanism of sound conversion into nerve impulses in the cochlea and the fact that different parts of the cochlea are sensitive to specific frequencies. The perceived sound quality of a tone is in fact determined by the particular values of the Fourier amplitudes so that changes in the Fourier amplitudes of a tone are perceived as a change in sound quality.

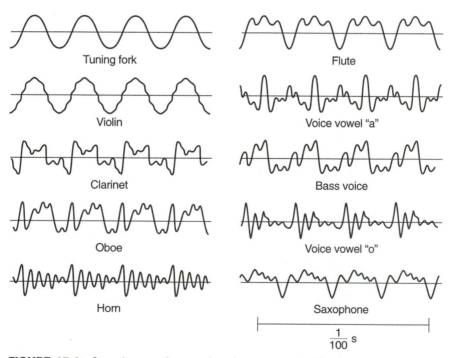

FIGURE 17.2 Sound waves from various instruments making the same note.

# TOPIC 17.3   Helmholtz Resonance Theory

We have already seen that the sound wave sets up an oscillation of the oval windows, and this will push the cochlea fluid back and forth. The basilar membrane contains some 23,000 hair cells which are connected to the nervous membrane. If a hair cell is moved back and forth by the fluid, then this sends a signal to the nervous system. The German physicist and physiologist Heinrich Helmholtz had a rather clever idea to explain this. He argued that the inner ear should have a structure which resembled a piano. To elaborate on his idea, suppose you take off the cover of an upright piano and expose the array of tense strings of varying length, $L$ say. Then each string will have its own fundamental frequency inversely proportional to $L$. Suppose now that the piano receives a sound wave produced by a pure tone. The piano will then produce an echo of the tone; that is, the strings will start to vibrate so that the resulting sound closely resembles the original tone of the impinging sound wave. To see this, we recognize that a pure tone corresponds to a periodic pressure oscillation. Now call to mind that a periodic signal $g(t)$ of period $T$ can be written as a Fourier series as

$$g(t) = A_1 \cos(\omega t + \phi_1) + A_2 \cos(2\omega t + \phi_2) + A_3 \cos(3\omega t + \phi_3) + \cdots \quad (17.4)$$

where the $A_n$ ($n = 1, 2, \ldots$) are the Fourier amplitudes. We apply this to the pressure oscillation $P(t)$ so that $g(t) = P(t)$ and produces a pure tone.

The $n$th term of the series in equation (17.4) is now a harmonic oscillator of frequency $n\omega$ that attempts to vibrate the strings of the piano. The result is that a tone from the piano can be represented by the Fourier series in equation (17.4). The piano may be thought of as a "Fourier analyzer." By that we mean that we could now measure the amplitude $A_n$ by measuring by how much that string, whose fundamental frequency equals or is close to $n\omega$, is excited. Helmholtz argued that the inner ear should work in the same manner.

It transpires that the inner ear does have such a structure, that is, the basilar membrane, and contains transverse fibers of increasing length and decreasing thickness. The formula for the fundamental frequency of a string is (see Chapter 10):

$$\omega_1 = \frac{\pi}{L} \sqrt{\frac{T}{\rho}} \quad (17.5)$$

where $T$ is the tension and $\rho$ is the linear density of the string.

From this equation we see that increasing $L$ and increasing the mass per unit length, $\rho$, decreases the fundamental frequency. We could therefore consider the basilar membrane and transverse fibers as a piano treating each transverse fiber of the membrane as an oscillator with its own natural frequency. That part of the basilar membrane close to the oval window contains oscillators with a higher fundamental frequency while the part of the membrane near the round window contains oscillators with a lower frequency. The basilar membrane is connected to nerve ends through the hair cells, and the Fourier amplitudes recorded by the hair cells could then be sent to the brain.

In 1960 von Bekesy tested the Helmholtz resonance theory by using the basilar membrane obtained from cadavers. The theory would predict that if the ear was exposed to a pressure oscillation of a certain frequency $\omega$, then only the fundamental frequency close to $\omega$ in that section of the basilar membrane will be excited. In Figure 17.3 we sketch the measured excitation pattern of the basilar membrane. The pulselike waveform in Figure 17.3 resembles a traveling wave a little, but a clear maximum of the pulse amplitude appears at a certain place along the membrane. The Helmholtz theory would lead us to

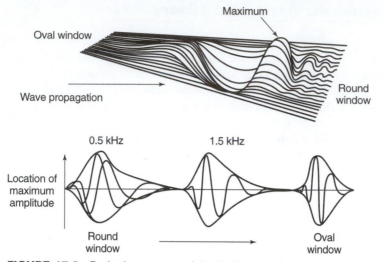

FIGURE 17.3 Excitation pattern of the basilar membrane.

expect that the maximum amplitude for a low-frequency sound pulse should be recorded near the round window and with increasing frequency the maximum should shift to the oval window. As shown in the measured waveform in Figure 17.3, this is precisely what happens.

A resonance curve records the amplitude of oscillation as a function of the driving frequency. When the resonance frequency is equal to the driving frequency, one should observe a sharp peak. An experiment of this type can be done on a basilar membrane. A position is chosen along the membrane, and the amplitude of oscillation is measured at that point as a function of sound frequency. In Figure 17.4 we display such a curve, measured with an elaborate method called the Mössbauer effect.

As predicted by Helmholtz, there is a clear peak, but it is very steep on one side and broad on the other, and at a different point on the membrane the curve is shifted to the left or to the right. Because of its asymmetry, this curve does not resemble the resonance curve of a driven harmonic oscillator very well, so we ask if the peak is sufficiently pronounced to tell different frequencies apart. Suppose we measure the frequency difference between two frequencies $\nu$ and $\nu + \delta\nu$ that a test subject can differentiate. The frequency difference $\Delta\nu$ is very small for low-frequency sound, that is, about 1.8 Hz for subjects with very good hearing. For higher frequencies it is about $(3 \times 10^{-3})\nu$. This ability to tell apart two frequencies differing by a few hertz does allow test subjects to identify two instruments of a large orchestra. Thus the amplitude of oscillation does change rap-

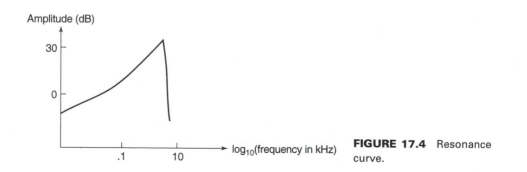

FIGURE 17.4 Resonance curve.

idly with frequency, provided you are at a point on the membrane which, for that frequency, is located on the steep side of the resonance curve.

## TOPIC 17.4    Hitting the Baseball

The task of hitting a baseball with the objective of achieving a home run is quite complex. It is interesting to show that it can be viewed as an application of wave propagation principles. A wooden baseball bat typically has a mass ranging from 0.85 to 1.7 kg with an average of about 1 kg. The batter tries to swing it quickly so that the baseball rebounds from the bat's surface at high speed, ideally at 100 km/h or more. It is also desirable to deliver most of the bat's energy to the ball, so one needs to find a compromise in terms of the bat's mass and hardness in order to have a bat which is both easy to swing and does not rebound when it hits the ball. Swinging a bat with a force of 500 N on the handle at approximately 1 m from the pivot point at the batter's shoulders produces a torque of 500 Nm and kinetic energy investment of 500 J. For each batter there is a particular optimum mass of the bat resulting in the best possible rebound of the ball. In the process of hitting the ball, most of the kinetic energy is transferred to the baseball (about 98%). Since the coefficient of restitution of the ball is between 0.514 and 0.578, the ball returns only about 30% of the energy going into its compression while the bat compresses by a factor of 50 less than the ball.

The bat's collision with a baseball causes two effects: (a) the bat's center of mass accelerates backward and (b) the bat begins to rotate around its center of mass. However, at a special point on the bat located about 17.5 cm from the broad end a collision with a ball will not make the handle accelerate. This is called a "sweet spot," and at this point the handle moves smoothly and the batter feels no jerking motion. It is interesting to note that the sweet spot also closely coincides with a vibrational node of the bat, and consequently, hitting this point wastes little energy on vibrations of the bat transferring a maximum possible kinetic energy to the rebound of the baseball.

A comparison made between wooden and aluminum bats showed visible differences between their vibrational frequencies: An aluminum bat of length 0.81 m and mass 0.825 kg registered a fundamental frequency of 27 Hz and a first harmonic of 317 Hz, both with a clamped handle. Under the same conditions a wooden bat of mass 0.846 kg and length 0.84 m possesses a fundamental frequency of 18 Hz and a first harmonic of 209 Hz.

## SOLVED PROBLEMS

17.1    The auditory canals of adults and children are 2.5 and 1.3 cm long, respectively. What are the frequencies of sounds to which they are maximally sensitive? The velocity of sound in air is 340 m/s.

**Solution**    In the open-ended auditory canal standing waves are produced from the interference of the incoming and the reflected sound waves. At the closed end, which is the skull surface of the middle ear, there must be a node. The fundamental resonant frequency occurs when the length of the canal is equal to one-quarter of the corresponding wavelength. The optimal wavelengths for resonance are 10 and 5.2 cm, and the corresponding sound frequencies are (340 m/s)/0.10 m = 3.40 kHz and (340 m/s)/0.052 m = 6.54 kHz for adults and children, respectively.

**17.2** Deep-sea divers breathe a mixture of helium, nitrogen, and oxygen to prevent nitrogen narcosis. Under these conditions, the speech is similar to "duck-talk." Why? Compare the resonant frequencies in the $L = 17$ cm long human vocal tract filled by air and helium. The velocities of sound in air and helium are $v_{air} = 353$ m/s and $v_{He} = 970$ m/s, respectively.

**Solution** The simplest acoustical model of the vocal tract is a pipe closed at one end (by the glottis) and open at the other end. As the motion of the gas particles is minimum at the closed end of the pipe (on the other hand, the pressure is maximum here), only odd-numbered harmonics can be established in the vocal tract. The resonant frequencies are

$$f_1 = \frac{v}{4L}, \ f_3 = 3f_1, \ f_5 = 5f_1, \ \ldots, \ f_n = nf_1$$

where $n = 1, 3, 5, \ldots$. For an $L = 17$ cm long pipe, the resonances occur at $f_1 = 519$ Hz, $f_3 = 1557$ Hz, $f_5 = 2596$ Hz, $\ldots$ in air and at $f_1 = 1426$ Hz, $f_3 = 4279$ Hz, $f_5 = 7132$ Hz, $\ldots$ in helium. Due to the higher velocity of sound in helium, the frequencies of the corresponding harmonics are higher in helium than in air.

# EXERCISE

**17.1** Using the figure, indicate at what times the larynx is opening at its maximum. When is the larynx closed? Is the time between subsequent closings of the larynx also the period $T$? Give the frequency and angular frequency of $x(t)$.

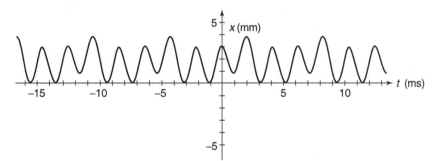

# CHAPTER 18

# ELECTRIC FORCES AND ELECTRIC FIELDS

## PHYSICAL BACKGROUND

Electric charge $q$ is a physical property of elementary particles of the atom. The SI unit of charge is the coulomb, abbreviated C. Charge can be either positive or negative, but the magnitude of charge is in units of the so-called elementary charge, $e = 1.60 \times 10^{-19}$ C. The charge of the proton is $+1.60 \times 10^{-19}$ C, and the charge of the electron is $-1.60 \times 10^{-19}$ C.

Conservation of charge means that the net charge of a system remains constant. In other words, the amount of positive charge equals the amount of negative charge regardless of its physical state.

Coulomb's law describes the electrostatic force $F$ in a vacuum between two charged particles $q_1$ and $q_2$ separated by a distance $r$:

$$F = \frac{1}{4\pi\varepsilon_0}\frac{q_1 q_2}{r^2} = k\frac{q_1 q_2}{r^2} \tag{18.1}$$

where $\varepsilon_0$ is the permittivity constant of free space, defined as $\varepsilon_0 = 8.854 \times 10^{-12}$ C$^2$/N $\cdot$ m$^2$. The value of the coefficient $k$ is

$$k = \frac{1}{4\pi\varepsilon_0} = 9.0 \times 10^9 \text{ N} \cdot \text{m}^2/\text{C}^2 \tag{18.2}$$

Electrostatic force is a vector quantity and is expressed in units of newtons. The direction of the electrostatic force is based on the charges involved. Opposite charges generate an attractive (negative) force, and the direction is toward the other charge; like charges generate a repulsive (positive) force, and the direction is away from the other charge. If the charges are placed not in a vacuum but in a medium with specific electrostatic properties, the force $F$ in equation (18.1) is divided by a factor, $\kappa$, which is called the dielectric constant. Its origin is discussed in Chapter 19.

Electric field $E$ defines the electric force exerted on a positive test charge positioned within any given space. A positive test charge $q_0$, is similar in most respects to a true charge except that it does not exert an electrostatic charge on any adjacent or nearby charges. Thus the electric field of a positive test charge provides an idealized distribution of electrostatic force generated by the test charge and is given by

$$E = \frac{F}{q_0} \tag{18.3}$$

The direction of $E$ depends on the identity of the charge. Since the test charge is positive, if the other charge is negative, an attractive force is generated and the direction of $E$ is toward the test charge. Likewise, if the other charge is positive, a repulsive force is generated and the direction of $E$ is away from the test charge. Electric field $E$ is expressed in units of force per charge (N/C). An electric field can be produced by one or more electric charges.

An electrical conductor is a material, such as copper, that conducts electric charges readily. An electrical insulator is a material, such as rubber, that conducts electric charge poorly. Excess negative or positive charge resides on the surface of a conductor at equilibrium under electrostatic conditions. In such a situation, the electric field at any point within the conducting material is zero, and the electric field just outside the surface of the conductor is perpendicular to the surface.

Gauss's law states that the electric flux $\Phi_E$ through a closed surface is equal to the net charge $Q$ enclosed by the surface divided by $\varepsilon_0$, the permittivity of free space:

$$\Phi_E = \frac{Q}{\varepsilon_0} \tag{18.4}$$

# TOPIC 18.1   Electric Forces in Molecular Biology: DNA Structure and Replication

An important area for the application of physics is the functioning of a living cell at the molecular level. The interior of a cell is mainly water, whose molecules possess various amounts of kinetic energy and often collide with one another. They interact with one another in a number of ways, one of which is via chemical reactions where bonds are made or broken between atoms. Another type of interaction is the electrostatic attraction between molecules executing random molecular motion. Thus, thermal motion is modified by ordering effects of electrostatic forces. All living things pass on genetic information from generation to generation. This is contained in the chromosomes, which are made up of genes. The information needed to produce a particular type of protein molecule is contained within each gene. Genetic information contained within a gene is built into the principal molecule of a chromosome, namely the DNA (deoxyribonucleic acid). A molecule of DNA is composed of a long chain of many small molecules known as nucleotide bases. There are only four types of DNA bases: adenine (A), cytosine (C), guanine (G), and thymine (T).

Deoxyribonucleic acid in a chromosome consists of two long DNA chains wrapped around each other in the form of a double helix. In Figure 18.1*a* we illustrate a section of a DNA double helix. In (*b*) a close-up view shows how the A on one strand is always opposite a T on the other strand. In a similar way G is always opposite C. The shapes of the molecules A, T, C, and G are such that a T always fits closely only onto A and G only onto C. In fact, only when the molecules of a pair are in close proximity is the electrostatic force arising from charged portions of each great enough to hold them together even for a short time. These are sometimes referred to as weak bonds or hydrogen bonds. The origin of these bonds is in the fact that the $H^+$ atom of adenine is attracted to the $O^-$ atom of thymine.

The particular helical arrangement occurs when the chromosome replicates itself just before cell division. The arrangement of A paired to T and G opposite C ensures that genetic information is passed to the next generation accurately. We show this process in Figure 18.2. The two strands or helices of the DNA separate with the help of enzymes leaving the charged parts of the bases fully exposed. The enzymes too operate via electrostatic forces.

To see how the correct order of bases occurs, we focus our attention on the G molecule, indicated by an arrow on the lowest strand in Figure 18.2. Unattached nucleotide bases of all four kinds move around in the cellular fluid. Of the four bases only one will

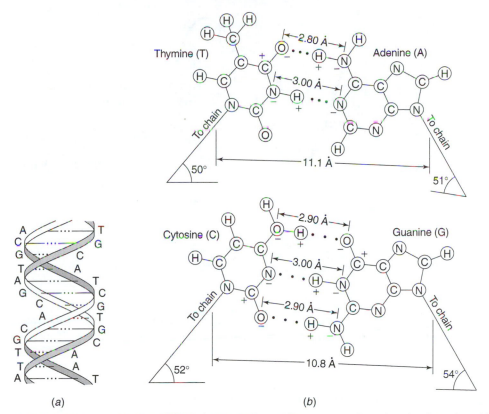

**FIGURE 18.1** (*a*) Section of DNA double helix and (*b*) close-up view showing how A and T and C and G are always paired (the distance unit is 1 Å = $10^{-10}$ m).

experience attraction to the G when it comes close to it, namely a C. The charges on the other three bases are arranged so they do not get into close proximity of those on G, and therefore there will be no significant attractive force exerted on them. These forces decrease rapidly with distance between molecules. Since A, T, and C are hardly attracted at all, they will tend to be knocked away by collisions with other molecules before enzymes can attach them to the growing chain. The electrostatic force will often ensure a C opposite our G long enough so that an enzyme can attach the C to the growing end of the new chain.

We see therefore that electrostatic forces not only hold the two helical chains together but also operate to select the bases in the right order during replication. In Figure 18.2 the new number 4 strand has the same order of bases as the old number 1 strand. Thus the two new helices, 1–3 and 2–4, are identical to the original 1–2 helix. If a T molecule was incorporated in a new chain opposite a G by accident, an error would occur. However, this would only occur very infrequently with an error rate of T being incorporated in a new chain opposite a G of 1 in $10^4$. Such an error may be deemed a spontaneous mutation resulting in a possible change in some characteristic of the organism. If the organism is to survive, this error must have a low rate, but it cannot be zero if evolution is to take place.

The process of DNA replication is often portrayed as occurring in a clockworklike fashion, that is, as if each molecule knew its role and went to its designated place. This is not the case. The forces of attraction between electric charges are rather weak and only

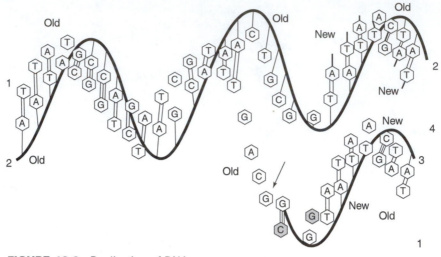

**FIGURE 18.2** Replication of DNA.

become significant when the molecules come close to one another. If the shapes of the molecules are not just right, there is almost no electrostatic attraction. This is why there are few errors.

## TOPIC 18.2   Electrophoresis of Proteins

The term *electrophoresis* describes the transport of charged macromolecules (e.g., proteins or DNA) through an electrolyte under an applied voltage difference and is a basic laboratory tool in biochemistry. The idea behind the technique is to segregate macromolecules according to size. Electrophoresis is the phenomenon of migration of small, electrically charged particles suspended in an aqueous electrolytic solution under the influence of an externally applied electric field. In biological applications electrophoresis is used to characterize various objects ranging from bacterial cells to viruses to globular proteins to DNA. This is possible due to these objects' net electric charge and has found important clinical and diagnostic applications intended to separate various components (e.g., proteins in the blood plasma). As a result, abnormal patterns of blood composition can be identified through electrophoresis. Proteins are folded polymers composed from nature's 20 amino acids arranged in a multitude of possible sequences. Sequence variations lead to different folding patterns and, consequently, functional properties in the body. The net charge on a protein may vary from −100 to +100 elementary charges and is largely determined by the pH value of the solution in which a given protein is suspended. A charged protein with a total electric charge $Q$ is subjected to the electric force $F_e = QE$, where $E$ is the applied electric field and the drag force $F_D = -6\pi\eta a v$, $\eta$ being the viscosity of the medium, $v$ the velocity of the migrating protein, and $a$ its radius. There is an additional complication due to the presence of counterions in the solution which shield the protein by surrounding it as if a cloud were created. The shielding "cloud" moves in the opposite direction under the influence of the field $E$. Without detailed exposition we simply summarize the results as follows. The mobility of a protein (velocity $v$ divided by the electric field intensity $E$) is proportional to the net charge on the protein and inversely proportional to the viscosity, the radius $a$ squared, and the ionic strength of the solution. For example, a protein of charge $10e$ placed in an aqueous solution of ionic strength 0.1 mol/L subjected to an electric field of 1 V/cm will attain a drift velocity of approximately $10^{-4}$ cm/s requiring a time of 3 h for a displacement over 1 cm. The distance traveled over a given time will be proportional to the charge on objects of the same size. Electrophoresis is useful in separating biological materials in a solution according to their charge and size. If some components possess no net charge but a net dipole moment, they, too, can be separated by an analogous method called dielectrophoresis, in which an electric field gradient is used providing a force on a dipole moment.

## SOLVED PROBLEM

**18.1**   Suppose we have $Cl^-$, $Na^+$, and $Ca^{2+}$ ions, as illustrated, in an aqueous solution (dielectric constant $\kappa = 80.4$, see Chapter 19). Find the electric force on the $Na^+$ ion due to the other two ions.

**Solution**   First, we find the magnitude of the force on the $Na^+$ ion due to the $Cl^-$ ion as

$$F_1 = \frac{k}{\kappa} \frac{q_1 q_2}{r_{12}^2} = \frac{(9.0 \times 10^9 \ \text{N} \cdot \text{m}^2)/\text{C}^2}{80.4} \frac{(1.60 \times 10^{-19} \ \text{C})(1.60 \times 10^{-19} \ \text{C})}{(1.50 \times 10^{-9} \ \text{m})^2}$$

$$= 1.27 \times 10^{-12} \ \text{N}$$

Since opposite charges attract, the direction of $\mathbf{F}_1$ on the $Na^+$ ion is to the left, that is, toward the $Cl^-$ ion.

Similarly, the magnitude of the force on the $Na^+$ ion due to the $Ca^{2+}$ ion is

$$F_2 = \frac{k}{\kappa} \frac{q_1 q_2}{r_{12}^2} = \frac{(9.0 \times 10^9 \ \text{N} \cdot \text{m}^2/\text{C}^2)}{80.4} \frac{(1.60 \times 10^{-19} \ \text{C})(2 \times 1.60 \times 10^{-19} \ \text{C})}{(3.00 \times 10^{-9} \ \text{m})^2}$$

$$= 6.37 \times 10^{-13} \ \text{N}$$

Since "like" charges repel, $\mathbf{F}_2$ on the $Na^+$ ion acts to the left, directly away from the $Ca^{2+}$ ion, as indicated in the diagram:

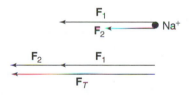

Finally, the total force $\mathbf{F}_T$ on the ion is the vector sum

$$\mathbf{F}_T = \mathbf{F}_1 + \mathbf{F}_2$$

Since $\mathbf{F}_1$ and $\mathbf{F}_2$ are parallel,

$$|\mathbf{F}_T| = |\mathbf{F}_1| + |\mathbf{F}_2| = 1.27 \times 10^{-12} + 0.64 \times 10^{-12} = 1.91 \times 10^{-12} \ \text{N}$$

and is directed to the left.

# EXERCISES

**18.1**   Suppose the plasma membrane of a cell has a surface charge per unit area of about $10 \ C/m^2$ on one side and an equal but opposite surface charge on the other side.
(a) Estimate the electrical field strength inside the plasma membrane.
(b) Estimate the force on an ion of $Ca^{2+}$ placed inside the plasma membrane of a cell.

**18.2**   Two DNA chains of length 3 $\mu$m in ion free water run parallel with a separation of 100 Å. What is the repulsive force between the chains? Assume the charge density on each chain to be 1 e per nm.

**18.3**   Give an expression for the electrostatic force between two 80-kD proteins in a 2 M salt solution, one with a positive charge $Q = 12e$ and one with a negative charge of $-Q$. Calculate the electrostatic force at a spacing of 20 and 100 Å. Compare with the force between the proteins in salt-free water and in a vacuum.

# CHAPTER 19

# ELECTRIC POTENTIAL ENERGY AND ELECTRIC POTENTIAL

## PHYSICAL BACKGROUND

Work $W_{AB}$ is done by an electric force as a positive test charge $+q_0$ moves from point $A$ to point $B$; the work equals the electric potential energy (EPE) at $A$ minus that at $B$:

$$W_{AB} = \text{EPE}_A - \text{EPE}_B \qquad (19.1)$$

The electric potential $V$ is the electric potential energy per unit charge, $V = \text{EPE}/q_0$. The electric potential difference between two points is

$$V_B - V_A = \frac{\text{EPE}_B - \text{EPE}_A}{q_0} = -\frac{W_{AB}}{q_0} \qquad (19.2)$$

A positive charge accelerates from a region of higher potential toward a region of lower potential. Conversely, a negative charge accelerates from a region of lower potential toward a region of higher potential. The electric potential at a distance $r$ from a point charge $q$ is given by

$$V = k\frac{q}{r} \qquad (19.3)$$

This expression for $V$ assumes the potential is zero at an infinite distance away from the charge.

An equipotential surface is a surface on which the electric potential is the same everywhere. The electric force does no work as a charge moves on an equipotential surface. The electric field created by any group of charges is always perpendicular to the associated equipotential surfaces and points in the direction of decreasing potential. The electric field is given by

$$E = -\frac{\Delta V}{\Delta d} \qquad (19.4)$$

where $\Delta V$ is the potential difference and $\Delta d$ is the displacement perpendicular to the equipotential surfaces.

The potential inside an electrolyte or a metal is constant if no electrical current flows. The potential of a constant electrical field varies linearly with position. The electrostatic work done when two ions move from infinity to a distance $d$ equals

$$\Delta W = -k\frac{q_1 q_2}{r} \qquad (19.5)$$

This leads to an ion–ion binding energy in vacuum of a few electron-volts.

A capacitor is a device that can store charge and consists of two conductors that are separated by a distance $d$. The magnitude $q$ of the charge on each plate is given by

$q = CV$, where $V$ is the magnitude of the potential difference between the plates and $C$ is the capacitance. The SI unit for capacitance is the farad (F), and one farad equals one coulomb per volt (C/V). The insulating material included between the plates is called a dielectric. The dielectric constant $\kappa$ of the material is $\kappa = E_0/E$, where $E_0$ and $E$ are, respectively, the magnitudes of the electric fields between the plates without and with a dielectric, assuming the charge on the plates is kept fixed. The capacitance of a parallel plate capacitor is

$$C = \frac{\kappa \varepsilon_0 A}{d} \tag{19.6}$$

where $\varepsilon_0 = 8.85 \times 10^{-12}$ C²/N · m² is the permittivity of free space and A is the area of each plate. The electric potential energy stored in a capacitor is

$$E = \tfrac{1}{2}CV^2 \tag{19.7}$$

The energy density or energy stored per unit volume is $\tfrac{1}{2}\kappa \varepsilon_0 E^2$. The voltage difference across a parallel plate capacitor with a uniform electric field is

$$\Delta V = E_0 d \tag{19.8}$$

with $E_0$ being the electrical field strength.

An electric dipole is formed when two equal but opposite charges $+Q$ and $-Q$ are brought together to a close distance $d$. The corresponding dipole moment is defined as $P = Qd$ and it is a vector quantity with its direction along the dipole axis. The electric potential of a dipole of magnitude $P$ is

$$V(r, \theta) = k\frac{P \cos \theta}{r^2} \tag{19.9}$$

where $r$ is the distance to the dipole's center and $\theta$ the angle between the position vector and the dipole axis. The force between the two dipoles $P_1$ and $P_2$ is then given by

$$F(r) = -3k\frac{P_1 P_2}{r^4}(2 \cos \theta_1 \cos \theta_2 - \frac{2}{3} \sin \theta_1 \sin \theta_2) \tag{19.10}$$

which can be both attractive and repulsive; see Figure 19.1 for an explanation.

Molecular dipole moments typically are on the order of one elementary charge $e$ times 1 Å, that is, about $10^{-29}$ C · m. The dipole moment of ethanol is, for instance, $5.6 \times 10^{-30}$ C · m. Water is polar, but it has a more complex charge distribution than ethanol: a negatively charged oxygen with two positively charged hydrogen atoms sticking out, making an angle of 104°.

The binding energy of charged particles such as Na⁺ and Cl⁻ is calculated from

$$\Delta W = -k\frac{q_1 q_2}{r} \tag{19.11}$$

where $r$ in this case is 2.4 Å, giving $\Delta W = 5.5$ eV. This is an example of an ionic bond. Another important form of chemical bonding is the covalent bond, the C–C bond for instance. It is responsible for maintaining the structural integrity of DNA, proteins, and polysachharides (chainlike sugar molecules). In a covalent bond, electrons shuttle back and forth constantly between positively charged nuclei, and hence the covalent bond is electrostatic in origin with a characteristic distance $d = 1$ Å. The C–C binding energy is 3.6 eV, which is of the same order of magnitude as the NaCl binding energy. Other covalent bonds are in the same range. The covalent bond thus has a binding energy compa-

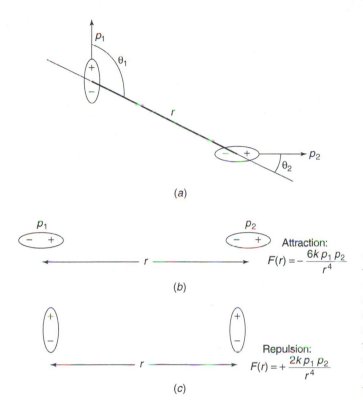

(a)

(b)

Attraction:

$$F(r) = -\frac{6k\,p_1\,p_2}{r^4}$$

(c)

Repulsion:

$$F(r) = +\frac{2k\,p_1\,p_2}{r^4}$$

**FIGURE 19.1** Force between two electric dipoles $P_1$ and $P_2$ separated by a distance $r$: (a) general orientation, (b) oriented along the lines joining them in the same sense, and (c) oriented perpendicular to the line joining them in the same sense

rable to that of the ionic bond in vacuum. Both types of bond have strength which at room temperature is large compared to thermal energy $k_B T$, where

$$k_B T = 4.2 \times 10^{-21} \text{ J} \tag{19.12}$$

Important biomolecules such as the carbohydrates are neither charged nor dipolar. They interact through a net induced dipole attraction. This is a result of the distortion of the electron distributions. This is called a van der Waals interaction, and it is always attractive. Two molecules separated by a distance $R$ attract each other by an attractive potential energy:

$$U_{vw}(R) = -\frac{A}{R^6} \tag{19.13}$$

The parameter $A$, whose units are $N \cdot m^7$, depends on the chemical structure of the two molecules but its value is generally close to $10^{-77} \text{ N} \cdot m^7$. Table 19.1 gives the values for

**TABLE 19.1  Coefficient *A* for Different Molecules**

| Molecule | $A$ $(10^{-77} \text{ N} \cdot m^7)$ |
|---|---|
| Water | 2.1 |
| Benzol | 4.3 |
| Phenol | 6.5 |
| Diphenylaniline | 14.4 |

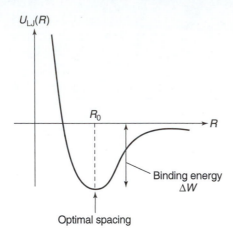

FIGURE 19.2 Lennard–Jones potential.

*A* between identical molecules in water. Note that the table goes from small to big molecules, and the bigger the molecule, the larger the *A* value. When two atoms or molecules get so close that the electron clouds overlap, the electrostatic interaction changes character and becomes highly repulsive, a force which is known as steric repulsion. The short-range repulsive force can be included by adding a repulsive potential energy to the van der Waals potential, giving

$$U_{\mathrm{VW}}(R) = -\frac{A}{R^6} + \frac{B}{R^{12}} \tag{19.14}$$

We call this combination formula the Lennard-Jones potential (see Figure 19.2). The depth of the potential energy curve $U_{\mathrm{LJ}}(R)$ at the minimum point, about 1 kcal/mol, is the strength of the van der Waals bond. Van der Waals bonds are somewhat weaker than polar and hydrogen bonds, but they play a similar role in molecular biology, namely as a source of weak, temporary links.

Often, the oxygen and hydrogens of different biological molecules (including water) are so close that they form a weak chemical bond called the hydrogen bond (see Figure 19.3 for illustration). The strength of the hydrogen bond again is about 5 kcal/mol. Hydrogen bonds play a very important role in the formation of macromolecules of biological importance such as the DNA and proteins.

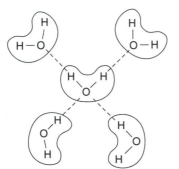

FIGURE 19.3 Hydration of $H_2O$ by $H_2O$: hydrogen bond.

## TOPIC 19.1   Discovery of Bioelectricity

Enormous interest in electric phenomena was generated by Benjamin Franklin's discovery in 1752 that lightning is an electric discharge phenomenon. Speculations concerning the electrical nature of nerve signals and muscle contraction can be traced back to the mid-eighteenth century and specifically to the work of the French scientists Dutay and Nollet as well as the German physiologist von Haller. However, a real epoch-making breakthrough took place in 1780 and is attributed to the Italian medical doctor Luigi Galvani, who serendipitously discovered the effect of electrical stimulation of animal muscle contraction. Galvani entered into a protracted controversy over the interpretation of his observations with another Italian discoverer, Alessandro Volta. Volta devised a method of generating constant electric currents from a source of electric power, the famous voltaic pile, a battery in which two different metals were stacked together with a moist felt pad. While Volta insisted on an external source of electricity in muscle contraction phenomena, Galvani attributed electricity to the muscle itself as its source. Both scientists were partly right and partly wrong since muscle cells are capable of establishing a potential difference of up to 100 mV but the mechanism of muscle contraction requires electrostimulation by nerve cells. Decisive contributions to the proper understanding of bioelectrochemical processes were made more than 50 years later by the German physicists von Helmholtz and Nernst. The explanation of the mechanism of electric pulse propagation in nerve axons was given by Hodgkin and Huxley in the mid-twentieth century but that is another story.

## TOPIC 19.2   Electrostatics in Water

Interactions between biomolecules take place in water. Water is the universal solvent for biomolecules. Water molecules carry a dipole moment and interact through the dipolar electrical fields they produce. As a consequence, the electric field of a charged protein polarizes the water molecules in the neighborhood of the protein which affects the nature of the electric field of a protein. In addition, both blood and the fluid inside living cells (cytoplasm) are electrolytes. The free ions of electrolytes further modify the electrical field. Since the dielectric constant of water is close to 80, the electrostatic interactions between ions in a water environment are reduced accordingly. This has important consequences, one of which is the fact that the ionic binding energy is reduced by a factor of 80. For instance, instead of $8.8 \times 10^{-19}$ J for the binding energy of NaCl in a vacuum, in water we find only about $10^{-20}$ J or 5 kcal/mol using again $d = 2.4$ Å and $q_1 = -q_2 = -e$. The ionic binding energy in water is only about 10 times the thermal energy $k_B T$. In many cases, biochemical reactions involve ATP-to-ADP conversion as a source of energy which releases about 7.3 kcal/mol (see Chapter 15). That is enough to break an ionic bond in water but not sufficient to break an ionic bond in vacuum.

   An important distance that relates the magnitude of the electrostatic energy in water to the thermal energy $k_B T$ is the so-called Bjerrum length. The Bjerrum length $l_B$ is the distance at which the electrostatic energy of two monovalent ions equals $k_B T$. It follows that

$$\frac{k}{\kappa} \frac{e^2}{l_B} = k_B T \tag{19.15}$$

For water at room temperature, the Bjerrum length is about 7.2 Å. The electrostatic interaction energy between two monovalent ions is thus negligible compared to the thermal energy if their separation is large compared to the Bjerrum length.

While the ionic (or polar) bond is greatly reduced by surrounding water molecules, this is not so true for the covalent bond. The strength of the covalent bond in water is about the same as in vacuum because the covalent bond involves the motion of an electron between two closely adjacent nuclei such that surrounding water molecules do not affect it too much. As a result, there are two very different energy scales available to biomolecules: (a) covalent bonds with energies in the range 50 to 100 kcal/mol and (b) ionic bonds in the 5-kcal/mol range. Covalent bonds cannot be broken by thermal fluctuations, and the covalent bonding energy is large compared to the chemical energy stored in one ATP-to-ADP conversion (see Fig. 19.4). Hence, covalent bonds are indeed used as permanent links keeping macromolecules together. The much weaker polar bonds can be broken by an ATP-to-ADP conversion, by thermal fluctuations, and also by light absorption. They are useful for temporary links, for example, to control the folding of proteins or the mutual binding of proteins or the binding of proteins to DNA.

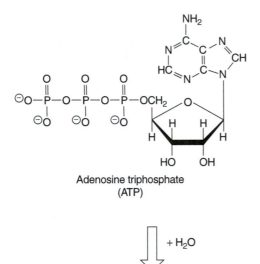

Adenosine triphosphate
(ATP)

$+ H_2O$

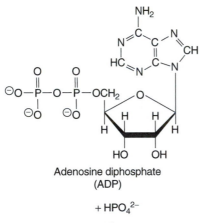

Adenosine diphosphate
(ADP)

$+ HPO_4^{2-}$

$+ H^+$

**FIGURE 19.4**   Conversion of ATP to ADP via hydrolysis.

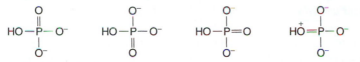

**FIGURE 19.5**   Significant resonance forms of orthophosphate with the same energy.

## TOPIC 19.3   ATP and ADP Molecules and Conversion of ATP to ADP

As already discussed in Topic 15.1, at a thermodynamic level the ATP molecule is a source of energy in biochemistry, and when it converts to ADP, energy is released. Approximately 7.3 kcal/mol of energy per ATP molecule is released in this conversion. The chemical reaction involved reads

$$ATP^{4-} + H_2O \rightarrow ADP^{3-} + HPO_4^{2-} + H^+ \tag{19.16}$$

In the reaction given in reaction (19.16) electrostatics plays a significant role. There are four negatively charged oxygen atoms in close proximity on the ATP molecules, and as these repel one another, the molecule is under considerable strain. From the right-hand side of reaction (19.16), as the ADP molecule contains only these negatively charged oxygen ions, we see that the result of the conversion is to spread out the negative charge over the ADP and $HPO_4^{2-}$ molecules. These latter molecules are called orthophosphate or Pi. Thus, in the conversion, the overall electrostatic energy is reduced. In Figure 19.4 we illustrate this conversion or hydrolysis, that is, the breakdown of a molecule by a water molecule, producing Pi and positively charged hydrogen. Electrostatics is not the only reason that the chemical reaction in reaction (19.16) is so favorable. If a molecule has a number of configurations of about the same energy and if the molecule is allowed to flip from one state to another, then the overall energy is lowered. The Pi molecule has four such resonance states which are illustrated in Figure 19.5.

If reaction (19.16) could proceed directly, then ATP molecules in water would be unstable; that is, they would immediately turn into ADP molecules, thus making ATP useless as a supplier of energy. Under physiological conditions the ATP molecule is metastable. When an ATP molecule binds to certain proteins called enzymes, however, the activation barrier is lowered and the molecule moves down the energy curve turning into ADP. We show this process in Figure 19.6.

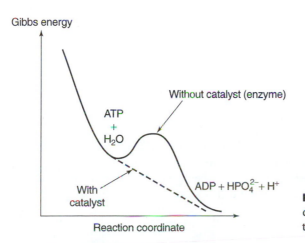
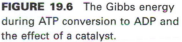

**FIGURE 19.6**   The Gibbs energy during ATP conversion to ADP and the effect of a catalyst.

## TOPIC 19.4 Electrostatic Potential of DNA

We have mentioned some properties of DNA before. Here, we examine its electrostatic potential. Deoxyribonucleic acid is a highly charged molecule and carries two fully ionized monovalent $PO_4^-$ groups on its outer surface per base pair. To model DNA, we assume the charge per unit length is $\lambda$, which is about one negative charge per 1.7 Å of length. For simplicity, we assume that DNA is a long, charged cylindrical molecule. The diameter of this cylinder is about 20 Å. The electrical field of a long strand of DNA in water will clearly depend on the perpendicular distance $r$ from the strand and on $\lambda$. The electric field lines must point inward because DNA is negatively charged, and because we have used a cylinder to model it, they must also have cylindrical symmetry. Denoting the electric field at a distance $r$ from the cylinder by $E(r)$, we observe that it is inversely proportional to the dielectric constant $\kappa$ and inversely proportional to some power, say $r^\alpha$, of $r$; hence we write

$$E(r) = \text{const } k\frac{\lambda}{\kappa r^\alpha} \tag{19.17}$$

In a vacuum, the electric field of a monopole of charge $Q$ a distance $R$ away is $kQ/R^2$, so as $\lambda$ is a charge per unit length, $\alpha = 1$; the constant is 2 and hence,

$$E(r) = 2k\frac{\lambda}{\kappa r} \tag{19.18}$$

Since $E(r) = -dV/dr$, we have an equivalent form of equation (19.18) as

$$\frac{dV(r)}{dr} = -2k\frac{\lambda}{\kappa r} \tag{19.19}$$

Integrating equation (19.19), we see that

$$V(r) = -2k\frac{\lambda}{\kappa}\ln r + k_0 \tag{19.20}$$

In equation (19.20), $k_0$ is a constant of integration which is conveniently chosen to be zero.
Now suppose a positive monovalent ion moves in from a distance of 1 $\mu$m to the surface of the DNA. The work done $\Delta W$ will be

$$\Delta W = |e|V(r = 1\ \mu m) - |e|V(r = 10\ \text{Å}) \tag{19.21}$$

Hence

$$\Delta W = -2k\frac{\lambda}{\kappa}|e|\ln\left(\frac{10^{-6}}{10^{-9}}\right) \tag{19.22}$$

since 1 Å $= 10^{-10}$ m. Using the values $k = 8.99 \times 10^9$ N · m$^2$/C$^2$, $\kappa = 80.4$, $\lambda = -|e|$ per $1.7 \times 10^{-10}$ m, and $e = 1.6 \times 10^{-19}$ C in equation (19.22), we find that $\Delta W$ exceeds by about 10 times the thermal energy $k_BT$ where $k_B$ is Boltzmann's constant and $T$ is the absolute temperature. This estimate suggests that a monovalent ion stays close to the surface of DNA and consequently in the cell we expect DNA to be surrounded by a cloud of Na$^+$ ions.

## TOPIC 19.5  Electrical Potentials of Cellular Membranes

All living cells are protected from their environment by a thin semipermeable wall called a cell membrane. The membrane possesses channels of pores which allow a selective passage of metabolites and ions in and out of the cell. The thickness of a cell membrane varies between 70 and 100 Å. First measurements of the electrical properties of cell membranes were made on red blood cells by H. Fricke and on sea urchin cells by K. S. Cole, and it was found that membranes act as capacitors maintaining a potential difference between oppositely charged surfaces composed mainly of phospholipids with proteins embedded in them. A typical value of the capacitance per unit area $C/A$ is about 1 $\mu F/cm^2$ for cell membranes. This relates to the membrane's dielectric constant $\kappa$ via

$$\frac{C}{A} = \frac{\kappa \varepsilon_0}{d} \tag{19.23}$$

where $\varepsilon_0 = 8.85 \times 10^{-12}\ C^2/N \cdot m^2$ giving a value of $\kappa \cong 10$ which is greater than $\kappa \cong 3$ for membrane phospholipids above resulting from the active presence of proteins.

The cellular membrane is much more permeable (the intercellular fluid contains primarily sodium chloride), in the normal resting state, to potassium ions than sodium ions, which results in an outward flow of potassium ions until the voltage inside the cell is $-85$ mV. This voltage is called the resting potential of the cell. If the cell is stimulated by mechanical, chemical, or electrical means, sodium ions diffuse more readily into the cell since the stimulus changes the permeability of the cellular membrane. The inward diffusion of a small amount of sodium ions increases the interior voltage to $+60$ mV, which is known as the action potential of the cell. The membrane again changes its permeability once the cell has achieved its action potential, and potassium ions then readily diffuse outward so the cell returns to its resting potential. Depending on the state of the cell, the interior voltage can therefore vary from its resting potential of $-85$ mV to its action potential of $+60$ mV. This results in a net voltage change of 145 mV in the cell interior. The voltage difference between the two sides of the membrane is fixed by the concentration difference. Having a salt concentration difference across a membrane and allowing only one kind of ion to pass the membrane produces a voltage difference given by the formula

$$V_L - V_R = \frac{k_B T}{e} \ln\left(\frac{c_R}{c_L}\right) \tag{19.24}$$

and called the Nernst potential. This is the basic mechanism whereby electrical potential differences are generated inside organisms (see Figure 19.7). Note that the Nernst potential difference only depends on the concentration ratio. In Table 19.2 are some concentration ratios for cellular membranes and the corresponding value of the Nernst potential computed from equation (19.24).

**TABLE 19.2  Concentration Ratios for Cellular Membranes and Corresponding Nernst Potential in Equation (19.24)**

| Ion | $c(R)/c(L)$ | Valency | $\Delta V$ (mV) |
| --- | --- | --- | --- |
| $Na^+$ | 10 | 1 | 62 |
| $K^+$ | 0.03 | 1 | $-88$ |
| $Cl^-$ | 14 | $-1$ | $-70$ |

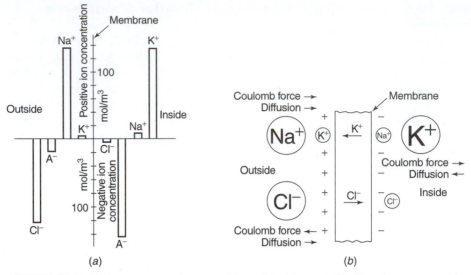

**FIGURE 19.7** (a) Ion concentrations outside and inside a cell. Positive ion concentrations are graphed above the line and negative ion concentrations below. (b) Membrane is permeable to $K^+$ and $Cl^-$ ions, which diffuse in opposite directions as shown. Small arrows indicate that the Coulomb force resists the continued diffusion of both $K^+$ and $Cl^-$ ions.

## TOPIC 19.6 Medical Diagnostic Techniques and Treatment

A number of medical diagnostic techniques rely on the fact that the surface of the human body is not an equipotential surface. In fact, between various points on the body there are tiny potential differences of the order of 30 to 500 $\mu$V. These provide the basis for electro-cardiography, electroencephalography, and electroretinography and can be traced to the electrical characteristics of nerve and muscle cells. The cells, to carry out their biological functions, utilize positively charged potassium and sodium ions and negatively charged chlorine ions that exist not only within the cells but on the outside of the cells in the in-tercellular fluid. Such charged particles generate electric fields which extend to the sur-face of the body and produce the small potential differences.

### Electrocardiography

The initial contraction and subsequent surge of pressure which the heart generates at the outset of a heartbeat are caused by an electrical signal. The potential difference between two points changes as the heart beats and forms a repetitive pattern. The recorded pattern of potential difference versus time is called an electrocardiogram (ECG or EKG). Record-ings are taken with electrodes attached to various positions on the body.

In Figure 19.8 we show a part of a normal ECG recording where the electrical ac-tivity of the heart is traced preceding one normal heartbeat. The vertical axis shows the current flowing toward the recording lead and the current is proportional to potential dif-ference. The rise at P occurs just before the atria—or upper heart chambers—begin to contract. The QRS spike occurs just before the ventricles, or lower chambers of the heart, begin to contract. The rise at T occurs as the electrical potential difference returns to zero. Any abnormality can be easily spotted from the recording by the doctor. In Figure 19.9 we illustrate the normal rhythm when the heart chambers are contracting regularly. In

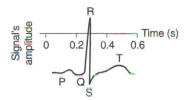

**FIGURE 19.8** Normal ECG recording.

**FIGURE 19.9** Normal rhythm.

Figure 19.10 we show a trace when the heart is undergoing ventricular fibrillation. Here the contractions of the lower heart chambers are extremely irregular. Another heart defect is a complete heart block, the corresponding trace of which is shown in Figure 19.11. When this occurs, the upper and lower heart chambers are beating independently. Lastly, in Figure 19.12 is a trace taken when the patient is suffering from atrial fibrillation when the upper heart chambers beat fast and irregularly.

A plot of the electrical field of a discharging nerve cell shows a dipole-type electric field. This is also true of the heart as a whole, and when a nervous signal passes across the heart, the dipoles of the depolarizing nerve cells add up to form one large, time-dependent dipolar electric field (see Figure 19.13). Denote by $P(t)$ the electric dipole moment of the heart. This changes direction and magnitude cyclically as nerve cells alternatingly polarize and depolarize various parts of the heart. Thus the time dependence is periodic, the period being equal to the time it takes for the duration of a heartbeat—the dipole moment being largest during stimulation of the ventricular cardiac muscle. The aim of an ECG measurement is a measurement of $P(t)$. Such a plot of $P(t)$ for a complete cycle of the heart is called a vector cardiogram. The dipole moment vector is found by measuring the electrical potential $V$ on a variety of sites on the skin. In equation (19.25), the relationship between the voltage as a function of $P$ and distance $r$, from the heart, is given as

$$k \frac{P \cos \theta}{r^2} = V_0 \tag{19.25}$$

where equation (19.9) has been applied to this case. Choosing a different $V_0$ gives another equipotential surface. The set of equipotential surfaces obtained this way are shown in Figure 19.14 and take the form of deformed spheres. A special case is $V_0 = 0$, which can be satisfied by choosing $\theta = \frac{1}{2}\pi$; that is, this corresponds to all points in the $x$–$y$ plane.

To discuss the generated potential in detail, we need a general formula in which the direction of the dipole moment varies during the cardiac cycle. This is

$$V(\pm) = k \frac{P(t) \cdot r}{r^3} \tag{19.26}$$

In equation (19.26), $r$ is the position vector on the skin, relative to an origin in the heart. Also, $P \cdot r = Pr \cos \theta$, where $\theta$ here is the angle between $r$ and the dipole moment $P(t)$. If the $z$ axis is along the dipole moment, then equation (19.26) becomes (19.25). During

**FIGURE 19.10** Ventricular fibrillation.

**FIGURE 19.11** Complete heart block.

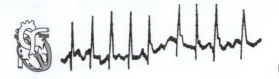

**FIGURE 19.12**   Atrial fibrillation.

an ECG we measure all these components of $\mathbf{P}(t)$ which we define as $P_x(t)$, $P_y(t)$, and $P_z(t)$ at a particular time $t$. This is done by measuring the electrical potential at three different points on the patient's body. Denoting these points by $R$, $L$, and $P$ and letting the potential at these points be $V_R(t)$, $V_L(t)$, and $V_P(t)$, respectively, assuming that the location of these points are known to be $\mathbf{r}_R$, $\mathbf{r}_L$, and $\mathbf{r}_P$, then equation (19.26) can be written as three equations, that is, in terms of components, as

$$V_R(t) = \frac{k}{r_R^3}[P_x(t)x_R + P_y(t)y_R + P_z(t)z_R] \tag{19.27}$$

$$V_L(t) = \frac{k}{r_L^3}[P_x(t)x_L + P_y(t)y_L + P_z(t)z_L] \tag{19.28}$$

$$V_P(t) = \frac{k}{r_P^3}[P_x(t)x_P + P_y(t)y_P + P_z(t)z_P] \tag{19.29}$$

This set of equations can then be solved on a computer to yield $\mathbf{P}(t)$ and the full vector cardiogram. The results can be checked by choosing additional points to measure the potential.

When the heart produces a rapid, irregular sequence of beats, the condition is called fibrillation. It can often be stopped by passing a fast discharge of electrical energy through the heart. A paddle is connected to each plate of a large capacitor. The paddles are placed on the chest of the patient close to the heart, and the capacitor is charged to a potential difference of about a thousand volts. In a few thousandths of a second the capacitor is discharged, and the discharge current passes through one paddle, the heart, and the other paddle. Within a few seconds the heart often returns to its normal beating pattern. A defibrillator is used to revive a person who has suffered a heart attack. The device uses the

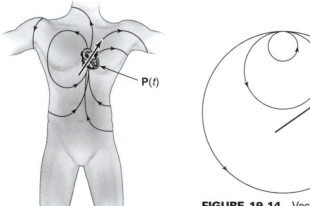

**FIGURE 19.13**   Heart as a dipole.

**FIGURE 19.14**   Vector cardiogram. Dipoles moment $P(t)$ moves over a closed loop during the cardiac cycle.

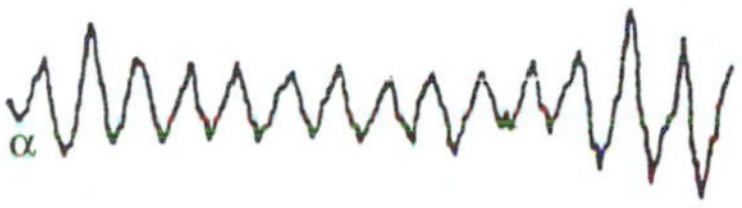

**FIGURE 19.15**  Alpha waves.

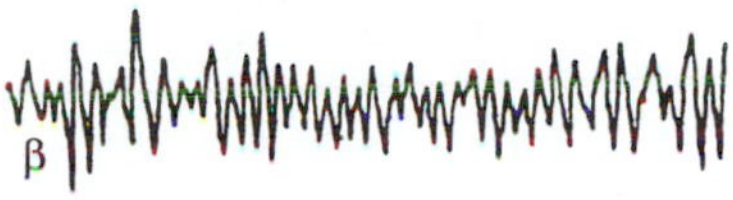

**FIGURE 19.16**  Beta waves.

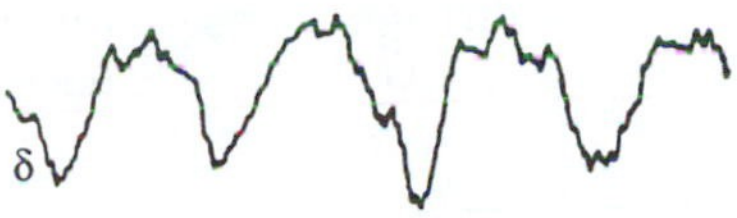

**FIGURE 19.17**  Delta waves.

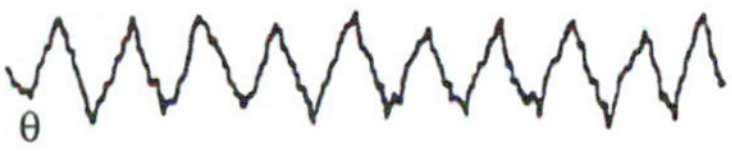

**FIGURE 19.18**  Theta waves.

electrical energy stored in the capacitor to deliver a controlled electric current which can restore the normal heart rhythm.

## Electroencephalography

Electrodes are placed at specific locations on the head in electroencephalography. The graph of potential difference versus time is known as an electroencephalogram (EEG). The various parts of the patterns produced in EEGs are referred to as waves or rhythms. The electrodes are connected to an instrument which measures the brain's impulses in microvolts and amplifies them for recording purposes. These recordings are taken with the subject at rest, with eyes open and eyes shut, during and after hyperventilation (an increase in the depth, duration, and rate of breathing) and while looking at a flashing light. It is helpful to record activity as a patient goes to sleep, particularly when epilepsy is suspected. In Figures 19.15 to 19.18 we illustrate the four main EEG wave patterns. The prominent patterns of an awake, relaxed adult whose eyes are closed are shown in Figure 19.15. They are termed alpha waves. In Figure 19.16 we illustrate beta waves, which are the lower faster oscillation from a person who is concentrating on an external stimulus. Delta waves are a characteristic pattern typical of sleep but also found in young infants. We show these in Figure 19.17. Sometimes, but rarely, beta waves are caused by a brain tumor. A trace example of theta waves is presented in Figure 19.18, and these are dominant in the waves of young children. In adults they may indicate an abnormality of the brain. Table 19.3 gives the brain wave frequency ranges.

## Electroretinography

Potential differences measured in electroretinography characterize the electrical signals of the retina of the eye. Figure 19.19 shows typical patterns of potential difference versus

**TABLE 19.3  Brain Wave Frequency Ranges**

| Band | Frequency range (Hz) |
| --- | --- |
| Alpha ($\alpha$) | 8–13 |
| Beta ($\beta$) | 13–22 |
| Gamma ($\gamma$) | 22–30 |
| Delta ($\delta$) | 0.5–4 |
| Theta ($\theta$) | 4–8 |

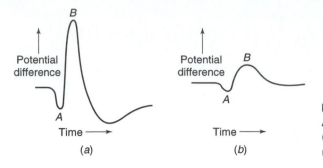

**FIGURE 19.19** ERGs showing A- and B-wave components for (*a*) normal and (*b*) abnormal retinas.

time when the eye is stimulated by a flash of light. One electrode is often placed on the forehead and the other on the contact lens. The recorded pattern is called an electro-retinogram (ERG). Parts of the pattern are referred to as "A wave" and "B wave." As can be seen in Figure 19.19, the ERGs of normal and diseased eyes can differ markedly.

## SOLVED PROBLEMS

**19.1** Let the electrical dipole moment of the heart of a patient be a vector $\mathbf{P}(t)$ which has a fixed length $|P| = 2 \times 10^{-13}$ C $\cdot$ m but which rotates around the $z$ direction, making a full turn every $T = 1.5$ s. Assume that at time $T = 0$, the dipole moment vector is along the $x$ direction.

**(a)** Give an expression for $\mathbf{P}(t)$.

**(b)** Compute the time-dependent voltage at the point $R$ on the surface of the patient with coordinates $x = 0$, $y = 50$ cm, and $z = 0$.

**Solution**

**(a)** A vector rotating in the $x$–$y$ plane can be represented as

$$x(t) = A \cos(\omega t + \varphi) \qquad y(t) = A \sin(\omega t + \varphi)$$

with $A$ the length of the vector, $\omega = 2\pi/T$ the angular frequency, and $\phi$ a constant (the phase). For our case, $A = |\mathbf{P}|$ and $\phi = 0$ since $y(t) = 0$ at time $t = 0$. So,

$$P_x(t) = |\mathbf{P}| \cos(\omega t) = 2 \times 10^{-13} \cos\left(\frac{2\pi t}{1.5}\right)$$

and

$$P_y(t) = |\mathbf{P}| \sin(\omega t) = 2 \times 10^{-13} \sin\left(\frac{2\pi t}{1.5}\right)$$

The vector cardiogram is a circle with radius $|\mathbf{P}|$.

**(b)** To compute the voltage at a point $x = 0$, $y = 50$ cm, $z = 0$, we use the equation for $V_R(t)$:

$$V_R(t) = k \frac{P_x(t)x_R + P_y(t)v_R + P_z(t)z_R}{r_R^3}$$

$$= 9 \times 10^9 \frac{P_x(t) \times 0 + P_y(t) \times 0.5 + P_z(t) \times 0}{0.5^3}$$

$$= 7.2 \sin\left(\frac{2\pi t}{1.5}\right) \text{ mV}$$

**19.2**   Calculate the interaction energy between the C=O dipole of thymine and the H–N dipole of adenine, assuming that $d_{CH} = 0.31$ nm, $d_{CN} = 0.41$ nm, $d_{OH} = 0.19$ nm, and $d_{ON} = 0.29$ nm. The charges involved are $q_H = -q_N = 0.19e = 3.0 \times 10^{-20}$ C and $q_C = -q_O = 0.41e = 6.6 \times 10^{-20}$ C.

**Solution**   The interaction energy $U$ consists of four contributions

$$U = U_{CH} + U_{CN} + U_{OH} + U_{ON}$$

where $U_{CH}$ means the potential energy of C in the presence of H, and similarly for the other terms. Since the potential energy $U = qV$, where $V$ is the electric potential, then for two point charges $U_{12} = kq_1q_2/r$, where $r$ is the distance between them. So

$$U = k\frac{q_C q_H}{r_{CH}} + k\frac{q_C q_N}{r_{CN}} + k\frac{q_O q_H}{r_{OH}} + k\frac{q_O q_N}{r_{ON}}$$

Using the distances given, we get

$$U = (9.0 \times 10^9 \text{ N} \cdot \text{m}^2/\text{C}^2)\left(\frac{(6.6)(3.0)}{0.31} + \frac{(6.6)(-3.0)}{0.41} + \frac{(-6.6)(3.0)}{0.19} + \frac{(-6.6)(-3.0)}{0.29}\right)\frac{(10^{-20} \text{ C})^2}{(10^{-9} \text{ m})}$$

$$= -1.83 \times 10^{-20} \text{ J} = -0.11 \text{ eV}$$

The potential energy is negative, meaning 0.11 eV of work is required to separate the molecules.

**19.3**   Estimate the capacity of the membrane in bacteria and mitochondria, where the average distance between adjacent electron transport proteins is $d = 10$ nm and a membrane potential of $\Delta V = 160$ mV is generated.

**Solution**   Consider a square section of the membrane of area $A$. Suppose that the membrane transport proteins form a two-dimensional grid with a characteristic lattice constant $d$. If the number of proteins along the side of the square is $N$, then the total number of proteins covering the square section of the membrane is $N^2$. Each of these transport proteins drives one electron with charge $e$ from one side of the membrane to the other side during one turnover. As the proteins are oriented, they all pump the electrons in one direction, and the electric charge $Q = eN^2$ will accumulate on the opposite side of the membrane, charging the membrane capacitor to electric potential $\Delta V = Q/C$ during one turnover. Thus, the electric capacity of the membrane is calculated as

$$C = \frac{eN^2}{\Delta V}$$

The membrane capacity per unit area is defined as $c = C/A$ and it is given by

$$c = \frac{e}{\Delta V\, d^2} = \frac{1.6 \times 10^{-19} \text{ C}}{(1.60 \times 10^{-3} \text{ V})(10 \times 10^{-9} \text{ m})^2} = 1 \text{ F/m}^2$$

**19.4**   Suppose the electric capacity of biomembranes of unit surface area is $c = 0.01$ F/m². How many H$^+$ ions should be pumped through the membrane of a spherical vesicle of diameter $d = 30$ nm to generate a membrane potential $\Delta V = 100$ mV?

**Solution**   The membrane potential $\Delta V = Q/C$, where $Q$ is the amount of electric charge transported through the total membrane surface area $A$ and $C$ denotes the total electric capacity of the membrane, which can be calculated from the specific capacity $c$ and the area of the membrane as $C = cA$. As $n$ protons are pumped across the membrane, the total electric charge $Q = ne$, which produces a membrane potential

$$\Delta V = \frac{Q}{C} = \frac{ne}{cA}$$

The surface area of a sphere with diameter $d$ is $A = \pi d^2$. This leads to an expression for $n$, the number of transported $H^+$ ions, as

$$n = \frac{\pi d^2 c\, \Delta V}{e} = \frac{\pi (30 \times 10^{-9}\ \text{m})^2 (10^{-2}\ \text{F/m}^2)(0.1\ \text{V})}{1.6 \times 10^{-19}\ \text{C}}$$

Substituting the numerical values for the vesicle, we find $n = 18$.

**19.5**   A number $Z$ ($>>1$) of discrete elementary negative charges are distributed uniformly in a globular protein of radius $R$. What is the formula for the electrostatic energy as a consequence of the interaction of the charges?

**Solution**   The total electrostatic energy of the protein is the sum of the pairwise interaction energies of the point charges. The electrostatic (Coulomb) energy of the charge pair $(i, j)$ is $e^2/4\pi\varepsilon_0 r_{ij}$, where $e$ denotes the negative elementary charge, $\varepsilon_0$ is the dielectric constant in a vacuum, and $r_{ij}$ is the distance between the charge pair $(i, j)$. It is assumed that the dielectric properties of the protein matrix are similar to those in vacuum. To calculate the total electrostatic energy of a large number of charge pairs, the mean value of $1/r_{ij}$ in the Coulomb expression must be determined. It can be proved that, if the charges are uniformly distributed in a sphere of radius $R$, the average value of $1/r_{ij}$ is $6/5R$. As there are $Z$ point charges in the sphere, the number of charge pairs is $Z(Z - 1)/2$. The average Coulomb interaction energy of the protein is then

$$W = \tfrac{3}{5} Z(Z - 1) \frac{e^2}{4\pi\varepsilon_0 R}$$

# EXERCISES

**19.1**   Suppose that the electric potential outside a living cell is higher than that inside the cell by 0.075 V. How much work is done by the electric force when a sodium ion (charge $+e$) moves from the outside to the inside?

**19.2**   A potential difference of about 0.090 V exists across a membrane. The thickness of the membrane is $8.5 \times 10^{-9}$ m. What is the magnitude of the electric field in the membrane?

**19.3**   The outer surface of a neuron's axon membrane ($\varepsilon = 5$, thickness $d = 1.2 \times 10^{-8}$ m) is charged positively, and the inner portion is charged negatively. Assuming that an axon can be treated like a parallel plate capacitor with a plate area of $6 \times 10^{-6}$ m$^2$, what is its capacitance?

**19.4**   The surface area of a cell membrane is $A = 5.0 \times 10^{-9}$ m$^2$ and its thickness $d = 1.2 \times 10^{-8}$ m. Assume that the membrane behaves like a parallel plate capacitor and has a dielectric constant $\varepsilon = 5.0$

**(a)** If the potential on the outer surface of the membrane is $+60$ mV greater than that on the inside surface, what is the amount of charge that resides on the outer surface?

**(b)** If the charge in part (a) is due to $K^+$ ions (charge $+e$), how many such ions are present on the outer surface?

**19.5**  What is the dipole–dipole force between two water molecules separated by a distance of 2 Å if one of the two dipoles is perpendicular to the line of separation and one is parallel? Is the force repulsive or attractive? The dipole moment of water is $P = 6.2 \times 10^{-30}$ C · m.

**19.6**  What is the voltage across a 15-nm-thick membrane of a cell if the electric field across it is $5 \times 10^6$ V/m?

**19.7**  A 10-$\mu$F capacitor is used to defibrillate the heart. It is charged to 10 V. Calculate the amount of energy and charge stored in it. If this energy is released over 10 ms, what is the power output of this defibrillator?

**19.8**  What is the membrane potential for $K^+$ ions given the concentrations of $K^+$ ions to be 120 mol/m$^3$ inside the cell and 10 mol/m$^3$ outside?

**19.9**  Show that the membrane potential for $Na^+$ ions is $+59$ mV, given the concentrations of $Na^+$ ions to be 15 mol/m$^3$ inside the cell and 140 mol/m$^3$ outside.

# ELECTRIC CIRCUITS

## PHYSICAL BACKGROUND

The rate of flow of charge in a conductor is called the electric current. If the rate is constant, the current $I$ is given by $I = \Delta q/\Delta t$, where $\Delta q$ is the magnitude of the charge crossing a surface in a time $\Delta t$, the surface being perpendicular to the motion of the charge, The SI unit for current is the coulomb per second (C/s), which is referred to as an ampere (A). When the charges flow only in one direction around a circuit, the current is called direct current (dc). When the direction of charge flow changes from moment to moment, the current is known as alternating current (ac). The definition of electrical resistance is

$$R = \frac{V}{I} \tag{20.1}$$

where $V$ is the voltage applied across a piece of material and $I$ is the current through the material. If the ratio $V/I$ is constant for all values of $V$ and $I$, the relation $R = V/I$ or $V = IR$ is referred to as Ohm's law. Resistance is measured in volts per ampere, a unit called an ohm ($\Omega$).

The resistance of a conductor depends on the resistivity $\rho$ unique to a material, its length $L$, and cross-sectional area $A$, according to

$$R = \rho\frac{L}{A} \tag{20.2}$$

Resistivity is given in units of ohm $\cdot$ meter ($\Omega \cdot$ m). The resistivity of a material depends on the temperature. For many materials and limited temperature ranges, the temperature dependence is given by

$$\rho = \rho_0[1 + \alpha(T - T_0)] \tag{20.3}$$

where $\rho$ and $\rho_0$ are the resistivities at temperatures $T$ and $T_0$, respectively, and $\alpha$ is the temperature coefficient of resistivity. We can rewrite the equation for resistance as an inverse relationship:

$$G = R^{-1} = \sigma\frac{A}{L} \tag{20.4}$$

The proportionality constant $\sigma$ incorporates the effect of different materials; $\sigma$ is called the conductivity of the material. Materials with a high conductivity will have a large conductance for given $A$ and $L$; they are good conductors. Since $\sigma = GL/A$, the SI unit of conductivity $\sigma$ is siemen per meter (1 S = 1 $\Omega^{-1}$). Some typical conductivities are given in Table 20.1.

As current flows through a resistor, electric energy or power $P$ is dissipated into the resistor according to the following equivalent expressions:

$$P = IV = I^2R = \frac{V^2}{R} \tag{20.5}$$

**TABLE 20.1  Typical Conductivities**

| Material | $\sigma$ (S/m) | Characteristic |
|---|---|---|
| Copper | $5.9 \times 10^7$ | Good conductor |
| Germanium | 2.2 | Fair conductor |
| Body fluids | 0.5–5 | Fair conductor |
| Glass | $10^{-12}$ | Good insulator |

In a dc circuit, electricity generated by a voltage source flows as current through an arrangement of circuit components connected by a conductor or a material that provides minimal resistance to the flow of electrons. In contrast to conductors, insulators are materials that permit minimal current flow. These circuit elements may be connected in series (see Figure 20.1) or in parallel (see Figure 20.2). In a series circuit, two or more circuit elements are connected in a direct line or sequence. In a parallel circuit, two or more circuit elements are connected in a branching arrangement. Current that flows through one circuit element connected in series must also flow through the remaining elements connected in the series circuit. Therefore,

$$I = I_1 = I_2 = I_3 = \cdots = I_n \tag{20.6}$$

where $n$ refers to the $n$th element in the series circuit. The equivalent resistance $R_S$ of a series combination of resistances ($R_1, R_2, R_3, \ldots$) is

$$R_S = R_1 + R_2 + R_3 + \cdots \tag{20.7}$$

The equivalent resistance dissipates the same total power as the series combination.

When devices are connected in parallel, the same voltage is applied across each device. In general, devices wired in parallel carry different currents. The reciprocal of the equivalent resistance $R_P$ of a parallel combination of resistances is

$$\frac{1}{R_P} = \frac{1}{R_1} + \frac{1}{R_2} + \frac{1}{R_3} + \cdots \tag{20.8}$$

Kirchhoff's junction rule states that the sum of the magnitudes of the currents directed into a junction equals the sum of the magnitudes of the currents directed out of the junction. Kirchhoff's loop rule states that around any closed circuit loop the sum of the potential drops equals the sum of the potential rises.

The equivalent capacitance $C_P$ for a parallel combination of capacitances ($C_1, C_2, C_3, \ldots$) is

$$C_P = C_1 + C_2 + C_3 + \cdots \tag{20.9}$$

In general, each capacitor in a parallel combination carries a different amount of charge. The equivalent capacitor carries the same total charge and stores the same total energy as the parallel combination.

**FIGURE 20.1**  Resistors in series.

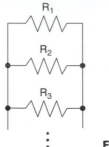

**FIGURE 20.2**   Resistors in parallel.

The reciprocal of the equivalent capacitance $C_S$ for a series combination of capacitances is

$$\frac{1}{C_S} = \frac{1}{C_1} + \frac{1}{C_2} + \frac{1}{C_3} + \cdots \tag{20.10}$$

Each capacitor in the combination carries the same amount of charge. The equivalent capacitor carries the same amount of charge as any one of the capacitors in the combination and stores the same total energy as the entire combination.

The charging or discharging of a capacitor in a dc series circuit (resistance $R$, capacitance $C$) does not occur instantaneously. Charge builds up gradually according to the relation

$$q = q_0(1 - e^{-t/RC}) \tag{20.11}$$

where $q$ is the charge on the capacitor at time $t$, $q_0$ is the equilibrium value of the charge, and $\tau = RC$ is the time constant of the circuit. The discharging of a capacitor through a resistor is described by

$$q = q_0 e^{-t/RC} \tag{20.12}$$

where $q_0$ is the charge on the capacitor at time $t = 0$.

## TOPIC 20.1   Electrical Current through Electrolytes

Electrical current in organisms is generally not carried by electrons. Instead, the electrical current is transported by the mobile ions of electrolytic solutions. It turns out that the relationship between the electromotive force $E$ and the electrolytic current $I$ is

$$E = IR \tag{20.13}$$

that is, Ohm's law remains valid for electrolytic conduction. Just as for metallic conductors, the resistance depends on the dimensions of the electrolytic cell. If two poles of the electrolytic cell are in the form of two parallel plates spaced a distance $L$ apart, then the resistance is again given by $R = \rho L/A$ with $A$ the cross section of the plates, $\rho$ the resistivity of the electrolytic cell, and the conductance $\sigma = 1/\rho$. The typical order of magnitude of the resistivity for body fluids is about $1\ \Omega \cdot m$. This is nine orders of magnitude larger than the resistivity of copper, so electrical conduction by ions is less effective than electrical conduction by electrons.

**TABLE 20.2  Molar Conductance at Infinite Dilution of Various Salt Solutions**

| Salt | $\Lambda_0$ [$(\Omega \cdot m)^{-1} M^{-1}$] |
| --- | --- |
| NaCl | 12.8 |
| KCl | 14.9 |
| $NaNO_3$ | 12.3 |
| $KNO_3$ | 14.5 |
| NaOH | 24.6 |
| KOH | 27.1 |

The conductivity $\sigma$ of the solution as a function of salt concentration $c$ obeys the relationship

$$\sigma(c) = \Lambda_0 c - k_c c^{3/2} + \cdots \qquad (20.14)$$

with both $\Lambda_0$ and $k_c$ positive and independent of concentration. The conductivity of an electrolytic solution is proportional to the ionic concentration for low salt concentrations. For a given temperature, the constant $\Lambda_0$, called the *molar conductance at infinite dilution,* only depends on the kind of salt used. Table 20.2 gives the molar conductance at infinite dilution. Here, the concentration is expressed in molars. According to the Kohlrausch law, the molar conductance of a salt is the sum of the conductivities of the ions comprising the salt. Table 20.3 lists the values of $\Lambda_0$ for several key ions.

For electrolytic conduction the electric force, $\mathbf{F} = q\mathbf{E}$, applied to the ion is balanced by the Stokes friction $\mathbf{F}_H = 6\pi\eta r \mathbf{v}$. Consequently, velocity can be computed as

$$\mathbf{v} = \mu q \mathbf{E} \qquad (20.15)$$

where $\mu$ is given by

$$\mu = \frac{1}{6\pi\eta r} \qquad (20.16)$$

and is called the *electrophoretic mobility.* Since the current passing through a cross-sectional area $A$ is proportional to $q$, $A$, $c$, and $v$, where $c$ is the ionic concentration, we obtain, combining positive and negative ionic mobilities $\mu^+$ and $\mu^-$,

$$I = Ac(\mu^+ + \mu^-)eE = \frac{A}{L}c(\mu^+ + \mu^-)e \,\Delta V \qquad (20.17)$$

**TABLE 20.3  Values of $\Lambda_0$ for Various Ions**

| Ion | $\Lambda_0$ [$(\Omega \cdot m)^{-1} M^{-1}$] |
| --- | --- |
| $H^+$ | 34.9 |
| $OH^-$ | 19.8 |
| $Na^+$ | 5.0 |
| $Cl^-$ | 7.6 |
| $K^+$ | 7.4 |

which is just another version of Ohm's law. Furthermore, we find that the molar conductance at low dilution is an additive quantity since:

$$\Lambda_0 = e(\mu^+ + \mu^-) \tag{20.18}$$

## TOPIC 20.2 Electrical Signal Transmission through Nerves

The human body is able to interpret, receive, and respond to sensory stimuli by the transmission of electric signals through the nervous system along nerve fibers of networks of individual cells. These latter are known as neurons. Figure 20.3 illustrates the structure of a neuron. It consists of a cell body with a nucleus connected to dendrites which serve as the input for a signal. A long conducting tail, which is attached to the neuron, is called an axon and propagates the electrical signal or impulse away from the neuron. The axon is made up of a series of connected segments about 2 mm in length and 20 $\mu$m in diameter. The segments of the axon are encased in a myelin sheath which acts as a good insulator. Segments of the axon are separated by small uninsulated gaps known as the Ranvier nodes. Nerve endings branch out at the end of the axon, and these transmit the signals across the small gaps to other neurons. Electrical signals travel along an axon and communicate with other neurons and so on, until they are eventually registered in the brain. When the electrical signals travel through the myelinated fibers of the axon, their propagation velocity ranges from 1 to 100 m/s. Simultaneously each nerve can accommodate many signals in an analogous way as a number of telephone calls transmitted along a telephone cable.

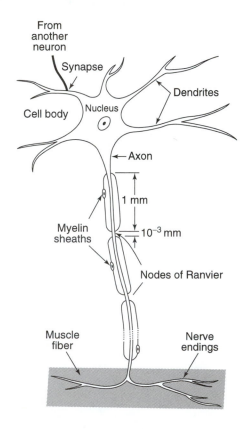

**FIGURE 20.3** Neuron and its structural elements.

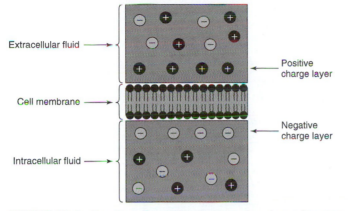

**FIGURE 20.4** Forming of positive and negative charge layers on the outside and inside surfaces, respectively, of a membrane during its resting state.

While the extracellular fluid is rich in sodium ions ($Na^+$) and $Cl^-$ ions, the intracellular fluid is rich in potassium ($K^+$) ions. The fluid inside the cell also contains negatively charged proteins. We have seen that a living cell has a selective permeable membrane and we have introduced the concept of a resting membrane potential. As a result of the selective membrane permeability, there is a small buildup of negative charges just inside the membrane and an equal amount of positive charges on the outside (see Figure 20.4). In neurons this ranges from $-40$ to $-90$ mV, it is typically $-70$ mV, and the minus sign indicates that the inside of the membrane is negative relative to the outside. When a neuron is in its "resting" state, it is not conducting an electrical signal. It is the change in the resting potential which is so important in the initiation and conduction of a signal. "Gates" in the membrane open when a sufficiently strong stimulus is applied to a given point on the neuron and sodium ions rush into the cell. This rush of $Na^+$ ions is illustrated in Figure 20.5.

The $Na^+$ ions are driven into the cell by attraction to the negative ions on the inside of the cell as well as by the relatively high concentration of positive sodium ions outside the cell. This large influx of $Na^+$ ions first neutralizes the negative ions on the interior of the membrane and then cause it to become positively charged. For a very short time, as a result, the membrane potential in this localized region goes from about $-70$ mV (the resting potential) to about $+30$ mV (see Figure 20.6). This change in potential, from $-70$ to $+30$ mV and back to $-70$ mV is known as the action potential. The action potential lasts for a few milliseconds. It is the electrical signal which propagates down the axon, at speeds between 0.5 and 130 m/s, to the next neuron or to a muscle cell.

The axon may be up to 1 m long and its diameter varies between 0.1 and 20 $\mu$m with the giant axon of the squid having the greatest diameter, up to 1 mm. The more prim-

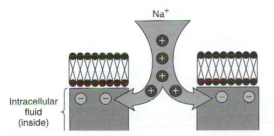

**FIGURE 20.5** As a stimulus is applied to the cell, sodium ions rush into the cell.

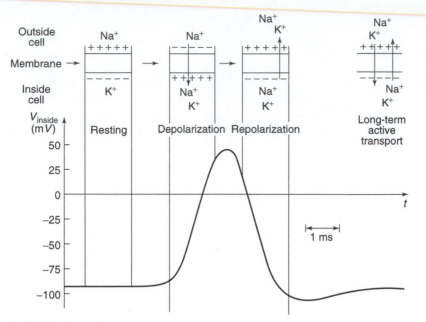

**FIGURE 20.6** Profile of an action potential at a fixed location on a nerve as a function of time.

itive unmyelinated axon such as those found in invertebrates is a cylindrical pipe whose walls consist of the cell membrane. The interior fluid is the *axoplasm*, whose ionic composition differs from that of the extracellular fluid surrounding the cable. Both these fluids are electric conductors whose resistivity is typically of the order of $\rho \cong 60 \ \Omega \cdot \text{cm}$. The cable wall forms a leaky capacitor with a typical value of its capacitance $C = 10^{-6} \ \text{F/cm}^2$ and a finite conductivity $\sigma_M = \rho_M^{-1}$, where $\rho_M$ is the membrane resistivity; typically, $\rho_M = 2 \times 10^3 \ \Omega \cdot \text{cm}^2$. The ability of ions to pass through a membrane depends, of course, on the size, length, shape, and chemical properties of any pores present in the membrane. All of this is lumped into a single parameter, average conductance of a pore, $g_P$, such that the membrane conductivity $\sigma_M$ is given by

$$\sigma_M = n g_P \tag{20.19}$$

where $n$ is the number of pores per unit area. For example, taking the radius of a squid axon as $r = 0.25 \times 10^{-3}$ m and the typical conductivity as $\sigma_M = 0.5 \times 10^{-3} \ \Omega^{-1} \cdot \text{cm}^{-2}$, we obtain the value of membrane conductance as $\sigma_M$ times the surface area of 1 cm of the axon's length; that is, $A = 2\pi r \ell = 1.6 \times 10^{-5} \ \text{m}^2$, and hence

$$g_M = \sigma_M A = 0.8 \times 10^{-4} \ \Omega^{-1} \tag{20.20}$$

Conversely, the *resistance* of 1 cm of a squid axon is found to be

$$R_M = g_M^{-1} = 12.5 \ \text{k}\Omega \tag{20.21}$$

With a potential difference $V = 80$ mV we find the current flow through 1 cm of the length of an axon to be

$$I = g_M V = (80 \times 10^{-3} \ \text{V})(0.8 \times 10^{-4} \ \Omega^{-1}) = 0.64 \times 10^{-5} \ \text{A} \tag{20.22}$$

which translates into a flow of approximately $0.4 \times 10^{11}$ elementary charges during the 1-ms duration of an *action potential* that travels at speeds on the order of 10 m/s in this case.

## TOPIC 20.3   Conduction across a Synapse: A Biological Computer Chip?

A synapse is the gap between one neuron and another or between a nerve terminal and a muscle fiber across which electrical signals pass in only one direction. Conduction across a synapse therefore cannot be the same as that along the neuron, in which the spike potential travels in both directions from the point of stimulation. It has been shown that, in the case of the crayfish, synaptic conduction is purely electrical. In most other cases of synaptic conduction studied, the transmission is chemical involving *neurotransmitter molecules* such as acetylcholine (ACh).

The end of the presynaptic fiber contains a dense accumulation of vesicles, each about 50 nm in diameter. The nerve impulse, on arriving at the synapse, causes the vesicles to move to the presynaptic nerve membrane, fuse with it, and eject the neurotransmitters they contain into the gap. On the order of $10^7$ molecules are released, which cross the gap by reaching the end of the postsynaptic receptor in about 1 ms. The gap is about 20 nm wide between neurons but perhaps four times larger between a neuron and a muscle fiber. Assuming that a given neurotransmitter has a net charge of several elementary charges, the current flow across the synapse is of the order of $10^{-8}$ A.

The neurotransmitter molecules are adsorbed once they cross the synapse and change the ion permeability of the membrane in the second fiber, causing a spike potential to start along it. The adsorbed molecules are destroyed by an enzyme; in the case of ACh this is acetyl-cholinesterase. If the enzyme is not present, the transmitter substance keeps on working, and the second fiber cannot receive a second impulse.

It is interesting to note that a synapse may act in the simple manner described above or it may perform more complicated "logical" operations such as adding, subtracting, dividing, and multiplying. "Addition" is performed if the synapse transmits only subthreshold responses, several therefore being necessary to produce a spike potential. Although ACh produces a spike potential at motor nerve endings, it inhibits heart muscle and is therefore "subtracting." At other synapses several arriving potentials may be necessary to produce a single response, and the synapse "divides." An impulse in one fiber may give rise to several impulses in a further fiber, and the original spike potential may be considered to be "multiplied." Since the response at a synapse may show a very complex time dependence, processes analogous to other mathematical operations can also take place. Indeed, the processes at synapses are essentially nonlinear and highly complex and display the logic circuitry of a modern computer.

## TOPIC 20.4   Resistance in the Human Body

The tissues and fluids beneath the skin conduct electricity almost as well as metals, and the conduction is principally ionic. If one attempts to measure the body's resistance by placing two electrodes at separate locations on the body, the resistance between them may appear large because of poor contact between the electrodes and the skin. This contact resistance is often reduced by placing a special conducting gel between the electrodes and the skin. Skin resistance does, in fact, vary markedly over the human body because the response is principally due to the sympathetic nervous system. When one electrode is fixed on one part of the body (e.g., the leg), and the other is moved over the body when a 50- to 100-V cell is connected between them, the reciprocal of the current, for a particular location of electrodes, is a measure of the resistance between them. For a normal subject the resistance varies in a predictable way. However, if the nerves in any region are damaged or if there is a carcinoma or other tumor compressing nerves, resistance increases appreciably near this position. For example, this can be very marked in the chest when lung cancer is present. Resistance measurements have proved very effective as a simple and rapid means of detecting the disease. The magnitude of the effect is an indication of the state of advancement of the cancer. The technique has also been useful in the diagnosis and location of regional nerve lesions and peripheral vascular disease.

## TOPIC 20.5 The Electrical Origin of the Heartbeat

The human heartbeat is synchronized and regulated by electrical impulses. These electrical stimuli are generated by unique muscle cells in the right atrium of the heart which act as a pacemaker for the heart or sinoatrial node. This electrical impulse is transferred over the atria to the atrioventricular node in the base of the right ventricle. Then it is relayed to the muscular ventricles by specialized conducting fibers called Purkinje fibers. The heart beats or contracts almost as a single entity or mass, assuming normal conditions. Figure 20.7 shows the location of the sinoatrial node and the atrioventricular node.

In normal action the heart pumps blood at about 5 L/min (see Chapter 11 for details). The purpose of keeping blood circulating is manifold. Carbon dioxide from the tissues is transported to the lungs for removal, and oxygen from the lungs is returned to the tissues. Food and metabolic products are transported to and from various sites, and endocrine secretions are transported from the ductless glands to the organs they control. Finally, the blood carries agents to fight any invading organisms. Because the blood performs so many essential functions, it occupies a key role. If the heart stops beating, the organism cannot survive for long.

The rhythmic pulsations of the heart are initiated at the SA (sinoatrial) node, which puts out about 76 impulses per minute. The spike potentials produced travel over the atria in all directions, canceling out on the opposite side when the two spikes meet. The auricles then contract, forcing blood through valves into the ventricles.

In the resting state a muscle cell maintains a constant resting potential of approximately 90 mV across its cell membrane. However, under the action of a stimulus from

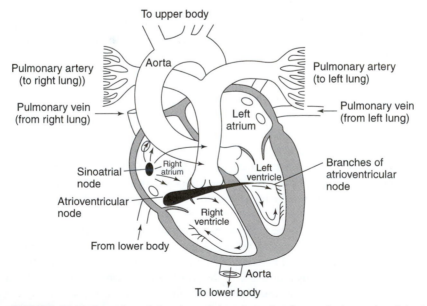

**FIGURE 20.7** Location of the sinoatrial node and atrioventricular node of the heart.

the nervous system, the cell membrane becomes locally and transiently permeable to the positively charged ions on the outside which mesh in depolarizing the cell in the form of a wave that propagates at a speed of 0.1 to 1 m/s. After approximately 250 ms in a de-polarized state, a muscle cell recovers and becomes polarized again. The passage of the depolarizing wave is associated with the contraction of the heart muscle, which is stimu-lated electrically by it. The electrical activity of the heart is controlled by the *pacemaker cells*, which have a characteristic oscillation rate for the resting state depolarization cy-cle. This rate also depends on the signals arriving from the surrounding nerve cells, which may tune the heart rate to the metabolic demands of the body.

## TOPIC 20.6  Health Hazards, Electrical Shock, and Physiological Effects of Current

The human body is a volume conductor whose combined resistance is approximately $1.5 \times 10^3$ $\Omega$ when dry. When the body becomes wet, as water is an excellent conductor, the resistance is reduced to about 500 $\Omega$. If we assume the voltage source is at 120 V, the current through the dry body is, via Ohm's law,

$$I = \frac{\text{voltage}}{\text{resistance}} = \frac{120 \text{ V}}{1500 \text{ }\Omega} = 80 \text{ mA} \tag{20.23}$$

whereas when the body is wet,

$$I = \frac{120 \text{ V}}{500 \text{ }\Omega} = 240 \text{ mA} \tag{20.24}$$

As we see below, ventricular fibrillation can occur at currents of 100 mA, so the harmful effects of exposure to electricity, particularly if the electrical contact is wet, become quite clear.

Electrical currents can be very hazardous, but to reduce the inherent danger when using circuits, proper electrical grounding is necessary. The third prong of a plug connects the metal casing directly to a copper rod driven into the ground or to a copper water pipe that is in the ground. When the dryer malfunctions, a person touching it receives no shock since the current flows through the third prong and into the ground via a copper rod, rather than through the person's body, which provides a much higher electrical resistance than does the metal.

Sometimes fatal or serious injuries can arise from electric shock. The magnitude of the current and the parts of the body through which moving charges pass determine the severity of the injury. A mild tingling sensation is caused by about 1 mA. Currents of or-der 10 to 20 mA can lead to muscular spasms in which a person "cannot let go" of the ob-ject causing the shock. However, currents of approximately 100 to 200 mA can be poten-tially fatal. This is because they can make the heart fibrillate, or beat in an uncontrolled manner, and larger currents stop the heart completely. Ventricular fibrillation prevents the heart from pumping properly, the oxygen supply to the brain is diminished, and permanent brain damage results if the process continues for more than a few minutes. Death by res-piratory paralysis may also result although artificial respiration may revive the victim if the effect has not lasted too long. The danger from static electricity is not so much that it could produce a large current but that it can cause fires or explosions by sparking. Table 20.4 summarizes the physiological effects of electrical shock as a function of the magnitude of the current assuming that an average male is shocked through intact skin for 1 s by 60 Hz ac. (Values for females are 60% to 80% of those listed in the table for male subjects).

**TABLE 20.4   Effects of Electrical Shock as a Function of Current**

| Current (mA) | Effect |
|---|---|
| 1 | Threshold of sensation |
| 5 | Maximum harmless current |
| 10–20 | Onset of sustained muscular contraction; cannot let go for duration of shock; contraction of chest muscles may cause breathing to cease during shock (fatal if continued) |
| 50 | Onset of pain; heart still unaffected |
| 100–300+ | Ventricular fibrillation possible; very often fatal |
| 300 | Onset of burns (thermal hazard); depends on concentration of current |
| 6000 (6 A) | Onset of sustained ventricular contraction and respiratory paralysis; both cease when shock is over; heartbeat often returns to normal |

There are two main methods of protection. First, all wires carrying electricity are surrounded by an insulating layer and, as we mentioned above, the piece of apparatus is generally grounded by joining it to a lead which is connected to a buried metal plate. Second, most electrical circuits include a fuse. This is essentially a piece of metal of low melting point which melts when a preset current passes through it. The current passing to ground is generally large if a fault develops in the circuit and the fuse melts. This will break the circuit and electricity will cease to flow. More modern pieces of apparatus have circuit breakers instead of fuses. When the current in the circuit exceeds a predetermined value, the circuit breaker is subjected to a force which moves it backward. This movement operates a switch and the electricity ceases to flow. The breaker will not go back into its former position as long as the fault in the circuit remains uncorrected.

A lightning bolt is one of the most spectacular examples of a charge transfer to a conducting medium. In violent storms a lightning discharge takes place between a cloud and the surface of the earth. A typical lightning discharge carries a current of about 25,000 A and lasts for a short period of about $4 \times 10^{-5}$ s, during which a usually negative charge $Q = It = 1$ C (or $\approx 0.6 \times 10^{19}$ elementary charges) is transferred to the earth's surface. It is worth noting that the electrical conductivity for wet soil is $\sigma \cong 10^{-2} \, (\Omega \cdot m)^{-1}$ and its dielectric constant $\kappa \cong 20$. It is instructive to calculate the potential difference experienced by a person standing some distance $r$ away from the point of impact where the lightning strikes. With the parameter values given above, it can be shown that with feet approximately 0.5 m apart, a person standing 1 m away from the impact point would experience a potential difference of approximately 20,000 V, while 10 m away only 200 V is felt since $\Delta V$ varies as an inverse square of the distance: $\Delta V \propto r^{-2}$. A voltage difference induces a current flow in the body which, when sufficiently large, can result in burns, or fibrillation of the heart as discussed above in this section. If the skin is not very well insulated from the earth's surface, this voltage difference may cause injury or even death.

# SOLVED PROBLEMS

**20.1**   Seawater has a salt concentration of about 35 g salt/L, predominantly NaCl. The molecular weight of NaCl is 58. What is the conductivity of seawater?

**Solution**   Since 1 mol of NaCl has a mass of 58 g, 1 L of seawater holds about 0.6 mol of NaCl. The conductivity of seawater should thus be roughly $0.6 \times 12.8 = 7.68 \, (\Omega \cdot m)^{-1}$.

**20.2**  Assume that in 3.0 s, $5.0 \times 10^{-14}$ g · m of $Na^+$ ions flow out of a biological cell. What current passes through the membrane? Avogadro's number $N_A = 6.02 \times 10^{23}$. Estimate the value of the current density for a cell whose surface area $A = 200$ $\mu m^2$.

**Solution**  The charge is calculated as

$$\Delta q = 5.0 \times 10^{-14} \text{ g} \cdot \text{mol} \times 6.02 \times 10^{23} \frac{\text{ions}}{\text{g} \cdot \text{mol}} \times \frac{1.60 \times 10^{-19} \text{ C}}{\text{ion}} = 4.82 \times 10^{-9} \text{ C}$$

The current is found to be

$$I = \frac{\Delta q}{\Delta t} = \frac{4.82 \times 10^{-9} \text{ C}}{3.0 \text{ s}} = 1.6 \times 10^{-9} \text{ A}$$

Since the above cell has a total surface or membrane area $A = 200$ $\mu m^2$, the current density through the membrane is

$$i = \frac{I}{A} = \frac{1.6 \times 10^{-9} \text{ A}}{200 \text{ } \mu m^2} = 8 \text{ A/m}^2$$

**20.3**  A pacemaker applies 72 stimulating square-pulses per minute to the heart. The pulse length $\tau = 0.5$ ms, and the amplitude $U = 5$ V. Two electrodes convey the pulses from the pulse generator to the heart wall. The resistance of the heart tissue between the connecting electrodes is $R = 600$ $\Omega$.

(a) What is the power of one pulse?

(b) What is the energy of one pulse?

(c) What is the average power of the pacemaker?

(d) For how many years can a pacemaker work if the total energy $W_1 = 15$ kJ of its power supply is used with an efficiency of 35%?

**Solution**

(a) The power of one pulse is

$$P = \frac{U^2}{R} = 41.7 \text{ mW}$$

(b) The energy of one pulse is

$$W_1 = P\tau = 20.8 \text{ } \mu J$$

(c) The average power $\overline{P}$ is given as the ratio of the energy of one pulse and the time period $T$ of the pulse series, which is $T = 60$ s/72 $= 0.833$ s. Thus,

$$\overline{P} = \frac{W_1}{T} = 25 \text{ } \mu W$$

(d) The energy $W$ available from the power supply is 35% of its total energy $W_t$, which means that

$$W = 0.35W_t = 5.25 \text{ kJ}$$

The pacemaker uses this energy during the time $t$, which can be calculated from the average power $\overline{P}$ as

$$t = \frac{W}{\overline{P}} = 2.1 \times 10^8 \text{ s} = 6.65 \text{ years}$$

**20.4** Following an electric shock and a subsequent fibrillation of the heart, an appropriate electric shock causing simultaneous contraction of all heart muscles can restore the function of the heart. This procedure utilizes a charged capacitor, which is discharged through the chest of the patient.

(a) What is the energy of the defibrillating pulse if the capacitance of the capacitor is $C = 30 \ \mu F$ and it is charged to a voltage of $U_0 = 5$ kV?

(b) What is the resistance of the electric circuit (capacitor–electrodes–body tissues) formed during defibrillation? The diameter of the large electrodes pressed onto the chest is $d = 10$ cm. The resistivity of the skin as a layer is $\rho^* = 0.2 \ \Omega \cdot m^2$, and the resistances of all other tissues and the connecting wires are negligible.

(c) What is the current at the beginning of the defibrillation pulse?

**Solution**

(a) The energy of the defibrillating pulse is equal to the total energy stored in the capacitor:

$$W_0 = \tfrac{1}{2}CU_0^2 = 375 \text{ J}$$

(b) The contact surface of one electrode is

$$A = \tfrac{1}{4}d^2\pi = 7.85 \times 10^{-3} \text{ m}^2$$

Thus, the resistance of the skin under one electrode is

$$R_1 = \frac{\rho^*}{A} = 25.5 \ \Omega$$

The total resistance of the circuit is the sum of the two resistances of the skin under the individual electrodes connected in series:

$$R = 2R_1 = 51 \ \Omega$$

(c) The current at the beginning of discharge is

$$I_0 = \frac{U_0}{R} = 98 \text{ A}$$

# EXERCISES

**20.1** A digital thermometer uses a thermistor as the temperature-sensing element. A thermistor is a kind of semiconductor and has a large negative temperature coefficient of resistivity $\alpha$. Suppose $\alpha = -0.055°C^{-1}$ for the thermistor in a digital thermometer used to measure the temperature of a patient. The resistance of the thermistor decreases to 80% of its value at the normal body temperature of 37.0°C. What is the patient's temperature?

**20.2** The current through a cardiac pacemaker is 0.015 A. Determine the number of electrons that flow during (a) 5 min and (b) 5 hours of operation.

**20.3** Below is a table of the ion concentrations of the most common ions in the cell interior and in the blood. Compute the conductivity of the cell interior and that of blood. Compare with seawater.

| Type of Ion | Ion Concentration in Cell Interior (mM) | Blood (mM) |
|---|---|---|
| $K^+$ | 140 | 4 |
| $Na^+$ | 10 | 145 |
| $Cl^-$ | 5 | 116 |

**20.4** Estimate the molar conductance of the $Cl^-$ ion using

$$\mu = \frac{1}{6\pi\eta r} \quad \text{and} \quad \Lambda_0 = e(\mu^+ + \mu^-)$$

Assume that the spherical ion is surrounded by a "hydration shell" of water molecules. Use the van der Waals radius for the radius of the $Cl^-$ ion and the dimensions of the water molecules.

**20.5** Assume that one nerve impulse must end before a new one can begin. Estimate the maximum firing rate of a nerve cell, expressing it in impulses per second.

**20.6** During open-heart surgery, a small current of 25 $\mu$A may cause ventricular fibrillation. Assuming the heart resistance to be 250 $\Omega$, calculate the voltage level posing a hazard to the patient's health.

**20.7** A heart defibrillator passes 12 A of current through a patient's torso for $3 \times 10^{-3}$ s to restore normal beating of the heart. (a) How much charge passed through the patient's body? (b) What voltage was used if a total energy of 300 J was dissipated by the current? (c) What was the resistance of the path through the person?

**20.8** Suppose a physician who is well insulated from the ground touches an appliance that has shorted to its hot wire and has a voltage of 120 V ac. She simultaneously touches the pacemaker lead of a patient who is grounded. Her resistance is 100,000 $\Omega$ and that of the patient is 1000 $\Omega$. Calculate the current through the two people assuming they are in series and identify the likely effect on each.

# CHAPTER 21

# MAGNETIC FORCES AND MAGNETIC FIELDS

## PHYSICAL BACKGROUND

A magnetic field **B** can be generated by a moving charged particle. Like magnetic poles repel and unlike magnetic poles are attractive. A standard bar magnet can be thought of as made up of a positive north pole where the lines of force begin and a south pole where the lines of force end. Magnetic field is a vector quantity. The SI unit of magnetic field is the tesla (T) or the newton per ampere-meter (N/A · m). Smaller magnetic fields are measured in gauss (G), where $1\text{ T} = 10^4$ G.

A magnetic field exists in the space around a magnet. The magnitude $B$ of the magnetic field at any point in space is defined as

$$B = \frac{F}{q_0 v \sin \theta} \tag{21.1}$$

where $F$ is the magnitude of the magnetic force that acts on a charge $q_0$, whose velocity **v** makes an angle $\theta$ with respect to the magnetic field. The direction of the magnetic force is perpendicular to both **v** and **B**, and for a positive charge the direction can be determined with the aid of the right-hand rule. The magnetic force on a moving negative charge is opposite to the force on a moving positive charge.

A magnetic force **F** exerted on a charged particle $q$ moving with a velocity **v** in a uniform magnetic field **B** is defined as

$$\mathbf{F} = q\mathbf{v} \times \mathbf{B} = q|\mathbf{v}||\mathbf{B}| \sin \theta \ \mathbf{n} \tag{21.2}$$

where **n** is the unit vector perpendicular to the plane of **v** and **B**, $\theta$ is the angle between the lines of the magnetic field **B**, and the direction of the velocity of the charged particle **v**.

If a particle of charge $q$ and mass $m$ moves with a speed $v$ perpendicular to a uniform magnetic field $B$, the magnetic force causes the charge to move on a circular path of radius

$$r = \frac{mv}{qB} \tag{21.3}$$

A constant magnetic force does not work on the charged particle, because the direction of the force is always perpendicular to the motion of the particle. Being unable to do work, the magnetic force cannot change the kinetic energy of the particle; however, the magnetic force does change the direction in which the particle moves.

An electric current, being composed of moving charges, can experience a magnetic force when placed in a magnetic field. For a straight wire that has a length $L$ and carries a current $I$, the magnetic force has a magnitude of

$$F = ILB \sin \theta \tag{21.4}$$

where $\theta$ is the angle between the directions of $I$ and $B$. The direction of the force is perpendicular to both $I$ and $B$.

Magnetic forces can exert a torque on a current-carrying loop of wire and thus cause the loop to rotate. If a current $I$ exists in a coil of wire with $N$ turns, each of area $A$, in the presence of a magnetic field $B$, the coil experiences a torque of magnitude

$$\tau = NIAB \sin \phi \qquad (21.5)$$

where $\phi$ is the angle between the direction of the magnetic field and the normal to the plane of the coil.

The magnetic field $B$ generated by a particle of charge $q$ moving at a speed $v$ is

$$B = \frac{\mu}{4\pi} \frac{qv \sin \theta}{r^2} \qquad (21.6)$$

where $r$ is the distance from the moving charge to the point where the magnetic field is to be calculated and $\mu$ is the permeability of the medium in which the magnetic field exists. In free space, $\mu_0 = 4\pi \times 10^{-7}$ T $\cdot$ m/A $= 1.257 \times 10^{-6}$ T $\cdot$ m/A.

An electric current produces a magnetic field, with different current geometries giving rise to different field patterns. For a long, straight wire, the magnetic field lines are circles centered on the wire. The magnitude of the field at a radial distance $r$ from the wire is

$$B = \frac{\mu_0 I}{2\pi r} \qquad (21.7)$$

Thus, the magnetic field at the center of a flat circular coil consisting of $N$ turns, each of radius $R$, is

$$B = \frac{N\mu_0 I}{2R} \qquad (21.8)$$

The coil has associated with it a north pole on one side and a south pole on the other side. A solenoid is a coil of wire wound in the shape of a helix. Inside a long solenoid the magnetic field is nearly constant and has the value

$$B = \mu_0 nI \qquad (21.9)$$

where $n$ is the number of turns per unit length of the solenoid. One end of a solenoid behaves like a north pole, and the other end like a south pole.

Ampere's law states that

$$\sum B_{\parallel} \Delta l = \mu_0 I \qquad (21.10)$$

where $\Delta l$ is a small segment of length along a closed path of arbitrary shape around the current $I$, $B_{\parallel}$ is the component of the magnetic field parallel to $\Delta l$, and it specifies the relationship between a current and its associated magnetic field. The Hall effect is the creation of voltage $\varepsilon$ across a current-carrying conductor by a magnetic field $B$. The Hall electromagnetic force (emf) is given by

$$\varepsilon = Bvl \qquad (21.11)$$

for a conductor of length $l$ with a charge moving at a velocity $v$.

Ferromagnetic materials such as iron are made up of tiny regions called domains, each of which behaves as a small magnet. In an unmagnetized ferromagnetic material, the domains are randomly aligned. In a permanent magnet, many of the domains are aligned, and a high degree of magnetism results. An unmagnetized ferromagnetic material can be induced into becoming magnetized by placing it in an external magnetic field.

## TOPIC 21.1 Nuclear Magnetic Resonance and Magnetic Resonance Imaging

A diagnostic technique utilizing magnetic fields exploits an effect called nuclear magnetic resonance (NMR) and is called magnetic resonance imaging (MRI). The nuclei of some atoms have small magnetic fields because of their spins (consult Chapter 29 for details), just as electrons do. Usually these spins are randomly oriented, but when placed in a strong magnetic field, they align themselves with that field. The directions in which these tiny magnets point can be altered by sending in a radio signal. By measuring the amount of radio waves absorbed and reemitted, it is possible to measure the location and abundance of certain elements. It is a type of resonance since only certain frequencies of radio waves work, depending on the type of nucleus and the strength of the magnetic field—hence the name NMR. One element that can easily be detected using NMR is hydrogen, which is found in great abundance throughout the body.

Unlike X-ray imaging, MRI does not use potentially harmful ionizing radiation. It exploits the behavior of protons, which are the nuclei of hydrogen atoms, when subjected to a very strong magnetic field and radio waves. The patient in MRI lies inside a large, hollow, cylindrical magnet in which the body is exposed to a magnetic field which is many times more powerful than that of the earth. The effective magnetic field acting on nuclei is slightly different in different locations in the body. The protons, whose spins normally point randomly in different directions, under the influence of the scanner's magnetic field, line up parallel to each other. A strong pulse of radio waves is then applied to knock out the protons from alignment. As the protons realign themselves, they produce radio signals characteristic of their location which can be picked up by radio receiver coils in the scanner. Computer processing of these radio signals then converts them into a two- or three-dimensional image based on the strength and location of the signals. In the body the most important sources of protons are the hydrogen atoms in water molecules. An MRI scan therefore reflects differences in the water content of tissues. The MRI can give a detailed picture of the structure of soft tissue because of the differences in water content within them. The MRI scans are extremely useful for studying the brain and spinal cord. A more recent application of MRI—called magnetic resonance spectroscopy—can detect other chemical elements like phosphorus and calcium. Figure 21.1 illustrates the principles of the scanning process.

## TOPIC 21.2 Biomagnetism

It is generally assumed in biology that migratory birds are able to orient themselves with respect to the earth's magnetic field. Experiments were conducted in which birds were captured and placed in cages surrounded by large coils which produced magnetic fields. These birds changed their direction of migration according to the field applied by the experimentalists. There is, however, no authoritative explanation to date of the physical mechanism of magnetic field detection by migratory birds as no ferromagnetic material has been found in their organs that could play the role of a compass. However, an explanation can be furnished on the basis of the law of electromagnetic induction. In particular, a potential difference $V$ is established between any two points on an object moving at a velocity $v$ through a magnetic field whose magnetic induction is $B$. The potential differ-

**The Scanning Process**
An MRI scanner consists of a powerful electromagnet, a radio-wave emitter, and a radio-wave detector. A plane of the body is selected for imaging and the electromagnet is turned on.

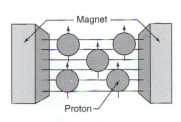
Magnet

Proton

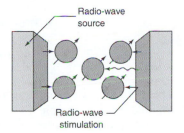
Radio-wave source

Radio-wave stimulation

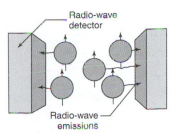
Radio-wave detector

Radio-wave emissions

1. Normally, the protons (nuclei) of the body's hydrogen atoms point randomly in different directions, but under the influence of the scanner's powerful magnetic field, they align themselves in the same direction.

2. Next, the radio-wave source emits a powerful pulse of radio waves, the effect of which is to knock the protons out of alignment.

3. However, milliseconds later, protons realign themselves, emitting faint radio signals as they do. These signals are picked up by the scanner's radio-wave detector.

**FIGURE 21.1**   Scanning process in an MRI scan.

ence $V$ is proportional to the product of $B$, $v$, and $l$, the distance separating these two points, such that

$$V = Bvl \qquad (21.12)$$

Taking $B = 0.4$ G $= 0.4 \cdot 10^{-4}$ T, $v = 10$ m/s, and $l = 2$ cm representing the small size of the bird's head, we obtain $V = 8.10^{-6}$ V, a small but not negligible value. There exist species of fish called electric fish which are capable of detecting an even smaller potential difference. Looking at these effects from the point of view of forces being generated on current flows in the brain's axons, we find using

$$F = B \cdot I \cdot l \qquad (21.13)$$

with $I$ of the order of $10^{-7}$ A for a single axon and $l$ on the order of 1 cm that the force is only a fraction of a piconewton ($10^{-12}$ N)—too small to make an effect. However, groups of hundreds of axons would experience a combined effect of tens of piconewtons. This topic is still under active investigation and remains a scientific puzzle even today.

# TOPIC 21.3   Magnetotactic Bacteria

Richard Blakemore, a graduate student in 1975, discovered that certain anaerobic bacteria (e.g., *Aquaspirillum magnetotacticum*) tended to swim toward the geographical north in his laboratory. He found that the body axis of the bacteria and swimming direction would align with a strong magnetic field which he applied. This behavior could not be due to some complex mechanism within or on each bacterium since dead bacteria also

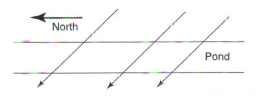
North

Pond

**FIGURE 21.2**   Magnetic field lines pointing downward in a pond in the Northern Hemisphere.

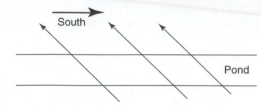

South

Pond

**FIGURE 21.3** Magnetic field lines pointing upward in a pond in the Southern Hemisphere.

aligned with the magnetic field. These bacteria possess a magnetic moment so each bacterium has a magnetic north pole and a south pole. In fact, electron micrographs of these bacteria reveal they have a small number of (roughly 5 to 20) magnetic particles, which are called magnetosomes, having an average diameter of about 500 Å.

The fact that the bacteria in a pond or stream swim to the north is accidental. Magnetotactic bacteria swim along the magnetic field lines. In the Northern Hemisphere these lines are pointing downward the more so the farther north one goes (see Figure 21.2). Hence, by swimming along the magnetic field lines, the bacteria swim downward to food-rich water and so avoid being poisoned by oxygen at the surface of the pond. In the Northern Hemisphere the earth's magnetic field has a downward component. A bacterium is oriented by the earth's field with its north pole pointing downward so it automatically swims to the bottom. If one goes to the Southern Hemisphere and collects the same bacteria, since the earth's magnetic field comes out of the earth's surface in the Southern Hemisphere, they swim toward the south geographical pole in order to swim down; that is, they swim in a direction opposite to the magnetic field lines (see Figure 21.3). In fact, in the Southern Hemisphere the magnetic dipole moment of these magnetotactic bacteria is flipped. If such bacteria were placed in a pond in the Northern Hemisphere, they would tend to swim toward the south pole, and therefore upward relative to the field lines, and die. If this experiment is done, they do die.

These magnetotactic bacteria use just enough magnetic material to overcome thermal noise. The question arises as to why they do not use one large magnetic particle instead of 20 small ones. If this were so, the larger magnetic particles break up into magnetic domains with the magnetic moment in different domains pointing in different directions, leading to cancellation. Why then do they not use 200 magnetosomes of one-tenth the volume? If the particles were made too small, thermal fluctuations would cause the magnetic moment of each particle to wander in different directions. Thus the 50-nm size is about the best size for the magnetic particles, and evolutionary pressure has solved, at a stroke, a very complex physical optimization problem.

# EXERCISES

**21.1** A patient's heart is being scanned by an MRI unit which utilizes the Hall effect. Assume the conducting path on the heart wall to be 0.10 m long and estimate the velocity of the charges to be 0.12 m/s perpendicular to the applied magnetic field of 1.2 T intensity. What is the value of the generated Hall voltage?

**21.2** Blood flow measurements are conducted using a Hall voltage probe. What is the emf value if the velocity of blood is 0.2 m/s, the vessel's diameter is 0.01 m, and the magnetic field intensity applied is 2 T?

# CHAPTER 22

# ELECTROMAGNETIC INDUCTION

## PHYSICAL BACKGROUND

Magnetic induction is the generation of a magnetic field due to the motion of charged particles either in free space or along a conductor. The magnetic flux $\Phi$ that passes through a surface is

$$\Phi = BA \cos \phi \tag{22.1}$$

where $A$ is the area of the surface, $B$ is the magnetic field at the surface, and $\phi$ is the angle between $B$ and the normal to the surface. Faraday's law of electromagnetic induction states that the average electromotive force (emf) $E$ induced in a coil of $N$ loops is

$$E = -N\left(\frac{\Phi - \Phi_0}{t - t_0}\right) = -N\frac{\Delta\Phi}{\Delta t} \tag{22.2}$$

where $\Delta\Phi$ is the change in magnetic flux through one turn and $\Delta t$ is the time interval during which the change occurs. For the special case of a conductor of length $L$ moving with speed $v$ perpendicular to a magnetic field $B$, the induced emf is called motional emf and its value is given by $E = vBL$.

Lenz's law provides a way to determine the polarity of an induced emf. Lenz's law is stated as follows: The induced emf resulting from a changing magnetic flux has a polarity that leads to an induced current whose direction is such that the induced magnetic field opposes the original flux change. This law is a consequence of the law of conservation of energy.

In its simplest form, an electric generator consists of a coil of $N$ turns that rotates in a uniform magnetic field $B$. The emf produced by this generator is

$$E = NAB\omega \sin \omega t = E_0 \sin \omega t \tag{22.3}$$

where $A$ is the area of the coil, $\omega$ is the angular speed (in radians per second) of the coil, and $E_0$ is the peak emf. The angular speed in radians per second is related to the frequency $f$ in cycles per second or hertz according to $\omega = 2\pi f$.

Mutual induction is the effect in which a changing current in the primary coil induces an emf in the secondary coil. The emf $E_2$ induced in the secondary coil by a change in current $\Delta I_1$ in the primary coil is

$$E_2 = -M\frac{\Delta I_1}{\Delta t} \tag{22.4}$$

where $\Delta t$ is the time interval during which the change occurs. The constant $M$ is the mutual inductance between the two coils and is measured in henries.

Self-induction is the effect in which a change in current $\Delta I$ in a coil induces an emf

$$E = -L\frac{\Delta I}{\Delta t} \tag{22.5}$$

in the same coil. The constant $L$ is the self-inductance or inductance of the coil and is measured in henries. To establish a current $I$ in an inductor, work must be done by an external agent. This work is stored as energy in the inductor, the amount being

$$\text{Energy} = \tfrac{1}{2}LI^2 \qquad (22.6)$$

The energy stored in an inductor can be regarded as being stored in its magnetic field. At any point in air or vacuum or in a nonmagnetic material where a magnetic field $B$ exists, the energy density, or the energy stored per unit volume, is

$$\text{Energy density} = \frac{B^2}{2\mu_0} \qquad (22.7)$$

where $\mu_0$ is the permeability of free space.

## TOPIC 22.1   Electromagnetic Flowmeter

An instrument which is used to qualitatively measure flow velocity is the electromagnetic flowmeter. It consists of a magnetic core or ring clamped around a vessel segment which produces a magnetic field. The principle on which it is based is that an electromotive force is induced in a conductor moving so as to cut the lines of force of the magnetic field. Signal electrodes detect the induced emf and are subsequently amplified. The conductor in this case is the blood. It moves through the field in a direction perpendicular to its own axis and to the lines of force. The potential difference in volts, $V$, measured at the ends of the vessel in terms of the average blood flow velocity, $v$, is given as

$$V = Bdv \times 10^{-8} \text{ V} \qquad (22.8)$$

where $B$ is the strength of the magnetic field (in gauss) and $d$ is the internal diameter of the vessel (in centimeters). The volumetric blood flow rate $Q$ is then determined from

$$Q = \frac{\pi d^2}{4} \frac{V}{Bd} \qquad (22.9)$$

The above equation for blood flow is true provided three criteria are met:

(a) The magnetic field is uniform.

(b) The conductor moves in a plane perpendicular to the magnetic field.

(c) The length of the conductor extends at right angles to both the magnetic field and the direction of motion.

Unfortunately, to use an electromagnetic flowmeter to determine accurate blood flow, direct exposure of the vessel concerned is required, limiting clinical applications during surgery.

## SOLVED PROBLEM

**22.1**   An electromagnetic flowmeter (see figure below) applied to an artery of radius 0.35 cm yields a blood flow velocity of 25 cm/s using a magnetic field intensity of 600 G. What is the expected measured voltage?

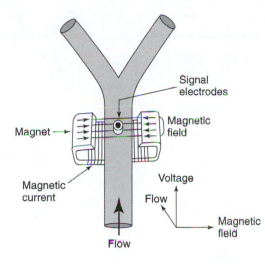

**Solution** From

$$V = Bdv \times 10^{-8} \text{ V}$$

where the diameter $d = 0.7$ cm, corresponding to the radius $r = 0.35$ cm, $B = 600$ G, and $v = 25$ cm/s, the voltage is

$$V = (0.7 \text{ cm})(600 \text{ G})(25 \text{ cm/s}) \times 10^{-8} = 1.05 \cdot 10^{-4} \text{ V}$$

# CHAPTER 23

# *ALTERNATING CURRENT CIRCUITS*

## PHYSICAL BACKGROUND

Voltage in ac circuits is produced by an electric power generator which generates an oscillating potential across its terminals. As a result, the voltage potential is sinusoidal, changing direction and consequently magnitude over time, as shown in Figure 23.1, and can be expressed mathematically by the formula

$$V = V_0 \sin \omega t = V_0 \sin 2\pi ft \qquad (23.1)$$

where $V$ is the instantaneous voltage in volts, $V_0$ is the amplitude or maximum value of the voltage in volts, $\omega$ is the angular velocity in radians per second, $f$ is the frequency in hertz, and $t$ is time in seconds. Because it continuously changes over time, ac voltage can be expressed as an effective, or root-mean-square (rms), voltage, given by

$$V = V_{rms} = \frac{V_0}{\sqrt{2}} \qquad (23.2)$$

Current in ac circuits, similar to voltage, is sinusoidal and is defined as

$$I = I_0 \sin \omega t = I_0 \sin 2\pi ft \qquad (23.3)$$

where the peak value of the current is $I_0$ and is related to the peak voltage via

$$I_0 = \frac{V_0}{R} \qquad (23.4)$$

The rms current is also related to the peak value according to

$$I_{rms} = \frac{I_0}{\sqrt{2}} \qquad (23.5)$$

The power in an ac circuit is the product of the current and the voltage and oscillates in time. The average power is

$$\overline{P} = I_{rms}V_{rms} \qquad (23.6)$$

Since $V_{rms} = I_{rms}R$, the average power can also be written as

$$\overline{P} = I_{rms}^2 R \quad \text{or} \quad \overline{P} = \frac{V_{rms}^2}{R} \qquad (23.7)$$

In an ac circuit the rms voltage across a capacitor is related to the rms current according to $V_{rms} = I_{rms}X_C$, where $X_C$ is the capacitive reactance. The capacitive reactance is measured in ohms and is given by

$$X_C = \frac{1}{2\pi fC} \qquad (23.8)$$

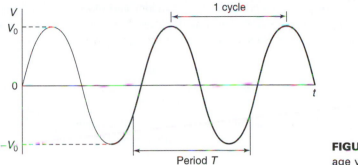

**FIGURE 23.1** An ac voltage versus time.

for a capacitance $C$ and frequency $f$. The current in a capacitor leads the voltage across the capacitor by a phase angle of 90°, and, as a result, a capacitor consumes no power, on average.

For an inductor the rms voltage and the rms current are related by $V_{rms} = I_{rms}X_L$, where $X_L$ is the inductive reactance. For an inductance $L$ and frequency $f$ the inductive reactance is given in ohms as

$$X_L = 2\pi fL \qquad (23.9)$$

The ac current in an inductor lags behind the voltage by a phase angle of 90°. Consequently, an inductor, like a capacitor, consumes no power, on average.

When a resistor, a capacitor, and an inductor are connected in series, the rms voltage across the combination is related to the rms current according to $V_{rms} = I_{rms}Z$, where $Z$ is the impedance of the combination. The impedance (in ohms) for the series combination is

$$Z = \sqrt{R^2 + (X_L - X_C)^2} \qquad (23.10)$$

The phase angle $\phi$ between current and voltage for a series $RCL$ combination is given by

$$\tan \phi = \frac{X_L - X_C}{R} \qquad (23.11)$$

Only the resistor in the combination dissipates power, on average, according to the relation

$$\overline{P} = I_{rms}V_{rms} \cos \phi \qquad (23.12)$$

The term $\cos \phi$ is the power factor of the circuit.

A series $RCL$ circuit has a resonant frequency $f_0$ that is given by

$$f_0 = \frac{1}{2\pi\sqrt{LC}} \qquad (23.13)$$

At resonance the impedance of the circuit has a minimum value equal to the resistance $R$, and the rms current has a maximum value.

## TOPIC 23.1 Electrical Current across Biomembranes

Nerve cells can be viewed as conducting cylinders surrounded by insulating sheets of biomembrane. In particular, from the viewpoint of electrostatics such a sheet behaves like a capacitor with a capacitance of about $10^{-2}$ F/m². However, a biomembrane sheet, compared with rubber or glass, is not much of an insulator and indeed leaks current. To discuss this situation quantitatively, we must assign a conductance and resistance to a sheet of biomembrane. Suppose $A$ is the area of a sheet of thickness $d$, and the resistance $R$ of the sheet is

$$R = \rho \frac{d}{A} \tag{23.14}$$

where $\rho$ is the resistivity of the membrane material. Thus the conductance $G = 1/R$ is proportional to the area of the membrane. We can therefore define a conductance per unit area by

$$g_m = \frac{G}{A} \tag{23.15}$$

which has units of reciprocal ohm-meters squared e.g. the conductance per square meter of a typical phospholipid membrane is about 9 $\Omega^{-1} \cdot m^{-2}$.

The current per unit area crossing a membrane, or current density $J$, is denoted by

$$J = \frac{I}{A} \tag{23.16}$$

where $I$ is the current flowing through the membrane in the direction perpendicular to it. From Ohm's law

$$I = \frac{\Delta V}{R} \quad \text{or} \quad I = \Delta V \, G \tag{23.17}$$

where $\Delta V$ is the voltage difference across the membrane. From equations (23.16) and (23.17), the current density is found to be

$$J = g_m \, \Delta V \tag{23.18}$$

Typical voltage differences across membranes are about 10 to 100 mV, so we expect the current density, for a conductance of 9 $\Omega^{-1} \cdot m^{-2}$, to be about 0.1 to 1.0 A/m². However, equation (23.16) is not entirely correct. Now suppose the concentrations of a particular ion inside and outside the cell membrane are $c(\text{in})$ and $c(\text{out})$, respectively. Then the equilibrium Nernst potential difference (see Chapter 14) for this particular ion is

$$V_c = V(\text{in}) - V(\text{out}) = \frac{k_B T}{e} \ln \frac{c(\text{out})}{c(\text{in})} \tag{23.19}$$

The effective potential difference across the membrane therefore becomes $\Delta V - \Delta V_c$, and the true current density is given by (see Figure 23.2)

$$J = g_m (\Delta V - \Delta V_c) \tag{23.20}$$

We may now set up an equivalent circuit where the resistor plays the role of leakage of current through the membrane; see Figure 23.3. We imagine that the sides of the capacitor represent the axoplasm and the exoplasm of a nerve cell. The axoplasm is outside the cell; the exoplasm is the layer of cytoplasm inside the cell but nearest the cell wall. If the section of membrane has an area $A$, then the capacitance in the diagram is $A \cdot c$, where $c$ is the capacitance per unit area of the membrane. The conductance $G$ is therefore $gA$, where $g$ is the conductance per unit area. The resistance $R = 1/G = 1/gA$. For a typical

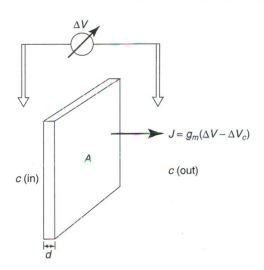

$$J = g_m(\Delta V - \Delta V_c)$$

$c$ (in)                $c$ (out)

**FIGURE 23.2**  Electrical current density $J$ across a membrane.

biomembrane, $c$ is about $10^{-2}$ F/m$^2$ and $g = 9\ \Omega^{-1} \cdot$ m$^{-2}$. The characteristic time for the discharge of a capacitor in an $RC$ circuit is therefore

$$\tau = RC = \frac{cA}{gA} = \frac{10^{-2}}{9} \cong 10^{-3}\ \text{s} \tag{23.21}$$

Left by itself, a membrane of a nerve cell would thus discharge in about a millisecond because of leakage. Thus ion pumps have to pump constantly to maintain the membrane voltage to compensate for the leakage. The RC time calculated above was shown by Nobel Prize winners Sir Alan Lloyd Hodgkin and Sir Andrew Fielding Huxley to be the characteristic switching time for nerve signals.

## TOPIC 23.2   Electrical Analogue of Nonpulsating Blood Flow

Blood flow through a vessel can be approximated by a steady fluid flow $Q$ through a rigid pipe of resistance $R$. This is generated by a steady pressure gradient $\Delta P$ in a similar way to Ohm's law in an electrical circuit:

$$\Delta P = QR \tag{23.22}$$

which is consistent with Poiseuille's law (see Chapter 11).

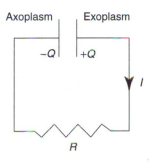

**FIGURE 23.3**  Equivalent circuit for current passing through a cell membrane.

**TABLE 23.1  Electrical Analogues of Hemodynamic Parameters and Their Units**

| Hemodynamic parameter | Electrical analogue |
| --- | --- |
| Volumetric blood flow rate $Q$, m³/s | Current $I$, A |
| Pressure gradient $\Delta P$, N/m² | Voltage $V$, V |
| Vascular resistance $R$, N · s/m⁵ | Resistance $R$, $\Omega$ |
| Compliance $C$, m⁵/N · s | Capacitance $C$, F |

This is probably a very simplistic representation of the human circulatory system. One physiological factor which is not easily accounted for by this oversimplified description is the type of fluid flow. The fluid flow through a rigid tube is analogous to current flow through a wire and thus an electrical analogy between the two may be drawn. In the electrical example, current is analogous to fluid flow rate, voltage is analogous to the pressure gradient, and resistance remains the same in both cases. The analogy of conductance is analogous to vascular compliance or elasticity of the blood vessel. Table 23.1 summarizes the fluid flow parameters and the corresponding electrical entities. Thus, by utilizing electrical principles, one can investigate flow through an isolated tube or vessel.

Electrical principles are more importantly applied to the calculation of current (flow), not through single wires necessarily but through networks of wires (or vessels) connected as circuits. In other words, the vessels in a vascular system resemble an electrical circuit such as the vessels presented in Figure 23.4. Referring to Figure 23.4, a voltage source $V$ (pressure gradient $\Delta P$) is driving a current (fluid flow) through a single wire (vessel) which branches into three vessels connected in parallel. The aim of investigating this network is to determine the total flow rate $Q_T$ and the individual flows $Q_1$, $Q_2$, and $Q_3$ through each vessel. The total flow entering the network must be equal to the sum of the flows in the three arms. Hence

$$Q_T = Q_1 + Q_2 + Q_3 \qquad (23.23)$$

The total pressure gradient, since the vessels are connected in parallel, must be identical to the pressure gradient in each vessel. Therefore,

$$\Delta P_T = \Delta P_1 = \Delta P_2 = \Delta P_3 \qquad (23.24)$$

Hence, from equations (23.23) and (23.24), if the effective resistance is $R_{\text{eff}}$, we obtain

$$\frac{Q_T}{\Delta P_T} = \frac{1}{R_{\text{eff}}} = \frac{Q_1}{\Delta P_1} + \frac{Q_2}{\Delta P_2} + \frac{Q_3}{\Delta P_3} = \frac{1}{R_1} + \frac{1}{R_2} + \frac{1}{R_3} \qquad (23.25)$$

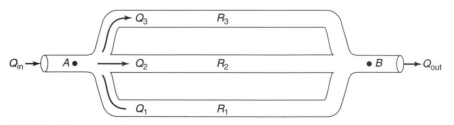

**FIGURE 23.4**  Fluid flow in a membrane.

which may be rearranged to give the effective resistance $R_{\text{eff}}$ as

$$R_{\text{eff}} = \frac{R_1 R_2 R_3}{R_2 R_3 + R_1 R_3 + R_1 R_2} \tag{23.26}$$

These relationships and their analogues can be used to approximate blood flow through any arrangement of vessels in the human circulatory system and are limited only by the complexity of the blood vessels.

## TOPIC 23.3    Pulsating Blood Flow

Steady fluid flow tends to hold for smaller blood vessels, but the oscillatory nature of the heartbeat introduces a pulsing flow through the circulatory system so blood flow should be better approximated by an ac circuit. The inductance represents the fluid inertia of the flowing blood, so if $Q$ is the analogue of current and $\Delta P$ of voltage, it is more accurate to write

$$\Delta P = L\frac{\Delta Q}{\Delta t} + QR \tag{23.27}$$

for a short time duration $\Delta t$. In terms of the geometry of a blood vessel, the inductance $L$ and resistance are defined by

$$R = \frac{8\pi\mu L}{A^2} \tag{23.28}$$

$$L = \rho\frac{\ell}{A} \tag{23.29}$$

where $\mu$ is the blood viscosity, $\ell$ is the vessel length, $A$ is the cross-sectional area of the vessel, and $\rho$ is the blood density.

## TOPIC 23.4    Diathermy

When a high-frequency ac generator is connected to two electrodes of large area which are attached to the body of a patient, heat is produced in the tissues. This heat is largely confined to tissues near the electrodes, and the heating effect is found to be very useful when treating many diseases (e.g., rheumatic and arthritic conditions and lesions in the vagina, urethra, and cervix of women or the prostate and vesicles of men). When very high frequencies are employed, the heating effect penetrates to a greater degree so that more deep-seated sites can be treated. This particular type of treatment has applications to bone and joint diseases, sinusitis, and axillary and cervical adenitis. Such short-wave medical diathermy, as it is called, is also very effective in speeding up the healing of wounds.

In surgery one electrode is reduced to a needle or button form, and the intensity of the current at that electrode can be made so great that cells are coagulated and killed by the concentrated heating effect of the needle (e.g., in the treatment of malignant and benign tumors). The frequency and power of the ac supply may be increased and the small electrode made into the form of a loop or needle. An intense arc may be produced and soft tissue cut instantly and cleanly by the moving electrode and at the same time the small blood vessels and lymphatics are sealed. The latter has been employed in extensive dissection such as the removal of a breast.

## TOPIC 23.5 Impedance Plethysmography

The measurement of the changes in volume of organs is called plethysmography. It utilizes the properties of tissue which behaves like a capacitor, and if alternating current is passed through it, the current leads the voltage by several degrees. In the conducting regions of the body, changes in the volume of organs can be followed by the resulting changes in electrical impedance of that region. This technique has been principally applied to changes in shape and content of the lungs on breathing and also to the changes which occur because of the cyclic flow of blood through arteries.

In an intensive care unit, lung functioning can be monitored by continual measurement of the changing current which is caused by the changing impedance across the patient's chest. If the current remains constant, some means of sounding the alarm is activated. One might achieve the same effect by putting thermistors or pressure transducers in the nostrils (or by some other means in the air flow from the lungs) but of course this is not convenient. In the case of small babies there is always the chance of accidental blockage, and if an insert became covered with mucus or vomit, the stoppage might go unnoticed. If leads were attached to the face, it would be uncomfortable and there would always be the danger of the leads being pulled off. The attachment of leads to the chest inside a special jacket is less restrictive.

For volume flow measurements it is normal to use four electrodes. These are attached to the legs or arms for blood flow measurements or to various points of the chest for air flow measurements. Impedance changes tend to be small, so to ensure they are maximal, a frequency range of 15 to 20 Hz is used for the ac supply. To ensure the patient is unaware of the current passed, it must be small, so amplifiers must be used to make the current change of a reasonable size.

The techniques of impedance plethysmography can also be used in diagnosis. For example, if fluid were to collect in the lungs, this allows easier passage of current through the region and the impedance is decreased. Such an accumulation can be extremely dangerous and may lead to acute pulmonary edema. This latter condition can pass undetected until well advanced and leakage into the patient's air passages occurs. This condition may be diagnosed early using impedance plethysmography before it becomes serious.

## TOPIC 23.6 Pacemakers

The electronic pacemaker is an interesting application of an *RC* circuit and can, by applying an electrical stimulus through electrodes attached to the chest, make a heart start beating again. If necessary, the stimulus can be repeated at the normal heartbeat rate. The heart itself already contains a pacemaker. It sends out electrical pulses at a rate of 60 to 80 per minute at rest, and these signals induce the start of each heartbeat. The pacemaker cells fail to function properly in some forms of heart disease; that is, the heart loses its rhythm. Those patients suffering from this ailment now commonly make use of electronic pacemakers which produce a regular voltage pulse which controls the frequency and triggers the heartbeat. There are two types, the fixed-rate type and the demand type. The fixed-rate variety produces signals continuously whereas the demand type operates only when the natural pacemaker fails. The electrodes of electronic pacemakers are implanted in or on the surface of the heart, the circuit usually containing a capacitor $C$ and a resistor $R$. The charge on the capacitor builds up to a certain point and then discharges. It then starts charging again. The pulsing rate depends on the values of $R$ and $C$, and the power source is usually a lithium battery which must be replaced after several years. Surprisingly, some pacemakers generate their energy from the heat produced by a radioactive element which is then converted to electricity using a thermocouple. There is also another form of pacemaker whose power is produced by the heart's own contractions which act on a piezoelectric crystal (the piezoelectric effect is the production of electricity by applying a mechanical stress to certain crystals). The crystal produces an alternating current which is converted to dc by a subminiature rectifier. The power is fed to and stored in a capacitor. At regular intervals the charge is switched to a pulse generator which powers the heart. One drawback of this device is that the piezoelectric crystal may eventually be encapsulated within a membrane by the body, and this would seriously affect its ability to respond to the heart's contractions.

## SOLVED PROBLEM

**23.1** Consider a heart defibrillator applied to an accident victim. This is basically an $RC$ circuit with $C = 10$ $\mu$F and $V_0 = 10^4$ V. Assuming that the resistance of the body is $R = 1000$ $\Omega$, calculate:

(a) The characteristic time $\tau$ of the discharge.

(b) The time it takes for the voltage to drop to 10 V.

**Solution**

(a) The characteristic time constant is simply

$$\tau = RC = (1000 \ \Omega)(10^{-5} \ \text{F}) = 0.01 \ \text{s} = 10 \ \text{ms}$$

(b) The time for the voltage to drop by a factor of 1000 is calculated from

$$V = V_0 e^{-t/\tau}$$

taking $V = 10$ V and $V_0 = 10^4$ V, so that

$$\ln\left(\frac{V}{V_0}\right) = -\frac{t}{\tau}$$

or

$$t = \tau \ln\left(\frac{V_0}{V}\right) = 3(10 \ \text{ms}) \ln 10 = 69 \ \text{ms}$$

## EXERCISES

**23.1** Consider a heart pacemaker as an $RC$ circuit. It fires 70 times a minute and uses a 20-$\mu$F capacitor. Calculate the resistance in the circuit.

**23.2** An ECG monitor uses an $RC$ circuit whose time constant must not exceed 100 $\mu$s in order to be able to measure rapid voltage fluctuations. Assuming the patient's resistance to be 1 k$\Omega$, determine the maximum suitable capacitance of the circuit used.

**23.3** An MRI machine employs a superconducting magnet with 60 H inductance. Find the required resistance of the circuit if a 1.2-s characteristic time constant is needed.

**23.4** An EEG machine records the activity of the brain by applying a 12-mV signal with a 0.6-Hz frequency to a capacitor in an *RLC* circuit with a negligible resistance. What is the circuit's capacitance?

# CHAPTER 24

# ELECTROMAGNETIC WAVES

## PHYSICAL BACKGROUND

An electromagnetic wave in a vacuum consists of mutually perpendicular and oscillating electric and magnetic fields. See Figure 24.1 for illustration.

The electromagnetic wave is a transverse wave, since the $\mathbf{E}$ and $\mathbf{B}$ fields are perpendicular to the direction of wave propagation. The frequency $f$ and wavelength $\lambda$ of an electromagnetic wave in a vacuum are related to its speed through the relation $c = f\lambda$. The series of electromagnetic waves, arranged in order of their frequencies or wavelengths, is called the electromagnetic spectrum. The spectrum includes radio waves, infrared radiation, visible light, ultraviolet radiation, X-rays, and gamma rays. Visible light has frequencies between about $4.0 \times 10^{14}$ and $7.9 \times 10^{14}$ Hz. The human eye and brain perceive different frequencies or wavelengths as different colors.

Maxwell showed that the speed of light in a vacuum is

$$c_0 = \frac{1}{\sqrt{\varepsilon_0 \mu_0}} = 3 \times 10^8 \text{ m/s} \tag{24.1}$$

where $\varepsilon_0$ is the (electric) permittivity of free space and $\mu_0$ is the (magnetic) permeability of free space. The velocity of electromagnetic waves through a medium of index of refraction $n$ (see Chapter 26) is

$$c = \frac{c_0}{n} \tag{24.2}$$

Electromagnetic waves transport energy and vary only in frequency. The total energy density $u$ of an electromagnetic wave is the total energy per unit volume of the wave and, in a vacuum, is given by

$$u = \tfrac{1}{2}\varepsilon_0 E^2 + \frac{B^2}{2\mu_0} \tag{24.3}$$

where $E$ and $B$ are the magnitudes of the electric and magnetic fields. In a vacuum, $E$ and $B$ are related by $E = cB$, and the electric and magnetic parts of the total energy density are equal. The intensity of an electromagnetic wave is the power that the wave carries perpendicularly through a surface divided by the area of the surface. In a vacuum, the intensity $S$ is related to the total energy density $u$ according to $S = cu$.

When the source of electromagnetic waves and their observer all travel along the same line in a vacuum, the Doppler effect takes place with a frequency shift which is given by

$$f' = f\left(1 \pm \frac{v}{c}\right) \tag{24.4}$$

where $f'$ and $f$ are, respectively, the observed and emitted wave frequencies. In this expression, $v$ is the relative speed of the source and observer and is assumed to be very small

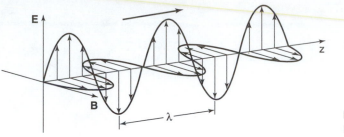

**FIGURE 24.1**
Electromagnetic wave.

with respect to the speed of light $c$. The plus sign is used when the source and observer approach each other, and the minus sign when they move apart.

A linearly polarized electromagnetic wave is one in which all oscillations of the electric field occur along one direction, which is taken to be the direction of polarization. In unpolarized light, the direction of polarization does not remain fixed but fluctuates randomly in time. Electromagnetic waves polarized along the $z$ direction are of the form

$$E_z(x, t) = E_0 \cos [kx - \omega(k)t] \tag{24.5}$$

with the dispersion relation given by

$$\omega(k) = ck \tag{24.6}$$

where $\omega = 2\pi f$ is the angular frequency and $k = 2\pi/\lambda$ is the so-called wave number.

Polarizing materials allow only the component of the wave's electric field along one direction to pass through them. The preferred transmission direction for the electric field is called the transmission axis of the material. When unpolarized light is incident on a piece of polarizing material, the transmitted polarized light has an intensity that is one-half that of the incident light. When two pieces of polarizing material are used one after the other, the first is called the polarizer, while the second is referred to as the analyzer. If the average intensity of polarized light falling on an analyzer is $\bar{S}_0$, the average intensity $\bar{S}$ of the light leaving the analyzer is given by Malus's law as

$$\bar{S} = \bar{S}_0 \cos^2 \theta \tag{24.7}$$

where $\theta$ is the angle between the transmission axes of the polarizer and analyzer. When $\theta = 90°$, the polarizer and analyzer are said to be "crossed," and no light passes through the analyzer.

## TOPIC 24.1 Cochlear Implants

Deaf people can sometimes recover part of their hearing with the help of a cochlea implant. The fact that electromagnetic waves can be broadcast and received is used to provide assistance to deaf people who have auditory nerves that are at least partially intact. To bypass the damaged part of the hearing mechanism, these implants utilize radio waves and gain access to the auditory nerve directly; see Figure 24.2. An external microphone, often set into an ear mold, detects sound waves and sends a corresponding electrical signal to a speech processor small enough to be carried in a pocket. The role of the speech processor is to encode the signals into a radio wave which is broadcast from an external transmitter coil placed over the site of a miniature receiver and antenna implanted under the skin. Acting much like a radio, the receiver detects the broadcasted wave and from

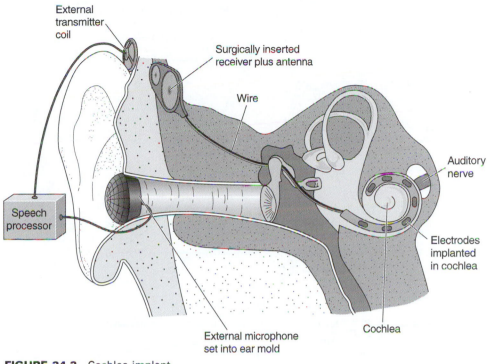

**FIGURE 24.2**   Cochlea implant.

the encoded audio information produces electrical signals which represent the sound wave. These signals are sent along a wire to electrodes which are implanted in the cochlea of the inner ear. The auditory nerves are stimulated by the electrodes and the corresponding signal in the nerves feeds directly between structures within the cochlea and the brain.

## TOPIC 24.2   Green Fluorescent Protein

Fluorescence provided by green fluorescent protein (GFP), which, as its name suggests, emits green light when excited, has found many uses in cell biology. This was first discovered in a small squid found in the North Pacific Ocean. Under sunlight the squid has a greenish color. The absorption spectrum of blue fluorescence protein (BFP), a variety of GFP, is shown in Figure 24.3. The molecule absorbs in the blue and emits in the green. The chemical structure of this molecule consists entirely of amino acids, which are building blocks of proteins. The sequence of amino acids of a protein is directly coded by the DNA sequence of an organism. Normally proteins which consist entirely of amino acids do not show fluorescence. The GFP, on the other hand, is fluorescent all by itself.

Suppose we wanted to know whether a certain protein is expressed in a cell under certain conditions. It is possible to insert the DNA code of the GFP protein into the DNA of the cell, by recombinant DNA methods and just alter the code of interest. When the cell expresses the protein of interest, the genetic machinery produces a new protein plus the smallest GFP protein attached. To see whether the cell is expressing the protein of interest, we simply shine a flashlight onto the cell. If the cell lights up in green, the protein is expressed; if it does not light up, the protein is not expressed. This observation also

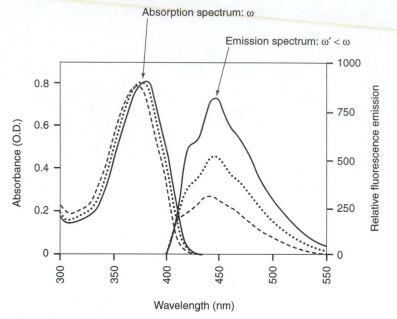

**FIGURE 24.3** Absorbance and relative fluorescence emission of intact BFP. The protein concentration was kept constant for all absorbance scans and also for emission scans. Dashed line, pH 5.0; dotted line, pH 6.0; solid line, pH 7.0.

tells where the protein is expressed. Thus GFP greatly facilitates a study of protein expression in the cell.

## TOPIC 24.3 Solar Radiation and Greenhouse Effect

Electromagnetic waves, with a wide range of frequencies, are generated by the sun. In the frequency range between ultraviolet and infrared, the solar radiation which reaches the outer atmosphere is intense. The intensity of the light, $I$, is defined as the amount of energy per unit area passing per second onto a surface. The maximum intensity lies right in the visible range. Electromagnetic waves with this frequency are able to reach sea level and can scatter there from various objects. This ultimately provides the basis of the sense of vision in animals, and by detecting scattered electromagnetic waves, an organism can obtain information about its environment. In a plot of $I$ against frequency, there are a number of dips which correspond to the fact that certain molecules can absorb electromagnetic radiation at certain special frequencies. This is an example of a resonance phenomenon; that is, only at certain frequencies will a molecule be excited resonantly by light. From the $I$-versus-frequency plot we find that $H_2O$, $CO_2$, and ozone remove certain frequencies from the solar radiation in the ultraviolet frequency range. In the last decade or so an ozone "hole" has appeared in the Southern Hemisphere—mostly in the Antarctic region—with a much reduced concentration of atmospheric ozone. The fluorocarbon compounds, which are powerful catalysts for the photochemical breakdown of ozone, are probably the culprits. If the hole continues to grow, it could have serious effects because the increased levels of ultraviolet radiation would lead to an even higher incidence of skin cancers.

Electromagnetic waves of particular importance are in the infrared region with wavelengths in the 800-nm range or longer. Infrared radiation is generated by heated objects and so is sometimes called heat radiation. When objects are heated to a modest 40°C, the infrared radiation is not visible to the human eye, although certain animals can detect it. Objects need to be heated to several hundred degrees Celsius before the radiation becomes visible (e.g., a spoon placed in a gas flame will start to glow with a red color). Solar radiation in the infrared range is shielded by the atmosphere to a considerable degree because atmospheric $H_2O$ and $CO_2$ reflect infrared radiation. In the wavelength interval around 1500 and 1900 nm none of the solar infrared radiation reaches the surface of the earth. However, solar radiation in the visible reaches the earth with a much smaller reduction.

Solar radiation in the visible, during daytime, heats up the surface of the earth. The heated surface then emits infrared radiation at night when the atmosphere cools. If this latter infrared radiation could reach the outer atmosphere, then the heat energy would be lost into space. However, just as the $H_2O$ and $CO_2$ molecules partially reflect the infrared radiation from the sun, they also partially reflect that trying to get out of the atmosphere, having been reflected from the earth's surface. This trapping of infrared radiation is called the greenhouse effect. The reason for the name is that greenhouses covered with sheets of glass absorb solar heat in the visible range but trap the infrared radiation inside.

The heating of the earth's surface is due to a combination of heat coming from the sun and heat from the earth's interior. Thus without the greenhouse effect the magnitude of the former contribution would be significantly reduced and the earth's surface would be almost 30°C colder. If, however, we increase the amount of $H_2O$ and $CO_2$ in the earth's atmosphere, we would trap more heat radiation, and an overall rise in temperature at the earth's surface would ensue.

For more than a century the amount of $CO_2$ in the atmosphere has been rising. This is probably due to the use of fossil fuels like oil and coal. Contributions to the greenhouse effect also arise from other gases like methane and nitrous oxide. It has been estimated that 4% of greenhouse gases in the atmosphere are due to man-made emissions. Assuming that the greenhouse heating effect is roughly proportional to the amount of greenhouse gas, we would estimate that the man-made rise in global temperature would be about 4% of 30 degrees, that is about 1.2°C. Based on various highly complex computer models of the dynamics of the earth's atmosphere, current estimates run between 0.3 and 1.0°C. One degree does not sound like a very great change in temperature until we realize that the mean global summer temperature drop during the last Ice Age was only about 2°C.

## Hazards and Benefits of Ultraviolet Radiation

The sun emits radiation over the entire electromagnetic spectrum, from the most energetic X-rays to the least energetic radio waves. However, most of the solar radiation's energy falls in the visible, infrared, and ultraviolet (UV) range which is created at the sun's surface. While most X-rays are blocked by the earth's atmosphere, ultraviolet radiation passes through it. Ultraviolet radiation has frequencies in excess of $8 \times 10^{14}$ Hz. In terms of wavelength, the part of the UV radiation with 280 nm $< \lambda <$ 320 nm is called UVB and that with 320 nm $< \lambda <$ 400 nm is referred to as UVA. The energy of UV radiation is high enough to shake the structure of many organic molecules and hence is readily absorbed by living matter. Especially dangerous is UVB, which can trigger skin cancer, kill cells, split molecules, and even cause genetic mutations. On the other hand, UVA is less harmful and triggers the body's production of vitamin D, which is essential for proper development. In recent years, the intensity of the sun's UV radiation reaching the earth's surface has been reported to have markedly increased as a result of the destruction of the ozone layer in the upper atmosphere. The latter effect has been blamed on the dramatic rise in industrial production of airborne pollutants, most specifically the so-called chlorofluorocarbons (CFCs). The 1995 Nobel Prize in Chemistry was awarded to Paul J. Crutzen, Mario J. Molina and F. Sherwood Rowland, who scientifically demonstrated the correlation between ozone depletion and CFC production.

Since the energy per photon of UV radiation is large enough to destroy organic compounds, UV radiation can be used as a germicide or sterilizer to kill microorganisms. It is most useful in sterilizing the surfaces of instruments such as combs and brushes in barber shops. Similarly, UV radiation causes sunburn but affects only the upper layers of tissue.

Ultraviolet radiation is used to treat many skin conditions, commonly with careful use of an ordinary sunlamp. The high photon energy of UV radiation apparently initiates chemical processes, such as the production of vitamin D, that alleviate certain skin conditions. There are definite long-term hazards to excessive UV exposure. The incidence of most types of skin cancer is strongly correlated with exposure to UV radiation.

## TOPIC 24.4  Medical Applications of X-rays and $\gamma$-Rays

X-ray radiation is emitted by the most energetic electrons in atoms. The wavelengths of X-rays are in the range of $10^{-10}$ m, which is comparable to atomic dimensions. For this reason, X-rays found applications in diffraction from crystalline solids whose atoms act like a grating (see Chapter 27). From the resultant diffraction patterns one can reconstruct the three-dimensional structure of the crystal (e.g., NaCl) from which X-rays have been diffracted. X-ray diffraction has also been applied with great success toward understanding the structure of biologically important molecules, such as proteins and nucleic acids. One of the most important breakthroughs of this century was the discovery in 1953 by J. D. Watson and F. H. C. Crick that the structure of the nucleic acid DNA is a double helix. It was achieved with the help of X-ray diffraction.

Furthermore, X-rays strongly interact with biological matter. They possess enough energy to ionize molecules within biological cells, which may lead to mutations and tumor growth. However, in controlled doses, X-rays found applications in medical diagnostics. Dense tissue, particularly bones, absorbs X-rays more strongly than other tissue. Thus, they form shadows of varying intensity on the screen or film. While X-rays cannot be seen by the human eye, fluorescent screens or special photographic films can make these shadow patterns visible. X-rays usually are used to detect relative tissue density, as in finding cavities in teeth and breaks in bones. Gamma-ray-emitting nuclei can be placed in various chemicals that are concentrated in specific organs, such as the thyroid. An image then can be made for medical diagnostic purposes by measuring the pattern of gamma ray emission.

Apart from their diagnostic use, X-rays are used in medicine for therapeutic purposes. Because their energy is quite large, these rays do considerable biological damage. A typical diagnostic X-ray photon has an energy of 50 keV, enough to disrupt about 10,000 molecules. Most cancer cells are more sensitive to this damage than are normal cells, and the radiation often can be localized to cancerous tissue. Since radiation kills cells, it can be employed to get rid of unwanted cells such as those in cancerous growth. This is possible only if the unwanted cells can be destroyed by a dose of radiation which does not permanently damage the surrounding healthy tissue. A collimated beam of X-rays can be directed to the target volume. Machines operating at under 250 kV are used for tumors.

X-rays pass right through the body, so that all the cells along the line of the beam are irradiated. Consult Chapter 31 for a more detailed exposition of the effects of ionizing radiation.

## TOPIC 24.5 Applications of Microwaves

The most common medical use of microwaves is for deep heating, called microwave diathermy. Water molecules in tissue absorb a range of microwave frequencies. The microwave energy is converted to thermal energy, increasing the temperature of the tissue. The energy per photon of microwaves is small, so very large numbers of them (high intensities) must be used. Microwaves penetrate much deeper into the body than the shorter wavelength infrared. The beneficial effects of microwave diathermy result from the increased temperature produced.

## TOPIC 24.6 Infrared Radiation

Infrared radiation is also used for deep heating. The common heat lamp puts out most of its energy in the infrared region. Infrared does not penetrate as deeply and does not present the same hazards as microwaves. About 95% of infrared that falls on the skin is absorbed in the upper layers of tissue. The interior of the body is heated by blood circulation. Infrared photon energies are too small to damage organic compounds directly.

## TOPIC 24.7 Applications of Polarimeters in Determination of Sugar Concentration

Polarized light can be used to measure the concentration of sugar in a liquid by merely shining a beam and determining its properties as it passes through the liquid. It turns out that the direction of polarization rotates as polarized light passes through the solution. Glucose, with a simplified formula $C_6H_{12}O_6$, exists in two main forms: (a) dextrose (also called $d$-glucose or corn sugar), which rotates the plane of polarization of light according to a right-handed screw orientation, and (b) fructose (also called $\ell$-glucose or levulose), which rotates the plane of polarization in the opposite direction, that is, according to the left-handed screw. While fructose is mainly found in fruits and honey, dextrose can be found in starches.

The polarimeter is a device which allows you to measure the angle of rotation of polarized light $\theta$ in radians, which depends on the length of path $\ell$ through which the light has traveled, the concentration of sugar $c$ in grams per cubic centimeter, and a characteristic parameter called specific rotation $\rho$. Knowing the value of $\rho$ and measuring $\theta$ with a polarimeter, one can determine the concentration $c$ from the formula

$$c = \frac{10\theta}{\ell\rho} \tag{24.8}$$

The total amount of glucose can be determined using methods of organic chemical analysis while the polarimeter will provide the food analyst with a way of determining the relative contribution of artificial sweeteners such as $d$-glucose.

# SOLVED PROBLEM

**24.1** High-frequency electromagnetic fields are used in medical practice for heat therapy. The usual frequencies are about $f_1 = 27$ MHz (short wave), $f_2 = 0.43$ GHz (decimeter wave), and $f_3 = 2.45$ GHz (microwave).
(a) Calculate the wavelengths that correspond to the above given frequencies in air.
(b) Calculate the corresponding wavelengths in muscle tissue. The relative dielectric constants of muscle are $\varepsilon_1 = 110$ (at 27 MHz), $\varepsilon_2 = 50$ (at 0.43 GHz), and $\varepsilon_3 = 43$ (at 2.45 GHz).

### Solution

(a) The velocity of electromagnetic waves in air is practically the same as in vacuum, that is, $c = 3 \times 10^8$ m/s. Hence, the wavelengths for the three frequencies can be calculated from the equation

$$\lambda_{air} = \frac{c}{f}$$

The results are $\lambda_{air,1} = 11.1$ m, $\lambda_{air,2} = 0.7$ m, and $\lambda_{air,3} = 0.12$ m.
(b) The wavelengths of an electromagnetic wave in a medium of relative dielectric constant $\varepsilon$ is

$$\lambda_{medium} = \frac{\lambda_{air}}{\sqrt{\varepsilon}}$$

Thus, the wavelengths in muscle tissue are $\lambda_{muscle,1} = 1.1$ m, $\lambda_{muscle,2} = 0.1$ m, and $\lambda_{muscle,3} = 0.018$ m

# EXERCISES

**24.1** Magnetic resonance imaging (MRI) and positron emission tomography (PET) scanning are two medical diagnostic techniques. Both employ electromagnetic waves. For these waves, find the ratio of the MRI scanning wavelength (frequency $6.35 \times 10^7$ Hz) to the PET scanning wavelength (frequency $1.25 \times 10^{20}$ Hz).

**24.2** Calculate the frequency of the 190-nm UV radiation used in laser eye surgery. Suppose that a 1-ms-duration laser burst deposits 0.5 mJ of energy on the cornea of a patient making a spot of 1 mm diameter. What are the intensities of the electric and magnetic fields used?

**24.3** The smallest details observable using electromagnetic radiation as a probe have a size of about one wavelength. (a) What is the smallest detail available with ultraviolet photons of energy 10 eV? (b) What is the smallest detail observable with X-ray photons of energy 100 keV?

**24.4** What would be the smallest possible detail observable with a microscope that uses ultraviolet light having a frequency of $1.5 \times 10^{15}$ Hz?

**24.5** Calculate the frequency range of visible light, given its wavelength range to be 380 to 770 nm. How does this compare with the ratio of high and low frequencies the ear can hear?

# REFLECTION OF LIGHT: MIRRORS

## PHYSICAL BACKGROUND

Geometric optics is a branch of physics concerned with the propagation of light utilized in the formation of images. In geometric optics, a light source impinges upon one of two basic types of optical components: mirrors and thin lenses. Wave fronts are surfaces on which all points of a wave are in the same phase of motion. If the wave fronts are flat surfaces, the wave is called a plane wave. Rays are lines that are perpendicular to the wave fronts and point in the direction of the velocity of the wave.

Mirrors are optical components that reflect light rays from a light source to form an image. Two types of mirror are plane mirrors and spherical mirrors. Spherical mirrors can further be subdivided into concave (curved-inward) mirrors and convex (curved-outward) mirrors.

When light reflects from a smooth surface, the reflected light obeys the law of reflection, which states that (a) the incident ray, the reflected ray, and the normal to the surface all lie in the same plane and (b) the angle of reflection $\theta_r$ equals the angle of incidence $\theta_i$:

$$\theta_i = \theta_r \tag{25.1}$$

The law of reflection explains how mirrors form images. A virtual image is one from which all the rays of light do not actually come out but only appear to do so. A real image is one from which all the rays of light actually do emanate. A plane mirror forms an upright, virtual image that is located as far behind the mirror as the object is in front of it. In addition, the heights of the image and the object are equal.

A spherical mirror has the shape of a section cut out from the surface of a sphere. The principal axis of a mirror is a straight line drawn through the center of curvature and the middle of the mirror's surface. Rays that lie close to the principal axis are known as paraxial rays. The radius of curvature of the mirror is the distance from the center of curvature of the mirror. The focal point of a concave spherical mirror is a point on the principal axis, in front of the mirror. Incident paraxial rays that are parallel to the principal axis converge to the focal point after being reflected from the concave mirror. The focal point of a convex spherical mirror is a point on the principal axis behind the mirror. For a convex mirror, paraxial rays that are parallel to the principal axis seem to diverge from the focal point after reflecting from the mirror. The focal length $f$ of a mirror is the distance along the principal axis between the focal point and the mirror. The focal point and the radius of curvature are related by $f = \frac{1}{2}R$ for a concave mirror and $f = -\frac{1}{2}R$ for a convex mirror. The image produced by a mirror can be located by a graphical method known as ray tracing.

The mirror equation can be used with either concave or convex mirrors and specifies the relation between the image distance $d_i$, the object distance $d_o$, and the focal length $f$ of the mirror:

$$\frac{1}{d_o} + \frac{1}{d_i} = \frac{1}{f} \tag{25.2}$$

The magnification $m$ of a mirror is the ratio of the image height $h_i$ to the object height $h_o$. The magnification is also related to $d_i$ and $d_o$ by the magnification equation:

$$m = \frac{h_i}{h_o} = -\frac{d_i}{d_o} \tag{25.3}$$

The intensity of light is the amount of energy per unit time passing through a unit area normal to the direction of flow and thus is power divided by unit area, or

$$I = \frac{P}{A} \tag{25.4}$$

As with sound, the intensity of light decreases as 1 over the square of the distance from a point source to the receiver:

$$I \propto \frac{1}{r^2} \tag{25.5}$$

## TOPIC 25.1  Lighting Devices and Their Intensity

The basic SI unit of light intensity is the lumen ($\ell$m). A typical value of the amount of light from the sun falling on 1 $m^2$ of the earth's surface is $4.4 \times 10^5$ $\ell$m. A comfortable illumination level for reading is about 1000 $\ell$m/$m^2$, and an ordinary 100-W light bulb provides more than an adequate level of lighting since its output is 1700 $\ell$m. The other kind of home and office lamp is the fluorescent lamp, which produces 2400 $\ell$m at only a 30-W power rating. The third type of lamp is the quartz halide lamp, commonly used in automobile headlamps, spotlights, and slide projectors. A typical 50-W quartz halide lamp is rated at 1600 $\ell$m. The efficacy of a lamp is the ratio of its light intensity per energy input in units of lumen per watts. Thus, the fluorescent lamp has the highest efficacy, followed by the quartz halide, with a regular incandescent bulb having the lowest efficacy value. For outdoor lighting such as street lights and highway intersection lights, the most commonly used are 400-W high-pressure sodium lamps at an efficacy of 125 $\ell$m/W and 400-W mercury lamps at 52 $\ell$m/W. They are typically mounted 11 m above the ground and for added efficiency have reflecting surfaces above them to divert the light to the street instead of wasting it by shining in all directions. It is generally assumed that the level of 9 $\ell$m/$m^2$ is considered adequate for highway use. The latest in light-producing technology is the laser, which produces a coherent (i.e., single-wavelength), light beam that is highly collimated. However, while lasers have found numerous applications from bar-code scanners to eye surgery, their efficacy as lighting devices is very low. For example, a helium–neon gas laser producing red light with the wavelength $\lambda = 632.8$ nm generates 0.35 $\ell$m of light requiring a 13-W electrical power supply!

## EXERCISE

**25.1**  A dentist uses a mirror that gives an upright image of a tooth with a magnification of 5.0 when held 0.025 m from the tooth. Calculate the focal length of the mirror.

# CHAPTER 26

# REFRACTION OF LIGHT, LENSES, AND OPTICAL INSTRUMENTS

## PHYSICAL BACKGROUND

When light strikes the interface between two media, part of the light is reflected and the remainder is transmitted across the interface. The change in the direction of travel as light passes from one medium into another is called refraction. The index of refraction $n$ of a material is the ratio of the speed of light $c$ in a vacuum to the speed of light $v$ in the material: $n = c/v$. Snell's law of refraction states that (1) the refracted ray, the incident ray, and the normal to the interface all lie in the same plane and (2) the angle of refraction $\theta_2$ is related to the angle of incidence $\theta_1$ by

$$n_1 \sin \theta_1 = n_2 \sin \theta_2 \qquad (26.1)$$

where $n_1$ and $n_2$ are the indices of refraction of the incident and refracting media, respectively. The angles are measured relative to the normal to the interface.

When light passes from a medium of larger refractive index $n_1$ to one of smaller refractive index $n_2$, the refracted ray is bent away from the normal. If the incident ray is at the critical angle $\theta_c$, the angle of refraction is 90°. The critical angle is determined from Snell's law and is given by

$$\sin \theta_c = \frac{n_2}{n_1} \qquad (26.2)$$

When the angle of incidence exceeds the critical angle, all the incident light is reflected back into the medium from which it came, a phenomenon known as total internal reflection.

When light is incident on a nonmetallic surface at the Brewster angle $\theta_B$, the reflected light is completely polarized parallel to the surface. The Brewster angle is given by

$$\tan \theta_B = \frac{n_2}{n_1} \qquad (26.3)$$

where $n_1$ and $n_2$ are the refractive indices of the incident and refracting media, respectively. When light is incident at the Brewster angle, the reflected and refracted rays are perpendicular to each other.

A glass prism can split a beam of sunlight into a spectrum of colors, because the index of refraction of the glass depends on the wavelength of the light. The splitting of light into its color components is known as dispersion.

Converging lenses and diverging lenses depend on the phenomenon of refraction in forming an image. With a converging lens, paraxial rays that are parallel to the principal axis are focused to a point on the axis by the lens. This point is called the focal point of the lens, and its distance from the lens is the focal length $f$. Paraxial light rays that are parallel to the principal axis of a diverging lens appear to originate from its focal point after passing through the lens. The image produced by a converging or diverging lens can be located with the help of a ray diagram.

The thin-lens equation can be used with either converging or diverging lenses that are thin, and it relates the object distance $d_o$, the image distance $d_i$, and the focal length $f$ of the lens:

$$\frac{1}{d_o} + \frac{1}{d_i} = \frac{1}{f} \tag{26.4}$$

The magnification $m$ of a lens is the ratio of the image height $h_i$ to the object height $h_o$. The magnification is also related to $d_i$ and $d_o$ by the magnification equation:

$$m = -\frac{d_i}{d_o} \tag{26.5}$$

When two or more lenses are used in combination, the image produced by one lens serves as the object for the next lens.

Thin lenses, like mirrors, exist as convex or concave lenses. However, lenses form images by refracting light rays, as opposed to mirrors that form images by reflecting light rays. Thin lenses have two surfaces and are thus characterized by two radii of curvature $R_1$ (side of lens closest to the object) and $R_2$ (side of lens opposite the object). A convex lens has a positive radius $R_1$ and a negative $R_2$, while a concave lens has a negative $R_1$ and a positive $R_2$. The lens maker's equation can be expressed as

$$\frac{1}{f} = (n-1)\left(\frac{1}{R_1} + \frac{1}{R_2}\right) \tag{26.6}$$

where $f$ is the focal length, $n$ is the index of refraction for the lens material, $R_1$ is the radius of the lens closest to the object, and $R_2$ is the radius of the lens farthest from the object. If the lens, characterized by an index of refraction $n_1$, is placed within a medium with an index of refraction $n_1$, then $n$ in the lens maker's equation is replaced by $n_1/n_2$.

The refractive power $P$ of a lens is defined by

$$P = \frac{1}{f} \tag{26.7}$$

where $P$ is expressed in units of diopters or inverse length $(1/L)$. The sign convention for $P$ is similar to that for $f$: positive for a converging lens and negative for a diverging lens.

In the human eye, a real, inverted image is formed on a light-sensitive surface, called the retina. Accommodation is the process by which the focal length of the eye is automatically adjusted, so that objects at different distances produce focused images on the retina. The near point of the eye is the point nearest the eye at which an object can be placed and still have a sharp image produced on the retina. The far point of the eye is the location of the farthest object on which the fully relaxed eye can focus. For a young and normal eye, the near point is located 25 cm from the eye, and the far point is located at infinity.

A nearsighted (myopic) eye is one that can focus on nearby objects but not on distant ones. Nearsightedness can be corrected by wearing eyeglasses or contacts made from diverging lenses. A farsighted (hyperopic) eye can see distant objects clearly but not those close up. Farsightedness can be corrected by using converging lenses.

The angular size of an object is the angle that it subtends at the eye of the viewer. For small angles, the angular size in radians is

$$\theta \approx \frac{h_o}{d_o} \tag{26.8}$$

where $h_o$ is the height of the object and $d_o$ is the object distance. The angular magnification $M$ of an optical instrument is the angular size $\theta'$ of the final image produced by the instrument divided by the reference angular size $\theta$ of the object, which is that seen without the instrument:

$$M = \frac{\theta'}{\theta} \tag{26.9}$$

A magnifying glass is usually a single converging lens that forms an enlarged, upright, and virtual image of an object placed at or inside the focal point of the lens. For a magnifying glass held close to the eye, the angular magnification is approximately

$$M \approx \left( \frac{1}{f} - \frac{1}{d_i} \right) N \tag{26.10}$$

where $f$ is the focal length of the lens, $d_i$ is the image distance, and $N$ is the distance of the viewer's near point from the eye.

A compound microscope usually consists of two lenses, an objective and an eyepiece. The final image is enlarged, inverted, and virtual. The angular magnification of such a microscope is

$$M \approx -\frac{(L - f_e)N}{f_o f_e} \tag{26.11}$$

where $f_o$ and $f_e$ are, respectively, the focal lengths of the objective and eyepiece, $L$ is the distance between the two lenses, and $N$ is the distance of the viewer's near point from the eye.

An astronomical telescope magnifies distant objects with the aid of an objective and eyepiece, and it produces a final image that is inverted and virtual. The angular magnification of a telescope is

$$M = -\frac{f_o}{f_e} \tag{26.12}$$

where $f_o$ and $f_e$ are the focal lengths of the objective and eyepiece.

Lens aberrations limit the formation of perfectly focused or sharp images by optical instruments. Spherical aberration occurs because rays that pass through the outer edge of a lens with spherical surfaces are not focused at the same point as those that pass through the center of the lens. Chromatic aberration arises because a lens focuses different colors at different points.

## TOPIC 26.1  Anatomy of the Human Eye

The human eye is one of the most remarkable of all optical devices which are found in nature or have been designed by humans. Figure 26.1 shows some of the main anatomical features of the eye. The eyeball itself is approximately spherical with a diameter of about 25 mm. Often the eye is likened to a camera, for its purpose is to form an image on the retina in the same way that a camera forms an image on a photographic plate.

Light enters the eye through a transparent membrane called the cornea. This latter membrane covers a clear liquid region—the aqueous humor—behind which is a diaphragm, the iris (or colored part of the eye). This adjusts automatically to control the amount of light entering the eye. Most of the refraction (bending) of light occurs at the

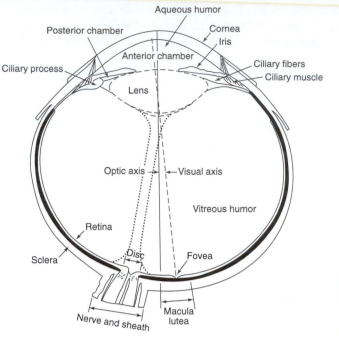

**FIGURE 26.1** Cross-sectional anatomy of the human eye seen from above.

cornea of the eye. The hole in the iris through which light passes—called the pupil—is black because no light is reflected back out from the interior of the eye. The pupil has a diameter which varies from about 2 to 7 mm, decreasing in bright light and increasing in dim light. Across this hole is the lens, which is connected to the ciliary muscles by the suspensory ligaments. The fact that the lens is flexible and its shape can be altered by the action of the ciliary muscle is of prime importance to the operation of the eye. The lens is a clear, flexible, cellular structure. The cells are long and hexagonal in cross section and are stacked in columns roughly parallel to the light path. The layers are ordered in such a way that relatively little light is scattered or absorbed. The lens consists of nested fibers which run from the front to the back of the lens (see Figure 26.2), creating an onion-like, gelatinous structure with an index of refraction of 1.40 at the center of the lens and 1.38 at the edge. These fibers are made of a protein called $\alpha$-crystallin. The presence of cataracts in the eye is due to the loss of structure of the $\alpha$-crystallin proteins, leading to a clouding of the lens. The lens is suspended by tense radial filaments called the zonules of Zimm that are connected to the surrounding ciliary muscles. The lens can adjust its shape by contraction of the ciliary muscle so that, for example, when the ciliary muscles are relaxed, the lens is flattened because the tension in the radial filaments stretches it out. Accommodation, like the adjustment of the pupil, is automatic via the autofocusing re-flex. This can be overridden, and it is possible to defocus when looking at an object. The change in refractive index from aqueous humor to $\alpha$-crystallin is rather small, so this ac-commodation of the lens is modest and only fine tunes the focusing. The main converg-ing action is performed by the cornea. Behind the lens is a region filled with a jellylike substance called the vitreous humor. The retina, which plays the role of a film in a cam-era, is on the curved rear surface of the eye. It is a very complex array of cells called rods and cones which act to change light energy into electrical signals which then pass along

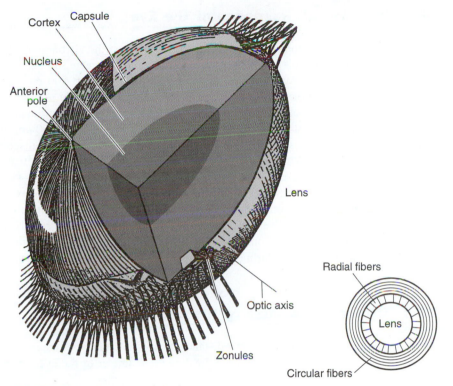

**FIGURE 26.2** Structure of the lens.

the optic nerve (see Figure 26.1) to the brain. The region of the retina where the optic nerve enters contains no photoreceptors and constitutes a blind spot on the eye. At the center of the retina is a small area, called the fovea, about 0.25 mm in diameter, where the cones are very closely packed and the sharpest image and best color discrimination are found. Unlike a camera, the eye contains no shutter. The equivalent operation is carried out by the nervous system, which analyzes the signals to form images at the rate of about 30 per second.

Tables 26.1 and 26.2 summarize the geometric and refractive properties of the eye.

**TABLE 26.1  Optical Constants of the Helmholtz Schematic Relaxed Eye**

| Optical value | Constant (mm) |
|---|---|
| Distance from the front surface of the cornea to the front of the lens | 3.6 |
| Lens thickness | 3.6 |
| Radii of curvature | |
|   Cornea | 8.0 |
|   Front of the lens | 10.0 |
|   Back of the lens | 6.0 |

**TABLE 26.2  Indices of Refraction for the Eye**

| Material | $n$ |
|---|---|
| Cornea | 1.38 |
| Aqueous humor | 1.33–1.34 |
| Lens | 1.41–1.45 |
| Vitreous humor | 1.34 |
| Air | 1.00 |
| Water | 1.33 |

## TOPIC 26.2   Wavelength Response of the Eye

The electromagnetic radiation reaching the eye extends beyond the visible range into both the infrared and the ultraviolet. However, only in the range between 380 and 700 nm is the eye responsive. The cornea absorbs most of the energy for $\lambda < 300$ nm, that is, in the UV range, and hence the risk of corneal damage is high for people not wearing protective sunglasses. On the other end of the spectrum, water molecules in the cornea and aqueous humor absorb most of the electromagnetic radiation with $\lambda > 1200$ nm. Furthermore, the eye pigments are unresponsive to $\lambda > 800$ nm and only slightly responsive to $800 \text{ nm} > \lambda > 700$ nm. What is not generally known is that in the visible region (i.e., for $380 < \lambda < 700$ nm) the eye's response depends rather strongly on the wavelength (i.e., color). Moreover, the response of the eye depends strongly on its intensity. If a person is kept in bright surroundings, the process is called photopic vision involving light adaptation. Conversely, in dark surroundings, when the person's eyesight is dark adapted, the process is called scotopic vision. Changing the light intensity of the surroundings requires a period of readaptation during which the person is temporarily blinded.

## TOPIC 26.3   Optical Properties of the Eye

When an object is being viewed and is a distance $d_o$ from the lens, the image distance always has the same value $d_i$ which is approximately the diameter of the eyeball. Thus the power of the lens, $P$, is given by

$$P = \frac{1}{f} = \frac{1}{d_o} + \frac{1}{d_i} \tag{26.13}$$

When the eye is fully relaxed, $d_o = \infty$ and $f = f_\infty$, so

$$P_\infty = \frac{1}{f_\infty} = \frac{1}{\infty} + \frac{1}{d_i} = \frac{1}{d_i} \tag{26.14}$$

On the other hand, if the eye is fully accommodated and the object is at the near point, the power $P_N$ is given by

$$P_N = \frac{1}{f_N} = \frac{1}{0.25 \text{ m}} + \frac{1}{d_i} \tag{26.15}$$

Thus,

$$P_N - P_\infty = \frac{1}{0.25 \text{ m}} = 4 \text{ diopters} \tag{26.16}$$

The power of accommodation of an eye is defined as the difference of the powers associated with the extremes of vision, and for a normal eye this is 4 diopters. The visual acuity of the eye is defined to be the minimum angular separation of two equidistant points of light which can be just resolved into two separate objects by the eye. If the points are closer than this, the eye sees light coming from a single point source. The acuity of scotopic vision (low-intensity light) is highest about 20° from the fovea (20° from the line through the center of the lens and the fovea), where the density of rods is greatest. For photopic vision acuity is highest on the fovea, where cones are most dense. The latter is always greater than the former.

# TOPIC 26.4   Light Absorption and Black–White Vision

An image is projected onto the retina after light from the exterior has passed through the lens of the eye. In this section we will discuss how this image is recorded and from where our color perception comes. On taking the eye of a frog out of a dark cupboard, in 1877, the German biologist Boll noticed a red color in the back of the eye which quickly faded away. If the eye was placed back in the cupboard and brought out a while later, the same results were observed. Boll interpreted the red color of the eye as a chemical change induced by the light. The color of certain pigments disappears under sustained exposure to light. This process is called bleaching and can be understood by assuming that the pigment alters its structure in the excited state in such a way as to make it difficult to return to the ground state (consult Chapter 29 for information on quantum mechanical aspects of this process). The color produced by the pigment must fade when all the pigment molecules have been altered photochemically.

The above suggests that a photochemical reaction is related to the process of vision. The retina contains two types of light-sensitive cells called photoreceptor (PR) cells: cones (about 6 to 7 million of them) and rods (about 120 million of them). The PR cells are inside a matrix of pigment cells which act as a nonreflecting background and are about $10^{-6}$ m in diameter. After light absorption PR cells send out an electrical signal. With a relatively high threshold of light intensity before they are triggered, on being fired, the cones send a signal to the brain and are responsible for color vision (see Topic 26.5 for details). On the other hand, the rods are responsible for low-intensity black–white vision, that is, for night time viewing, and have low firing thresholds. Cats and other nocturnal animals have a relatively large number of rods.

Shortly after the observation of Boll, the visual pigment was isolated by Wilhelm Kuhne. It was found to be a light-absorbing protein called rhodopsin, or visual purple, lo-

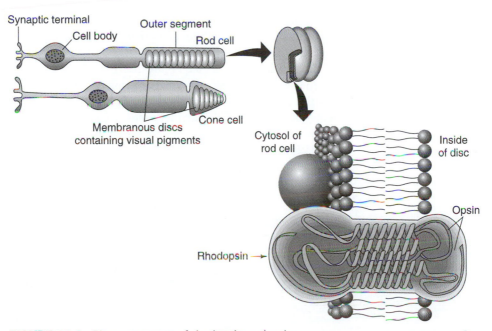

**FIGURE 26.3**   Planar structure of rhodopsin molecules.

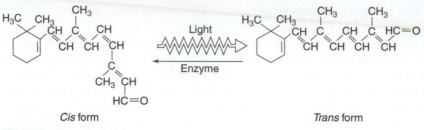

*Cis* form                                                                    *Trans* form

**FIGURE 26.4** Retinal isomers.

cated in the rods. Rhodopsin absorbs strongly visible light in certain frequency ranges which can be in the rose or purple. The molecules of this protein are arranged in planar structures anchored in a membrane which can transmit electrical signals in response to the stimuli received from the rhodopsin molecules (see Figure 26.3). The light-active group is called 11-*cis*-retinol. After retinol absorbs a quantum of light, it isomerizes (see Figure 26.4). The new structure is called 11-*trans*-retinol, which is basically a straightened-out version of *cis*-retinol (see Figure 26.4). This isomerization takes place in about $10^{-2}$ s. The energy of the light quantum has been transferred into a restructuring of the molecule. The isomerization blocks the excited molecule from rapidly falling back into its ground state and reemitting the light quantum. Thus rhodopsin is not a fluorescent molecule. Furthermore, the isomerization alters the charge of the trans-membrane part of the rhodopsin protein, and this starts a chemical reaction and produces electrical activity by closing ion channels in the membrane. This process is then the origin of the electrical signal which will eventually reach the brain as a part of black–white vision.

## TOPIC 26.5 Color Vision

Sir Isaac Newton was the first to demonstrate that white light is composed of several colors. He did this by shining a beam of sunlight onto a prism and examining the colors obtained. Based on the electromagnetic theory of light, we now know that visible light is a small part of the electromagnetic spectrum (see Figure 26.5).

As mentioned earlier in this chapter, there are two major types of light-sensitive cells in the retina: rods and cones. The rods are found throughout the retina, except for a small region in the center of the fovea. Rods are more sensitive than cones and are solely responsible for vision in very dark environments and for peripheral vision. Unlike the cones, rods do not yield color information. Cones are concentrated in the fovea and are associated with color vision. There are three types of cones, each type is sensitive to a different range of wavelengths (see Figure 26.6).

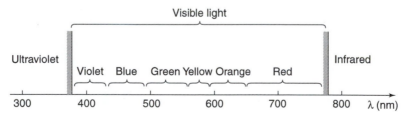

**FIGURE 26.5** A small part of the electromagnetic spectrum, with the type of electromagnetic radiation identified as a function of wavelength (in nanometers).

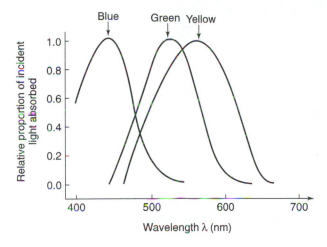

**FIGURE 26.6**  The relative sensitivity of the three types of cones is indicated by their relative absorption of light as a function of frequency.

The molecular basis of color vision was difficult to demonstrate experimentally but was predicted by the Young–Helmholtz theory of color vision, which states that there are only three different color pigments R, G, and B. It is assumed that pigment R absorbs predominantly in the red, pigment G in the green, and pigment B in the blue. Now suppose a light beam containing many different wavelengths falls on the retina and then excites only a fraction of the pigment molecules in the cones. What will be recorded is how many R pigment molecules have been excited, how many G molecules, and how many B molecules. The brain then receives this information. For example, suppose a light beam consists of 50% red and 50% green. The signal sent to the brain will correspond to excitation of the R and G receptors at half the light intensity. Alternatively, suppose yellow light excites R and G receptors but with reduced light intensity. Since yellow light lies between red and green in frequency, it is easy to imagine that the signal sent to the brain is similar in the two cases. It has indeed been found that cones have three kinds of pigments absorbing predominantly in the red, blue, and green, respectively. It is now known, however, that the Young–Helmholtz theory of color vision is too simple.

Assuming there are three types of pigment in the eye, the normal person being termed trichromat in consequence, one might anticipate that among all the various types of color blindness there would be three types corresponding to eyes having only two of the three pigments. It is found, in fact, that there are three types of dichromats, or persons whose sensations of color are limited to what can be reproduced by mixing only two primary colors. Of these three types two are very common and the third is somewhat rare. It is possible to deduce the response curves of the three pigments assumed to be present in the human eye by comparing the response to color of a normal person and each type of dichromat. Figure 26.7 illustrates these curves. None of these curves is similar to those of rhodopsin or iodopsin. These pigments are presumably concerned merely with the gross detection of light and therefore are present in large quantity, particularly rhodopsin. Only if the light intensity is great enough would color sensation be present and be produced by pigments which may be present in small amounts.

## TOPIC 26.6    Common Visual Defects

The human eye is capable of bringing objects a very large distance away and as close as 20 cm in proximity into focus. It can resolve particles of matter as small as 0.1 mm and

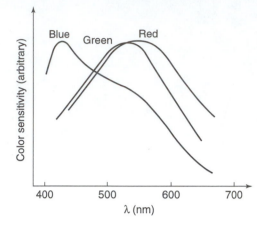

**FIGURE 26.7** Theoretical response curves for the three color-sensitive pigments in the human eye.

even detect one single quantum of light, although it will register this fact only if there are sufficient quanta present to make each quantum appear part of a pattern (see Topic 29.4 for an explanation). The ability to focus occurs as a result of the ciliary muscles which control the curvature of the cornea.

The most common visual defects are myopia, or nearsightedness, and hyperopia, or farsightedness. Both of these defects may be corrected with a proper choice of spectacle or contact lens. Focused vision is achieved, in myopia, for objects which are near while vision becomes blurred for distant objects. The opposite is the case in hyperopia, in which distant objects can be clearly seen and close objects become blurred. In the normal eye parallel rays enter through the lens and converge on the retina where an image is formed. The myopic eye focuses light to a point just in front of the retina due in part to a lens with too short a focal length. The ability of the normal eye to focus on objects over a wide range is called accommodation. In myopia the shortened focal length occurs as a result of either an elongated eyeball or excessive curvature of the cornea. In treating this complaint the objective is to reduce the converging power of the lens, which can be achieved by placing a diverging lens in front of the eye (see Figure 26.8). The hyperopic eye, on the other hand, focuses light behind the retina and is due in part to inadequate curvature or too flat a lens or too short an eyeball. This last condition is accomplished by placing a converging lens in front of the eye. We illustrate these common defects and their cures in Figure 26.8.

In the treatment of myopia or nearsightedness, an alternative to eyeglasses is a surgical procedure called radial keratotomy (RK). Myopia is caused by an elongated eyeball or lens which is too convex and RK involves 8 to 16 hairline incisions, $10^{-3}$ mm in depth, performed by a laser in the curved surface of the cornea. The idea of RK is to flatten the cornea and thus reduce the converging power of the eye to focus distant objects on the retina, as opposed to just in front of it.

The ability to accommodate decreases with age. The onset of presbyopia ("old eye") usually is noticed when reading materials must be held at arm's length to be legible. Distance vision usually is not affected, just the ability of the lens of the eye to increase its strength. The loss of accommodation ability is attributed primarily to stiffening of the lens. Presbyopia eventually affects nearly everyone, whether their eyes are normal, myopic, or hyperopic. One correction for presbyopia is the use of bifocals. A bifocal lens is ground to two different curvatures. The bottom part of the lens has a greater strength than the top since we generally look downward at close objects and need greater strength to see them.

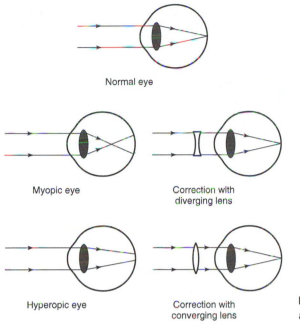

Normal eye

Myopic eye       Correction with<br>diverging lens

Hyperopic eye       Correction with<br>converging lens

**FIGURE 26.8** Defects of the eye and their correction.

In astigmatism, another very common vision defect, the cornea or lens of the eye is not symmetric. To a person with astigmatism, some lines appear darker and sharper because they are properly focused while other lines are not. Astigmatism can be corrected by using a spectacle lens with the opposite symmetry of the astigmatic eye. Astigmatism corrections are accomplished by adding a cylindrical lens to the front of the normal lens of the spectacle. Cylindrical lenses look as if they were cut from a can (cylinder) rather than from a sphere.

A cataract is an opacity of the lens of the eye. Cataracts are common in the elderly, but their advent can be hastened by exposure to ultraviolet radiation, microwaves, nuclear radiation, and certain chemicals. The cloudy lens must be removed, as it is not possible to make it clear again. Once the lens is removed, the patient needs glasses with a large positive strength. In addition, the eye then totally lacks accommodation ability, and bifocals are routinely prescribed.

## TOPIC 26.7  Pinhole Vision

In the invertebrate animals pinhole vision is a common form of vision. If this pinhole is too wide—as in the planarians—the animal will not be able to form a sharp image. One exception to this rule is the pinhole eye of the nautilus, which has a diameter of about 1 mm and can produce a sharp image on the retina. In the curious case of the viper the pinhole vision is formed by the infrared "eyes." In addition to its normal eyes, a viper has two pits located in the upper jaws. Theodore Bullock discovered that these pits act like pinhole eyes and can detect infrared radiation. This radiation consists of invisible electromagnetic waves with the wavelength range $10^{-6}$ to $10^{-3}$ m and is emitted by heated objects. The hotter the object, the relatively more radiation it emits. For example, the body temperature of a rabbit on a cold desert night is higher than its environment. The snake can detect the rabbit's infrared radiation against that of the background or environment.

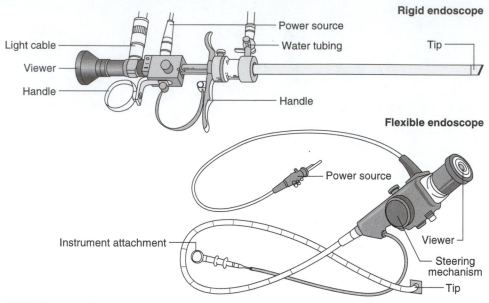

**FIGURE 26.9**   Endoscope.

The information from normal and infrared eyes, it is believed, is integrated in the snake's brain to provide a composite image.

## TOPIC 26.8   Endoscopes

The aim of an endoscope is to see inside body cavities. The instrument typically consists of flexible fiber-optic bundles, a channel or pipe through which air or water can be passed, and other channels through which attachments can be passed (see Figure 26.9). At one end is the head, with a viewing lens, steering device, and power source. The tip has a light to illuminate the area to be observed. For diagnostic purposes it may only be necessary to look down the tube or, if the tip contains a camera which transmits a picture electronically, to view it on a screen. Table 26.3 provides a summary of some common types of endoscopes.

A bronchoscope is a kind of endoscope which is inserted through the mouth or nose, down the bronchial tubes, and into the lungs. It consists of two fiber-optic cables. Light is provided by one to illuminate interior body parts and the other sends back an image for

**TABLE 26.3   Common Types of Endoscopes**

| Instrument | Region observed | Nature |
| --- | --- | --- |
| Cystoscope | Bladder | Rigid |
| Bronchoscope | Bronchi (main airways of the lungs) | Flexible or rigid |
| Gastroscope | Esophagus, stomach, and duodenum | Flexible |
| Colonoscope | Colon (large intestine) | Flexible |
| Laparoscope | Abdominal cavity | Rigid |
| Arthroscope | Knee joint | Rigid |

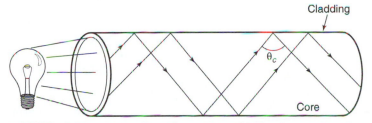

**FIGURE 26.10**  Light being reflected from the cladding of an optical fiber.

viewing. The diagnosis of pulmonary disease is greatly simplified by using a broncho-scope. In fact, even tissue samples can be collected by some bronchoscopes. Another type of endoscope, the colonoscope, has a similar design to the bronchoscope and is inserted through the rectum and utilized to examine the interior of the colon. This is one of the best instruments for diagnosing cancer in its early stages, when it can be treated.

Optical fibers have revolutionized surgical techniques. A small surgical instrument, in arthroscopic surgery, is mounted at the end of a fiber-optic cable so that the surgeon can insert it into a joint (e.g., the knee) with only a tiny incision, causing minimal dam-age to the surrounding tissue. Thus recovery from this procedure is relatively rapid com-pared to more traditional surgical techniques. In the above instruments a light source il-luminates the fiber-optic bundle, the light being emitted in all directions and subsequently reflected from the surrounding structures. Each optical fiber consists of a core, whose in-dex of refraction is $n_{core}$, encased in a cladding of refractive index $n_{clad}$ (see Figure 26.10). The refractive indices have values such that $n_{core} > n_{clad}$ when light strikes the cladding of the fiber (see Figure 26.10). Total internal reflection occurs for angles of incidence greater than the critical angle, $\theta_c$, whose value is determined by

$$\sin \theta_c = \frac{n_{clad}}{n_{core}} \tag{26.17}$$

## TOPIC 26.9  Polarizing Microscope

Some cellular components such as proteins and nucleic acids are strong absorbers of ul-traviolet radiation. They may be quite difficult to distinguish if viewed through an opti-cal microscope, but one can obtain strong image contrast in an ultraviolet microscope even without staining. Before, during, and after cell division the roles of DNA and RNA in liv-ing cells were unraveled using an ultraviolet microscope.

Many biological substances of interest contain birefringent (doubly refracting) re-gions. When viewed in ordinary light, they may well appear transparent and almost in-visible against their surroundings. However, if plane-polarized light illuminates them, in-tensity differences can be produced between those locations which are birefringent and those which are not. This effect is used in the polarizing microscope. The object is illu-minated by light which has passed through a polarizer so that vibrations of the electro-magnetic field are confined to one plane. In the microscope, an analyzer, which is crossed with respect to the polarizer, is inserted between the eyepiece and the objective. When ei-ther no object is viewed or the specimen has the same properties throughout, the field of view is dark. A material specimen which has birefringent regions, however, will appear bright where these regions are located.

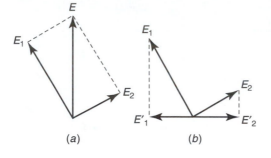

(a)  (b)

**FIGURE 26.11** (a) Plane-polarized light from a polarizer is split into two mutually perpendicular components by the birefringent material in the specimen. (b) When these waves arrive at the analyzer, a component in the permitted vibration direction is present. These two components in the same direction but out of phase interfere with one another.

To discuss the birefringent regions, suppose a light wave striking the birefringent material is split into two components, $E_1$ and $E_2$, vibrating at right angles to one another (see Figure 26.11). The material has different refractive indices, $n_1$ and $n_2$, with respect to two light waves. The phase difference $\theta$ between these two waves will be

$$\theta = \frac{2\pi(n_1 - n_2)d}{\lambda_0} \tag{26.18}$$

where $d$ is the thickness of the material and $\lambda_0$ is the wavelength in vacuum of the light employed. When they arrive at the analyzer, each wave will have a component along the permitted direction of vibration, that is, perpendicular to the initial direction of the wave $E$. In Figure 26.11$b$ these correspond to $E_1'$ and $E_2'$. These components add in the image plane, the amplitude and phase depending on the phase difference $\theta$ and hence the thickness $d$. If $\theta = 0$ or $\theta = 2\pi$, then $E_1'$ and $E_2'$ are vibrating in opposite directions. Thus the amplitude, and hence intensity, of the resulting wave will be a minimum. Similarly, if $\theta = \pi$, the amplitude will be a maximum and for any other value of $\theta$ the amplitude will have an intermediate value.

The value of the thickness $d$ may be deduced by inserting into the beam from the object a wedge of birefringent material (e.g., quartz or mica) whose characteristics are known. By rotating this and sliding the wedge until a suitable thickness is obtained, one can introduce a further phase difference, $-\theta$, which will just cancel out the phase difference $\theta$ introduced by the birefringent specimen material. This maneuver will make the intensity contrast of the image as small as possible, and from the known properties of quartz or mica, the value of $d$ may be deduced. Such a wedge is often used with a polarizing microscope even when $d$ is not required. Many biological samples are thin and the corresponding $\theta$ are small, so the image of the biological material may not show up. In this scenario a wedge is used to improve the contrast. The wedge is so oriented in the microscope that any phase difference it introduces between $E_1'$ and $E_2'$ enhances the detail of the image of the biological material.

Many facts obtained by staining and using an ordinary microscope have been confirmed with the polarizing microscope and others have been corrected. Cells need not be killed as they are with staining, so it has been possible to view changes as they occur in living cells (e.g., the process of cell division).

## SOLVED PROBLEM

**26.1** Calculate the image position produced by the cornea alone for the extremes of object position: (a) 25 cm, the so-called near point of the eye, and (b) at infinity, at the so-called far point of the eye.

**Solution**  (a) For an object at the near point,

$$n_1 = 1.00 \quad \text{(for air)} \qquad n_2 = 1.34 \qquad d_o = 25 \text{ cm} \qquad r = 0.8 \text{ cm}$$

Thus from equations (26.4) and (26.6)

$$d_i = 3.48 \text{ cm}$$

(b) For an object at the far point,

$$n_1 = 1.00 \quad \text{(for air)} \qquad n_2 = 1.34 \qquad d_o = \infty \qquad r = 0.8 \text{ cm}$$

thus from equations (26.4) and (26.6)

$$d_i = 3.15 \text{ cm}$$

Note that both image points are beyond the retina, which is 2.5 cm behind the cornea.

# EXERCISES

**26.1**  A farsighted person has a near point that is 48.0 cm from her eyes. She wears eyeglasses that are designed to enable her to read a newspaper held at a distance of 25.0 cm from her eyes. Find the focal length of the eyeglasses, assuming that they are worn (a) 2.0 cm from the eyes and (b) 3.0 cm from the eyes.

**26.2**  Suppose a person wears contact lenses that have a focal length of 35.1 cm. The lenses are designed so that this person can read a book held as close as 25.0 cm. Where is the near point of the unaided eyes?

**26.3**  A nearsighted person has a far point located only 250 cm from his eyes. Determine the focal length of contact lenses that will enable him to see distant objects clearly.

**26.4**  An optometrist prescribes contact lenses that have a focal length of 50.0 cm. (a) Are the lenses converging or diverging? (b) Is the person who wears them nearsighted or far-sighted? (c) Where is the unaided near point of the person located if the lenses are de-signed so that objects no closer than 30.0 cm can be seen clearly?

**26.5**  A nearsighted person wears contacts to correct for a far point that is only 3.50 m from his eyes. The near point of his unaided eye is 25.0 cm from the eyes. If he does not re-move the lenses while reading, how close can he hold a book and see it clearly?

**26.6**  John is farsighted and has a near point located 120 cm from his eyes. Mary is also far-sighted, but her near point is 60.0 cm from her eyes. Both have glasses that correct their vision to a normal near point (25.0 cm from the eyes), and both wear the glasses 2.0 cm from the eyes. Relative to the eyes, what is the closest object that can be seen clearly (a) by Mary when she wears John's glasses and (b) by John when he wears Mary's glasses?

**26.7**  A person has far points of 5.0 m from the right eye and 7.5 m from the left eye. Write a prescription for the refractive power of each corrective contact lens.

**26.8**  A woman can read the large print in a newspaper only when it is at a distance of 75 cm or more from her eyes. (a) Is she myopic or hyperopic? (b) What should be the refractive power of her glasses (worn 2.0 cm from the eyes) so she can read the newspaper at a dis-tance of 25 cm from the eyes?

**26.9**  A person holds a book 25 cm in front of the effective lens of her eye; the print in the book is 2.5 mm high. If the effective lens of the eye is located 2.5 cm from the retina, what is the size of the print image on the retina?

**26.10** A nearsighted person cannot read a sign that is more than 5.0 m from his eyes. To deal with this problem, he wears contact lenses that do not correct his vision completely but do allow him to read signs located up to distances of 10.0 m from his eyes. What is the focal length of the contacts?

**26.11** At some stage of his life, a man requires contact lenses ($f = 75.0$ cm) to read a book held 25.0 cm from his eyes. Five years later, he finds that while wearing these contacts he must now hold a book 30.0 cm from his eyes. (a) By what distance has his near point changed? (b) What focal length lenses does he now require to read a book at 25.0 cm?

**26.12** The far point of a nearsighted person is 5.0 m from her eyes, and she wears contacts that enable her to see distant objects clearly. A road sign is 20.0 m away and 2.5 m high. (a) When she looks through the contacts at the sign, what is its image distance? (b) How high is the image formed by the contacts?

**26.13** The contacts worn by a farsighted person allow her to see objects clearly that are as close as 25.0 cm, even though her uncorrected near point is 80.0 cm from her eyes. When she is looking at a poster, the contacts form an image of the poster at a distance of 200 cm from her eyes. (a) How far away is the poster actually located? (b) If the poster is 0.550 m tall, how tall is the image formed by the contacts?

**26.14** A person with myopia is unable to focus on objects which are beyond 100 cm from the eye. Determine the power of the diverging lens required to correct the person's vision to normal standards.

**26.15** A person with hyperopia is unable to clearly focus on objects closer than 90 cm from the eye. Determine the power of the converging lens required to allow the person to see the same object at a distance of 25 cm.

**26.16** Compare the range of light intensities to which the eye responds and the range of sound intensities to which the ear responds.

**26.17** A compound microscope consists of an objective lens with a focal length of 12 mm and an eyepiece lens with a focal length of 25 mm. Determine the distance between the lenses when the microscope is completely focused on a sample 10.0 mm from the objective lens.

**26.18** Would any inconvenience be caused if the spectacle lenses of a nearsighted person were rotated in their frames?

**26.19** One eye with myopia has a far point of 100 cm, and another eye with myopia has a far point of 200 cm. Which eye requires a corrective spectacle lens with focal length of greater magnitude?

**26.20** A myopic eye has a far point of 2.50 m and a near point of 0.150 m. Calculate the power of the spectacle lens required to correct the myopia (i.e., to make the far point infinity). Where should an object be placed so that the image produced by the spectacle lens would lie at the near point of the eye itself?

**26.21** A doctor is examining a mole using a magnifying glass of focal length 0.16 m. The lens is held 0.145 m from the mole. (a) Calculate the position of the image formed. (b) What is the magnification? (c) What is the height of the image if the mole is 0.007 m in diameter?

**26.22** The lens-to-retina distance of a patient is 0.0201 m and the totally relaxed strength of her eye is 50.0 D. (a) What spectacle lens strength will correct her distant vision? (b) If the eye has an 8% ability to accommodate, what will be the closest object she can see clearly with her glasses on?

**26.23** An amoeba is 0.00515 m away from the objective lens of a microscope. (a) If the objective lens has a focal length of 0.00500 m, where is the image formed by the objective? (b) What is the magnification of this image? (c) The eyepiece of the microscope has a focal length of 0.025 m and is 0.25 m away from the objective. Where is the final image? (d) What is the magnification produced by the eyepiece? (e) What is the overall magnification?

# CHAPTER 27

# INTERFERENCE AND WAVE NATURE OF LIGHT

## PHYSICAL BACKGROUND

The principle of linear superposition states that when two or more waves are present simultaneously in the same region of space, the resultant wave is the sum of the individual waves. According to this principle, two or more light waves can interfere constructively or destructively when they exist in the same place at the same time provided they originate from coherent sources. Two sources are coherent if they emit waves that have a constant phase relationship.

In Young's double-slit experiment, light passes through a pair of closely spaced narrow slits and produces a pattern of alternating bright and dark fringes on a viewing screen. The fringes arise because of constructive and destructive interference. The angle $\theta$ for the $m$th higher order bright fringe is given by

$$\sin \theta = \frac{m\lambda}{d} \qquad (27.1)$$

where $d$ is the spacing between the narrow slits, $\lambda$ is the wavelength of the light, and $m = 0, \pm 1, \pm 2, \pm 3, \ldots$ . Similarly, the angle for the dark fringes is given by

$$\sin \theta = \frac{(m + 1/2)\lambda}{d} \qquad (27.2)$$

Constructive and destructive interference of light waves can occur with thin films of transparent materials. The interference occurs between light waves that reflect from the top and bottom surfaces of the film. One important factor in thin-film interference is the thickness of the film relative to the wavelength of light within the film. The wavelength within the film is

$$\lambda_{\text{film}} = \frac{\lambda_{\text{vacuum}}}{n} \qquad (27.3)$$

where $n$ is the refractive index of the film. A second important factor is the phase change that can occur when light undergoes reflection at each surface of the film.

Diffraction is a bending of waves around obstacles or the edges of an opening. Diffraction is an interference effect that can be explained with the aid of Huygens's principle. This principle states that every point on a wave front acts as a source of tiny spherical wavelets that move forward with the same speed as the wave; the wave front at a later instant is the surface that is tangent to the wavelets. When light passes through a single narrow slit and falls on a viewing screen, a pattern of light and dark fringes is formed because of the superposition of such wavelets. The angle $\theta$ for the $m$th dark fringe on either side of the central bright fringe is given by

$$\sin \theta = \frac{m\lambda}{W} \tag{27.4}$$

where $W$ is the slit width and $m = \pm1, \pm2, \pm3, \ldots$.

The resolving power of an optical instrument is the ability of the instrument to distinguish between two closely spaced objects. Resolving power is limited by the diffraction that occurs when light waves enter an instrument, often through a circular opening. Consideration of the diffraction fringes for a circular opening leads to the Rayleigh criterion for resolution. This criterion specifies that two point objects are just resolved when the first dark fringe in the diffraction pattern of one falls directly on the central bright fringe in the diffraction pattern of the other. According to this specification, the minimum angle (in radians) that two point objects can subtend at an aperture of diameter $D$ and still be resolved as separate objects is

$$\theta_{\min} = 1.22\frac{\lambda}{D} \tag{27.5}$$

where $\lambda$ is the wavelength of the light.

A diffraction grating consists of a large number of parallel, closely spaced slits. When light passes through a diffraction grating and falls on a viewing screen, the light forms a pattern of bright and dark fringes. The bright fringes are referred to as principal maxima and are found at an angle $\theta$ such that

$$\sin \theta = \frac{m\lambda}{d} \tag{27.6}$$

where $d$ is the separation between two successive slits and $m = 0, \pm1, \pm2, \pm3, \ldots$.

## TOPIC 27.1 Resolution of the Human Eye

The criterion for resolvability of images produced by light passing through an aperture was formulated by Lord Rayleigh and can be expressed similarly to equation (27.5) as

$$\theta_{\min} = 1.22\frac{\lambda}{a} \tag{27.7}$$

where $\theta_{\min}$ is the minimum angular separation in radians, $\lambda$ is the wavelength of light in the medium, and $a$ is the diameter of the aperture.

The pupil of the eye is a variable aperture with a minimum diameter of about 2 mm. The eye is most sensitive to light of $\lambda = 500$ nm. Assuming that most of the refraction takes place in the cornea, we take $n = 1.33$ and find the value of $\theta_{\min}$ to be

$$\theta_{\min} = \frac{(1.22)(500 \times 10^{-9} \text{ m})}{(1.33)(2 \times 10^{-3} \text{ m})} = 2.3 \times 10^{-4} \text{ rad} \tag{27.8}$$

Since $\ell = 0.025$ m is the typical corneal–retinal distance, angular separation translates into linear separation according to

$$\theta_{\min} = \frac{d}{\ell} \tag{27.9}$$

giving $d = (2.3 \times 10^{-4} \text{ rad})(0.025 \text{ m}) \cong 6 \ \mu\text{m}$. Since the average cone separation in the eye's fovea is about 2 $\mu$m, the minimum distance of 6 $\mu$m calculated above corresponds to the span of three cones. In other words, two active cones must be separated by an inactive cone in order to create the sensation of an image.

Note that $\theta_{\min}$ depends inversely on the diameter of the pupil, $a$. Eagles, which are famous for their sharp eyesight, have their pupils twice as big as humans, leading to better resolution. It is also interesting to note that placing two objects at the near point of the eye ($N = 25$ cm) gives their linear distance as

$$x = \theta_{\min}N = 57 \ \mu\text{m} \tag{27.10}$$

As shown in the first part of this discussion, the smallest discernible detail makes an image about 6 $\mu$m in diameter. This is two or three times the size of retinal cells. Outside the fovea, light-sensitive cells are not as closely spaced, and peripheral visual acuity is consequently less by a factor of about 10. The eye cannot detect detail that creates an image smaller than the spacing of light-sensitive cells. For the eye to resolve two adjacent dots, their images must fall on two nonadjacent cells. If the two dots are images of small light sources and fall on adjacent cells, then they are interpreted as a single larger dot.

In addition to the separation between retinal cells, which limits the eye's resolution, there also exists a diffraction limit. In the eye, diffraction occurs because light passes through the pupil. The diameter of the pupil varies with light intensity, averaging about 3 mm. The outcome of the effect in the eye is such that a point source of light will not make a point image on the retina: It will make a spot about 2 $\mu$m in diameter, very nearly the size of a retinal cell.

## SOLVED PROBLEMS

**27.1**  A biological sample contains birefringent spindles of thickness 5.00 $\mu$m, and light of wavelength 690 nm in vacuum splits into beams which traverse the spindles with velocities of $2.257 \times 10^8$ and $2.262 \times 10^8$ m/s. What phase difference is introduced between the two beams in passage through the spindles?

**Solution**  The refractive indices of the spindles for the two beams traversing them are $n_1 = c/v_1$ and $n_2 = c/v_2$, where $v_1$ and $v_2$ are the respective velocities. The thickness $d$ of the spindles represents $d/\lambda_1$ and $d/\lambda_2$ wavelengths, respectively, for the two beams. Therefore, the phase difference is expressed in radians as

$$\Delta\phi = 2\pi\,\frac{dc}{\lambda}\left(\frac{1}{v_1} - \frac{1}{v_2}\right) = 0.134 \text{ rad}$$

**27.2**  (a) A hang glider is flying at an altitude $H = 120$ m. Green light (wavelength 555 nm in vacuum) enters the pilot's eye through a pupil that has a diameter $d = 2.5$ mm. The average index of refraction of the material in the eye is approximately $n = 1.36$. Determine how far apart two point objects must be on the ground if the pilot is to have any hope of distinguishing between them.
(b) An eagle's eye has a pupil with a diameter $d = 6.2$ mm and has about the same refractive index as does a human eye. Repeat part (a) for an eagle flying at the same altitude as the glider.

**Solution**

**(a)** The wavelength $\lambda$ within the eye takes into account the refractive index of the eye and is given by $\lambda = \lambda_{vacuum}/n = 555$ nm/1.36 $= 408$ nm. Therefore,

$$\theta_{min} \approx 1.22\frac{\lambda}{d} = 1.22\left(\frac{408 \times 10^{-9} \text{ m}}{2.5 \times 10^{-3} \text{ m}}\right) = 2.0 \times 10^{-4} \text{ rad}$$

According to equation (27.5), $\theta_{min}$ in radians is $\theta_{min} \approx s/H$, so

$$s \approx \theta_{min}H = (2.0 \times 10^{-4} \text{ rad})(120 \text{ m}) = 0.024 \text{ m}$$

**(b)** Since the pupil of an eagle's eye is larger than that of the human eye, diffraction creates less of a limitation for the eagle. A calculation like that above using $d = 6.2$ mm reveals that the diffraction limit for the eagle is 0.0096 m.

# EXERCISES

**27.1** Suppose you are looking down at the earth from inside a jetliner flying at a cruising altitude of 10,000 m. The pupil of your eye has a diameter of 2.00 mm, and the average refractive index of the material in the eye is 1.36. Determine how far apart two cars must be on the ground to distinguish between them in (a) red light (wavelength 665 nm in vacuum) and (b) violet light (wavelength 405 nm in a vacuum).

**27.2** It is claimed that some professional baseball players can see which way the ball is spinning as it travels toward home plate. One way to judge this claim is to estimate the distance at which a batter can first hope to resolve two points on opposite sides of a baseball which has a diameter of 0.0738 m. (a) Estimate the distance, assuming that the pupil of the eye has a diameter of 2.0 mm, the material within the eye has a refractive index of 1.36, and the wavelength of the light is 550 nm in vacuum. (b) Considering that the distance between the pitcher's mound and home plate is 18.4 m, can you rule out or verify the claim based on your answer to part (a)?

**27.3** Late one night on a highway, a car speeds by you and fades into the distance. Under these conditions the pupils of your eyes (average refractive index 1.36) have diameters of about 7.0 mm. The tail lights of this car are separated by a distance of 1.5 m and emit red light (wavelength 660 nm in vacuum). How far away from you is this car when its tail lights appear to merge into a single spot of light because of the effects of diffraction?

**27.4** The pupil of an eagle's eye has a diameter of 6.0 mm. The average refractive index in the eye is 1.36. Two field mice are separated by 0.012 m. From a distance of 200 m, the eagle sees them as one unresolved object and dives toward them at a speed of 20 m/s. Assume that the eagle's eye detects light that has a wavelength of 550 nm in a vacuum. How much time passes until the eagle sees the mice as separate objects?

**27.5** In a dot matrix printer, an array of dots is used to form printed characters. If the dots are close enough together, they cannot be resolved individually by the eye and therefore appear to form solid lines. Suppose that the pupil of the eye has a diameter of 2.0 mm in bright yellow-green light (wavelength 563 nm in vacuum), that the material in the eye has an average refractive index $n = 1.36$, and that the printed page is to be read at a distance of 0.50 m. Considering the limit created by diffraction, find the smallest separation between the dots that the eye can see.

**27.6**   Suppose the pupil of your eye were elliptical instead of circular in shape, with the long axis of the ellipse oriented in the vertical direction. (a) Would the resolving power of your eye be the same in the horizontal and vertical directions? (b) In which direction would the resolving power be greatest? Justify your answer by discussing how the diffraction of light waves would differ in the two directions.

**27.7**   Light enters the eye through the pupil. On the inner side of the pupil is material with an average index of refraction $n = 1.36$. Suppose, for comparison, that a sheet of opaque material has a hole cut into it. Light passes through this hole into the air on the other side. Assuming that the pupil and the hole have the same diameter, in which case does green light ($\lambda_{vacuum} = 550$ nm) diffract more, when it enters the eye or when it passes through the hole in the opaque material?

# CHAPTER *28*

## PARTICLES AND WAVES

## PHYSICAL BACKGROUND

At a constant temperature, a perfect blackbody absorbs and reemits all the electromagnetic radiation that falls on it. Max Planck calculated the emitted radiation intensity per unit wavelength as a function of wavelength. In his theory, Planck assumed that a blackbody consists of atomic oscillators that can have only quantized energies. Planck's quantized energies are given by $E = nhf$, where $n = 0, 1, 2, 3, \ldots$, $h$ is Planck's constant equal to $6.63 \times 10^{-34}$ J $\cdot$ s, and $f$ is the vibration frequency. It is therefore deduced that all electromagnetic radiation consists of photons, which are packets of energy. The energy of a photon is given as

$$E = hf \tag{28.1}$$

where $f$ is the frequency of the light. A photon in a vacuum always travels at the speed of light $c$ and has no mass.

The photoelectric effect is the phenomenon in which light shining on a metal surface causes electrons to be ejected from the surface. The work function $W_0$ of a metal is the minimum work that must be done to eject an electron from the metal. In accordance with the law of conservation of energy, the electrons ejected from a metal have a maximum kinetic energy $KE_{max}$ that is related to the energy $hf$ of the incident photon by

$$hf = KE_{max} + W_o \tag{28.2}$$

The Compton effect is the scattering of a photon by an electron in a material, the scattered photon having a smaller frequency than the incident photon. The magnitude $p$ of the photon's momentum is

$$p = \frac{h}{\lambda} \tag{28.3}$$

and in the Compton effect, part of it is transferred to the recoiling electron. The difference between the wavelength $\lambda'$ of the scattered photon and the wavelength $\lambda$ of the incident photon is related to the scattering angle $\theta$ by

$$\lambda' - \lambda = \frac{h}{mc}(1 - \cos \theta) \tag{28.4}$$

where $m$ is the mass of the electron and the quantity $h/mc$ is known as the Compton wavelength of the electron. Thus, the above effects demonstrate that photons have some particlelike properties.

The wave–particle duality refers to the fact that a wave can exhibit particlelike characteristics and a particle can exhibit wavelike characteristics. The de Broglie wavelength of a particle is defined as

$$\lambda = \frac{h}{p} \tag{28.5}$$

**TABLE 28.1  Comparison of Electromagnetic and Particle Waves**

| Electromagnetic waves | Particle waves |
|---|---|
| Magnetic and electric fields oscillate | Particle's wave function oscillates |
| Travel in packets called photons | Particles travel singly or in beams |
| Produced by accelerating a charge | Particle beams can be produced by accelerators, reactors, or potential gradients |
| Energy of a photon $E = h\upsilon$ | Energy of a particle $E = \frac{1}{2}mv^2$ |
| Travels with speed of light $c$ | Can travel at any velocity $(v < c)$ |
| Wavelength $\lambda = c/\upsilon = hc/E$ | Wavelength $\lambda = h/p = h/mv$ |

where $p$ is the magnitude of the momentum of the particle. Because of its de Broglie wavelength, a particle can exhibit wavelike characteristics. The wave associated with a particle is a wave of probability.

The Heisenberg uncertainty principle places limits on our knowledge about the behavior of a particle. The uncertainty principle states that

$$\Delta p_y \, \Delta y \geq \frac{h}{2\pi} \tag{28.6}$$

where $\Delta y$ and $\Delta p_y$ are, respectively, the uncertainties in the position and momentum of the particle in the $y$-direction. The uncertainty principle also states that

$$\Delta E \, \Delta t \geq \frac{h}{2\pi} \tag{28.7}$$

where $\Delta E$ is the uncertainty in the energy of a particle when the particle is in a certain state and $\Delta t$ is the time interval during which the particle is in that state. Table 28.1 summarizes the analogies between electromagnetic waves and particle waves in quantum theory.

# EXERCISES

**28.1**  Suppose a biological membrane is examined using an electron microscope. The position of the electron is determined to an accuracy of 0.5 $\mu$m. What is the electron's minimum uncertainty in velocity? What is the kinetic energy of an electron with this velocity?

**28.2**  The mass of a chlorine ion is $5.86 \times 10^{-26}$ kg. It has been located in a cell membrane to an accuracy of 1.2 $\mu$m. Calculate the associated uncertainty in its velocity. What is the corresponding kinetic energy of the ion?

# CHAPTER 29

# NATURE OF THE ATOM

## PHYSICAL BACKGROUND

The smallest building block of matter is the atom. All atoms consist of three elementary particles: the proton, the neutron, and the electron. The atom can be further subdivided into (a) a positively charged nucleus which is a densely packed core of protons and neutrons and (b) negatively charged electrons that occupy orbits around the nucleus.

The nucleus is composed of protons and neutrons, collectively known as nucleons, and it accounts for the vast majority of the atomic mass. The proton charge is identical in magnitude to that of the electron but is positive. The neutron is uncharged and is slightly heavier than the proton. The mass and charge of the proton, neutron, and electron are given in Table 29.1.

The following quantities describe the identity and structure of an atom:

Mass number $A$ is the number of protons and neutrons.

Atomic number $Z$ is the number of protons.

Neutron number $N$ is the number of neutrons and is equal to $A - Z$.

$Z$ is equal to the number of electrons if the atom is neutral since the number of protons must always equal the number of electrons so that charge is conserved. In symbolic form, the atom can be represented by $_{Z}^{A}X$, where $X$ symbolizes the chemical element.

The Bohr model applies to atoms or ions that have only a single electron orbiting a nucleus containing $Z$ protons. This model assumes that the electron exists in circular orbits that are called stationary orbits. According to this model, a photon is emitted only when an electron changes from a higher energy orbit to a lower energy orbit. The model also assumes that the orbital angular momentum $L_n$ of the electron can only have values that are integer multiples of Planck's constant divided by $2\pi$: $L_n = nh/2\pi$, $n = 1, 2, 3, \ldots$. With the assumptions above, it can be shown that the $n$th Bohr orbit of an electron has a radius (in meters) given by the formula

$$r_n = (5.29 \times 10^{-11})\frac{n^2}{Z} \tag{29.1}$$

and that the total energy associated with this orbit is

$$E_n = -(13.6 \text{ eV})\frac{Z^2}{n^2} \tag{29.2}$$

The ionization energy is the energy needed to remove an electron completely from an atom. The Bohr model predicts that the wavelengths comprising the line spectrum emitted by a hydrogen atom are given by

$$\frac{1}{\lambda} = \frac{2\pi^2 mk^2 e^4}{h^3 c} Z^2 \left( \frac{1}{n_f^2} - \frac{1}{n_i^2} \right) \tag{29.3}$$

where $m$ is the electron mass, $e$ is the electronic charge, and $k = 8.988 \times 10^9 \text{ Nm}^2/C^2$, $n_f$ is the principal quantum number of the final state to which a transition is made and

**TABLE 29.1 Mass and Charge of Proton, Neutron, and Electron**

|  | Mass (kg) | Charge (C) |
| --- | --- | --- |
| Proton | $1.67 \times 10^{-27}$ | $1.6 \times 10^{-19}$ |
| Neutron | $1.68 \times 10^{-27}$ | 0 |
| Electron | $9.11 \times 10^{-31}$ | $-1.6 \times 10^{-19}$ |

$n_i$ is the principal quantum number of the initial state. Thus, $n_i$, $n_f = 1, 2, 3, \ldots$, with $n_i > n_f$. A line spectrum is a series of discrete electromagnetic wavelengths emitted by the atoms of a low-pressure gas that is subjected to a sufficiently high potential difference. Certain groups of discrete wavelengths are referred to as "series." The line spectrum of atomic hydrogen includes, among others, the Lyman series, the Balmer series, and the Paschen series of wavelengths.

Quantum mechanics describes the hydrogen atom in terms of four quantum numbers: (a) the principal quantum number $n$, which can have the integer values $n = 1, 2, 3, \ldots$; (b) the orbital quantum number $\ell$, which can have the integer values $\ell = 0, 1, 2, \ldots, n - 1$; (c) the magnetic quantum number $m_l$, which can have the positive and negative integer values $m_l = -\ell, \ldots, -2, -1, 0, +1, +2, \ldots, +\ell$; and (d) the spin quantum number $m_s$, which, for an electron, can be $m_s = \pm\frac{1}{2}$. The Pauli exclusion principle states that no two electrons in an atom can have the same set of values for the four quantum numbers $n$, $\ell$, $m_l$, and $m_s$. This principle determines the way in which the electrons in multiple-electron atoms are distributed into shells (defined by the value of $n$) and subshells (defined by the values of $n$ and $\ell$). The arrangement of the periodic table of the elements is related to the exclusion principle.

A laser is a device that generates electromagnetic waves via a process known as stimulated emission. In this process, one photon stimulates the production of another photon by causing an electron in an atom to fall from a higher energy level to a lower energy level. Because of this mechanism of photon production, the electromagnetic waves generated by a laser are coherent and may be confined to a very narrow beam.

## TOPIC 29.1   Fluorescence in Biomolecules

In many plant materials the phenomenon of fluorescence has been known for more than a century (e.g., in quinone). For many years fluorescence methods have been used to check the presence of trace elements, drugs, and vitamins. However, it is surprising that fluorescent assays have only recently been used extensively for identification and location studies, particularly in view of the sensitivity of fluorescence detection and that, under suitable chemical treatment, almost all molecules are capable of fluorescing. This method affords many of the possibilities which radioactive techniques also offer but without any of the associated hazards. In many radioactive assays, studies *in vivo* require laboriously synthesized substances. The same investigations can be undertaken more simply and with equal accuracy using fluorescence. When it is combined with techniques for separation like electrophoresis, the method can detect impurities at concentrations of less than 1 in $10^{10}$.

When a substance to be studied is in solution, the container, the solvent, and other substances may also fluoresce. Clearly, no solvent should be used which fluoresces in the same general region, and the light used as the stimulant should be strictly monochromatic, even using filters if necessary. In some elaborate studies it may be necessary to use a spectrometer to analyze the fluorescent light and the wavelengths of the various components measured before identification becomes possible.

Fluorescence is routinely used to assay vitamin contents of food after various types of preparation and processing. It is also utilized to follow flow rates and pool formation in sewage systems. One can also follow metabolic activity by the excretion rates of substances in urine. In biochemistry, because of its sensitivity, it is much in use to investigate the location of enzymes and coenzymes in organisms and to follow the fate of a drug and its metabolic products. Fluorescence can also be used to determine the amino acid sequence in proteins. The detection of drug abuse, for example, the absorption and excretion of quite small amounts of lysergic acid diethylamide (LSD), also now uses fluorescence techniques.

To understand fluorescence, we must augment our elementary theory of quantum energy levels for electrons in atoms by vibrational degrees of freedom which are present in the energy structure of molecules. Vibrational energy levels arise because the molecular bond acts like a system of two masses connected by an elastic spring which has a natural frequency of vibration $\omega$. The energies are quantized according to the formula

$$E_n = \hbar\omega(n + \tfrac{1}{2}) \tag{29.4}$$

where $n = 0, 1, 2, \ldots$ This is illustrated in Figure 29.1. Transitions between vibrational energy levels can be induced by photons provided $\Delta n = \mp 1$. It should be mentioned that each electronic state may have a number of vibrational states. At room temperature most of the molecules will occupy the ground electronic state and its lowest vibrational energy level.

If a molecule does find itself in a high vibrational level, it will, after several collisions, drop to a lower level and give up its excess energy to other molecules in its vicinity. This process is called vibrational relaxation. A molecule in an excited electronic and vibrational state will undergo vibrational relaxation followed by an emission of a photon of energy $h\nu'$ taking it down to the ground state. When the emitted photon is in the visible range, the process is called fluorescence. The cycle of excitation by a photon of energy $h\nu$, vibrational relaxation, and eventual fluorescence takes typically $10^{-7}$ s. Figure 29.2 illustrates the quantum mechanical origin of fluorescence and $\nu$ labels vibrational states.

When we reincorporate the presence of spin in the electronic states, we find out that the ground state is usually a so-called singlet state; that is, it has a net spin of zero for the two electrons occupying it. Some excited states are singlet states also. Some other excited states may involve a net spin of unity, which is called a triplet state. We show the dis-

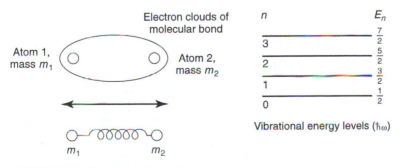

Molecular bond, mass–spring analogy

**FIGURE 29.1**    Molecular bond–mass spring analogy.

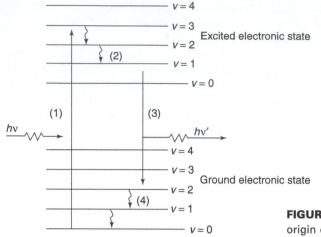

**FIGURE 29.2** Quantum mechanical origin of fluorescence.

tinction in Figure 29.3. This gives rise to the phenomenon of phosphorescence in which the return of the photoexcited electron to the ground state involves an intermediate transition to an excited triplet state. The distinguishing feature of phosphorescence is that it takes place, at times, even several seconds after the photon absorption process. In addition, the phosphorescent photon has an energy $h\nu''$ which is always lower than the absorbed photon $h\nu$ or the fluorescent photon $h\nu'$ (See Figure 29.4 for illustration.)

## TOPIC 29.2   Bioluminescence and Marine Organisms

Bioluminescence is the ability of glowing in the dark and is exhibited by several types of marine organism such as the sea anemone, jellyfish, and squid. Chemically, bioluminescence is a result of oxygen added to luciferin, which is a substance capable of emitting light. To promote this reaction mechanism, an enzyme called luciferase is required. The reaction generates energy, which is released as light in the blue-green range. The production of light energy from chemical energy can be nearly 100% efficient with very little heat produced. This contrasts strongly with the household light bulb, which is only 10% efficient, the other 90% of its energy being changed into heat.

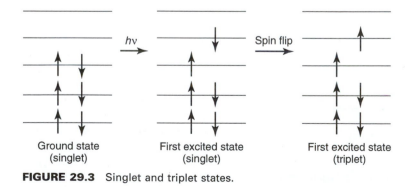

**FIGURE 29.3**   Singlet and triplet states.

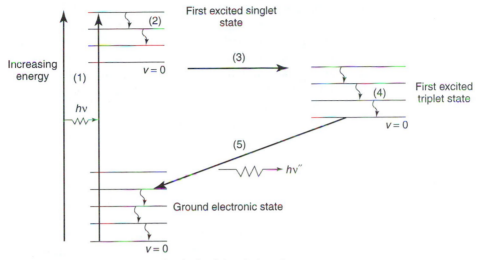

First excited singlet
state

Increasing
energy

First excited
triplet state

Ground electronic state

**FIGURE 29.4** Quantum mechanical origin of phosphorescence.

## TOPIC 29.3   DNA: Information and Damage

Genetic coding can be discussed using information theory. Deoxyribonucleic acid (DNA), in the living cell, is the primary genetic material and therefore specifies the information which can be passed on from one generation of cells to the next. It is also intimately involved with the synthesis of new material. For a long time it was not known how information from DNA was transferred to other parts of the cell so that the production of protein can take place. As we have seen earlier (see Chapter 18), DNA is a long, thin molecule made up of two strands, each of which consists of sequences of nucleotides. The strands are coiled around each other in the form of a helix and are held together by hydrogen bonds between pairs of nucleotide bases. The two strands may come apart when the nucleotides are exposed. When this happens, it is believed, specific sequences of the nucleotides are coded to form the amino acids which make up proteins. We also saw in an earlier chapter that there are only four distinct nucleotides in DNA. Thus, discovering that any one of these is in a specific location along a strand increases the information by $\log_2 4 = 2$ bits. As there are 20 amino acids, the identification of one particular amino acid requires $\log_2 20 = 4.32$ bits of information. We therefore conclude that coding must be done by at least three nucleotides arranged in order.

In a cell the DNA is too valuable to do the work of synthesizing proteins directly. Transcription of the coded DNA is accomplished by producing from it what are called messenger ribonucleic acid (RNA) molecules. The production of proteins is directed by the RNA, when they travel to the ribosomes, by specifying the sequence in which amino acids will be linked. Four nucleotides can be coded onto the messenger RNA by the DNA strand, namely adenine (A), cytosine (C), guanine (G), and uracil (U). To identify the groups of nucleotides that are coded for specific amino acids, scientists produced synthetic messenger RNA and examined the protein that was synthesized as a result. What was found was that if the RNA contains only uracil, for instance, the protein is made up only of the amino acid phenylalinine chain containing only lysine. As expected, the code is based on combinations of three nucleotides and there are duplicate ways in which any amino acid can be called up.

Three of the 64 possible combinations do not code for any amino acid and are believed to act as terminators. When the ribosome reaches this portion of the messenger RNA chain, the growth of the protein is halted. It is also believed that the code is universal and applies to all organisms. If a mutation or error occurs in the DNA, all copies will carry it and incorrect synthesis will take place at the ribosomes. As an example, GAA and GAG code for glutamic acid while CAT and CAC for histidine. If, as in this example, an error of U for A occurs in the appropriate portion of a strand of DNA which provides information for the production of hemoglobin, then it is believed that valine is built at that point instead of glutamic acid. The hemoglobin which results is called sickle-cell hemoglobin, the name arising from the shape of the red blood cells. Those people with this disease suffer severely from anemia. Many other hereditary diseases are very likely to be caused by minor coding errors of this sort.

As discussed in a previous chapter (see Chapter 24), electromagnetic radiation, especially in the ultraviolet range (200 to 350 nm), can be very harmful to living systems. Cellular injury, mutation, and lethality have been demonstrated to result from high-intensity UV exposure. One effect seen in samples of DNA extracted from UV-irradiated cells is the damage in the form of bond breaks called local denaturation and entanglement with proteins called protein cross-linking (see Figure 29.5).

## TOPIC 29.4  Quantum Response of the Eye

It is interesting to ask how many quanta are reaching each receptor on the retina when an eye sees a flash of light of wavelength 505 nm (a dark-adapted eye is most sensitive to this wavelength) at the threshold intensity. Materials of which the cornea, lens, and aqueous and vitreous humor (see Chapter 26) are made completely absorb wavelengths below 380 nm and also absorb, to a lesser extent, light with wavelengths in the visible region.

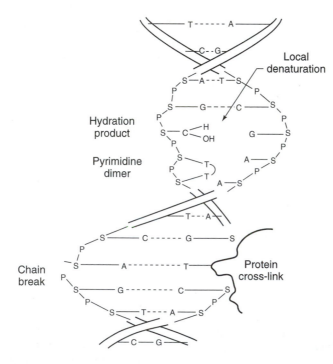

**FIGURE 29.5**  Schematic illustration of the various defects found in DNA that has been irradiated with ultraviolet light.

Only a few percent of quanta striking the cornea reach the retina. The rest are absorbed by materials along the path. Thus a measurement of the intensity of the light, $I$, striking the cornea does not allow a direct calculation of the average number, $\bar{n}$, of photons activating a single receptor on the retina. However, the two are linearly related; that is, doubling the intensity $I$ will automatically double the value of $\bar{n}$. Therefore,

$$\bar{n} = kI \tag{29.5}$$

where $k$ is a constant related to the absorbing ability of the materials of the eye and also depends on many other factors.

When the eye is near the threshold of the vision process, $\bar{n}$ and therefore $I$ are small numbers. Although the average number of quanta reaching each receptor on the retina is $\bar{n}$, some receptors will receive more and some less because of the random nature of the process. The statistical distribution of the frequency of occurrence of events of small probability around a mean value is well known. Thus we can work out the proportion of receptors receiving $\bar{n} - 1, \bar{n} - 2, \bar{n} - 3, \ldots$ quanta and the proportion of receptors receiving $\bar{n} + 1, \bar{n} + 2, \bar{n} + 3, \ldots$. We can also assume that a particular receptor will not pass on an electrical signal to its associated nerve fiber unless a minimum number $n$ of quanta stimulate it. One can then work out the proportion of receptors which will record a signal for various values of $\bar{n}$ and $n$.

Obviously, the smaller is $n$ and the larger is $\bar{n}$, the greater is the probability that any particular flash of light will be observed. Figure 29.6 shows the result of theoretical calculations (for $n = 1, 2, 3, 4$) of the variation of the frequency $\nu$ of response of an observer to a light flash as the number of quanta reaching the retina is increased. For a subject under test, the variation of the frequency of response to a light flash cannot be plotted as a function of log $a$ as in Figure 29.6 since the value of $\bar{n}$ is not known a priori. However, it can be plotted as a function of log $I$, because as $\bar{n} = kI$, we have

$$\log \bar{n} = \log(kI) = \log I + \log k \tag{29.6}$$

where $k$ and hence log $k$ are constant. An experimentally obtained response curve, if plotted on the same scale as in Figure 29.6, should be identical with the theoretical curves except that it is displaced along the abscissa axis by log $k$. The curves for different values of $n$ are of different shape, so it is easy to decide to which theoretical curve the experimental curve corresponds. Hence the number of quanta $n$ needed to cause a receptor

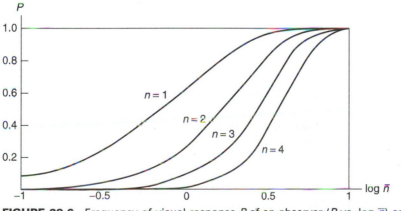

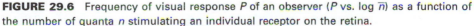

**FIGURE 29.6** Frequency of visual response $P$ of an observer ($P$ vs. log $\bar{n}$) as a function of the number of quanta $n$ stimulating an individual receptor on the retina.

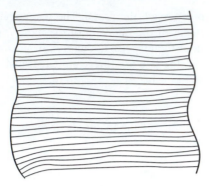

**FIGURE 29.7** Electron micrograph of a rod outer segment of a rat retina (magnification approximately 100,000 times).

to respond can be found. The parameter $n$ may be as high as 8 or as low as 1 in humans. In the most favorable cases a single photon is all that is necessary in any individual receptor. As these experiments have been performed in the main on dark-adapted eyes, the result can be said to be established only for rods, although we have no reason to believe that cone response is any different.

For many years it has been known that in the eyes of all vertebrates there are pigments which play some part in the vision process. All such pigments consist of a specific type of protein, called an opsin, with a molecular mass of about 40,000. This appears as a color group or chromophore, a particular configuration of vitamin A aldehyde called retinene. Biochemical reactions subsequently split off the retinene from the opsin and reduce it to vitamin A, ingestion of which is necessary for the continued resynthesis of the original pigment. One of the most studied pigments, rhodopsin, is a combination of retinene and opsin from rods. This is also known as visual purple because of its characteristic color. When white light is passed through a cell containing rhodopsin, one finds that the absorption spectrum agrees exactly with the luminosity curve for scotopic vision when corrected for absorption in the lens or taken from a lensless eye. Thus scotopic vision takes place by the absorption of light quanta by rhodopsin. Iodopsin, the pigment from cones, has an absorption spectrum different from rhodopsin; however, suitably corrected, it is different from the luminosity curve for photopic vision. This supports the idea that both rods and cones play a part in the photopic vision process.

When a section of a cone or rod cell is examined under an electron microscope, one finds that it is composed of a stack of approximately 1000 flattened sacs (see Figure 29.7). In all cells which employ the energy from light quanta, this layered arrangement is found (e.g., the chloroplasts which are responsible for photosynthesis in plants). The membranes which surround the sacs are about 50 Å thick. This is the diameter of the opsin molecule if it were spherical.

## TOPIC 29.5    Spectrophotometry

The amount of light absorbed by a solution or a solid can be expressed by defining the absorbance $A$ according to

$$A = \log \frac{I_0}{I} \tag{29.7}$$

where $I_0$ is the intensity of the beam before it passes through the sample and $I$ is the intensity after passing through the substance or the sample (see Figure 29.8). The Beer–

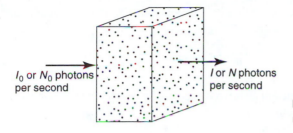
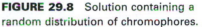

$I_0$ or $N_0$ photons per second

$I$ or $N$ photons per second

**FIGURE 29.8** Solution containing a random distribution of chromophores.

Lambert law states that the absorbance of a sample depends upon its overall thickness ($\ell$ in centimeters) and the concentration ($c$ in moles per cubic decimeter) of the light-absorbing species, the chromophore, according to

$$A = \log \frac{I_0}{I} = \varepsilon c \ell \qquad (29.8)$$

where $\varepsilon$ is a proportionality constant termed the extinction coefficient in units of cubic decimeters per mole centimeter. The value of $\varepsilon$ differs for every wavelength and every chromophore (see Figure 29.9).

The origin of the Beer–Lambert law is related to the random distribution of chromophores in the solution. If a photon passes very close to a chromophore, the chances are very good that it will be absorbed. Whether or not the photon will be absorbed depends on the area presented by the chromophore ($a$) and its ability ($p$) to capture the photon. The product of these two terms is called the effective cross section $s$: $s = ap$. Hence the average number of photons absorbed can be written as

$$\overline{N} = sc\ell \qquad (29.9)$$

The probability for a photon to pass through the solution is then given by

$$P(0) = e^{-\overline{N}} = e^{-sc\ell} \qquad (29.10)$$

but also

$$P(0) = \frac{\overline{N}}{N_0} = \frac{I}{I_0} \qquad (29.11)$$

Thus we deduce that

$$\frac{I}{I_0} = e^{-sc\ell} \qquad (29.12)$$

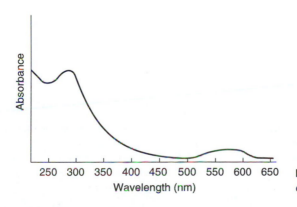

**FIGURE 29.9** Absorbance spectrum of metmyoglobin.

Using the base 10 logarithm on both sides of equation (29.12), we obtain the Beer–Lambert law:

$$\log \frac{I_0}{I} = \varepsilon c \ell \tag{29.13}$$

where $\varepsilon$ is the extinction coefficient equal to 0.4343 s. The device that measures the amount of light is called a spectrophotometer.

## TOPIC 29.6 Applications of Lasers in Medicine and Biology

The unique properties of lasers derive from the very distinctive way laser light is produced in contrast to ordinary light. Laser light is generated from atoms, ions, or molecules which are in an excited state through a process of stimulated emission of radiation (see Figure 29.10). The active laser medium is contained in a cavity. Here the emission process is organized into an intense directional, monochromatic coherent wave. Laser light has a well-defined phase and is coherent and monochromatic. These properties enable us to use lasers in a wide variety of applications based on interference or wave modulation.

Laser systems make use of a number of atomic gases, solids, or liquids as the working substance. They emit either continuous or pulsed monochromatic beams and can operate over a broad range of optical spectra with a single wavelength in, for example, the ultraviolet, visible, or infrared region. The output powers can vary between about a milliwatt to megawatts. There are a variety of laser systems in general use today. These include the 1-mW helium–neon laser producing a red beam at 632.8 nm (yellow and green beams are also available). The argon ion laser operates in the green or blue up to 10 W. The carbon dioxide gas laser emits in the infrared at 10 mm and can produce several hundred watts. A powerful solid-state optically pumped system, the neodymium YAG laser, emits either a continuous or a pulsed beam at 1.06 mm. By passing current through a semiconductor material, the diode junction laser emits in the near infrared. Diodes can emit up to 5 W and can be used to energize other laser materials.

A lens may be used to focus laser light to a very tiny area so that the energy per unit area in the focal spot can be made enormous. For example, small regions of an or-

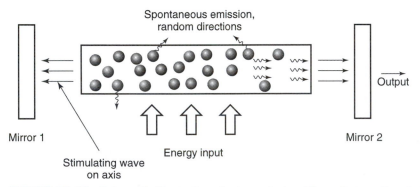

**FIGURE 29.10** Schematic illustration of a laser design. The active medium is the tube containing atoms. An external source of energy is needed to "pump" the atoms up to excited energy states. The parallel end mirrors, one of which is partially transparent, provide the feedback of the stimulating wave.

gan may be vaporized without harming surrounding tissues. This general property is utilized in most applications of the laser to biology or medicine. The functions of the various portions of cells and how the genetic materials function occupy much modern cell biology research. By employing a focused laser beam, damage is confined to the particular component required. The same technique has been used to study developing embryos, specific cells being destroyed at a particular stage of development. It is often important to isolate and collect unusual cells for study and growth, and a laser cell separator exploits the fact that specific cells can be tagged with fluorescent dyes. After all cells are dropped from a tiny charged nozzle, they may be laser scanned for the dye tag. When triggered by the correct light-emitting tag, a small voltage applied to parallel plates deflects the falling electrically charged cells into a collector beaker.

More sophisticated applications may be undertaken where only a small portion of a component is vaporized. An electric arc is set up between a pair of plates and the vapor passed through them. The spectrum of emitted light will contain spectral lines which are characteristic of the constituents of the vapor and which can be identified. This technique has been used extensively to study the elements in teeth. Lasers have been used to remove dental cavities. Irradiation of teeth at intervals appears to inhibit the production of cavities, and research is continuing into the use of lasers in preventive dentistry.

## TOPIC 29.7    Photorefractive Keratectomy and Applications of Lasers to Eye Surgery

The laser has had a substantial impact in the medical area of opthalmology, where the structure, function, and diseases of the eye are investigated. Photorefractive keratectomy (PRK), a laser-based procedure, offers alternative treatment for near-sightedness and far-sightedness that does not rely on lenses. This procedure involves using a laser to remove small amounts of tissue from the cornea of the eye, which changes its curvature. It is at the air–cornea boundary that most refraction of light occurs, and this curvature change of that boundary can correct deficiencies in the way the eye refracts light. This will cause the image to become focused onto the retina, where the focus should be. Laser light in the PRK technique is pulsed and is generated by an ultraviolet excimer laser that produces a wavelength of 193 nm. Weak pulses can be used since the cornea absorbs this wavelength extremely well, and this leads to a highly precise and controllable removal of corneal tissue. Typically from 0.1 to 0.5 $\mu$m of tissue is removed by each pulse without damaging adjacent layers. A major problem during the procedure is eye movement. However, this can be overcome with the aid of a mechanical device that stabilizes the eye. More recently this problem is tackled with the aid of systems capable of tracking eye movement. These movements occur every 15 ms and the tracking system follows them and retargets the laser.

In the eye the retina may become detached due to disease, injury, or degenerative changes. Treatment must be prompt because blindness can result. The alteration in nutrition and metabolism of the retina causes rapid degeneration. One attempts to fasten the retina back to the choroid by damaging small portions so that the resultant scar tissue will bond the two together at that point. It has been found that a 1-ms flash of light from a laser focused onto the retina will efficiently weld the retina back onto the choroid. No anaesthetic is required and the patient feels no pain. The exposure time is so short that the patient does not react to the light and move the eye. Hence mechanical fixing of the eye and an operating theater are not required.

A widespread condition called glaucoma, which is manifested by a high fluid pressure in the eye, may lead to damage to or destruction of the optic nerve. Iridectomy is a simple laser operation which can burn a small hole in a clogged membrane, relieving the pressure. Along similar lines a serious side effect of diabetes is neovascularization, the formation of weak blood vessels which can often leak into extremities. When this condition occurs in the eye (called diabetic retinopathy), vision deteriorates, leading to blindness. It is possible to direct green light from the argon ion laser through the eye lens and eye fluid, focus on the retina edges, and photocoagulate the leaky vessels.

## TOPIC 29.8   Photodynamic Therapy for Cancer

In conjunction with light-activated drugs, the laser is being used in the treatment of cancer. The procedure adopted is to administer the drug intravenously so the tumor can absorb it from the blood stream. The drug is located close to the cancer cells. Laser light then activates the drug, and a chemical reaction ensues that disintegrates the cancer cells and the small vessels that feed them. Therapy of this kind works best with small tumors in their early stages.

It is not possible to treat deep-seated tumors using a laser. Skin tumor therapy has been attempted by using high-energy focused laser beams. There is evidence that not only the treated cells but others in the tumor and adjacent secondaries regress and heal over a period of weeks. This may be due to either thermal denaturation of key enzymes in the structure or the disruption of the molecular structure of the tumor cells by the electric field of the beam. This latter results in rejection of the cells by the host body.

## TOPIC 29.9   Removal of Birthmarks

Approximately 0.3% of children at birth have port-wine birthmarks caused by congenital capillary malformations, which are usually found on the head and the neck. The preferred laser treatment for port-wine marks utilizes a pulsed dye laser with a laser light wavelength of 585 nm in the form of pulses lasting 0.45 ms and occurring every 3 s. The laser beam is focused onto a 5-mm-diameter spot, and the light is absorbed by oxyhemoglobin in the malformed capillaries, which are destroyed in the process without damaging adjacent tissue. The port-wine stain eventually fades; then the destroyed capillaries are replaced by normal blood vessels.

High-power continuous lasers are also used in surgery. Surgery on vascular organs such as the liver produces a large amount of bleeding, which can obscure the field of view of the surgeon. When a focused beam of laser light is used, it tends to seal off vessels as they are incised. As a consequence, much less bleeding occurs. When the laser technique is used in the removal of cancerous growths, there is no physical contact with the growth or surrounding tissue, so the risk of spreading the growth is considerably reduced.

## SOLVED PROBLEMS

**29.1**   What wavelength of light is sufficient for the photochemical reaction with an activation energy $U = 300$ kJ/mol?

**Solution**  A photochemical reaction is an interaction between one molecule and one photon. The activation energy is the energy needed by 1 mol of material; thus, for one molecule, the sufficient energy is

$$\varepsilon = \frac{U}{N_A} = \frac{300 \text{ kJ/mol}}{6 \times 10^{23} \text{ mol}^{-1}} = 5 \times 10^{-19} \text{ J}$$

where $N_A$ is the Avogadro constant. This energy is given by a single photon, and therefore

$$\varepsilon = h\frac{c}{\lambda}$$

where $h$ is the Planck constant, $c$ is the speed of light in a vacuum, and $\lambda$ is the wavelength. The wavelength of the photon is calculated as

$$\lambda = \frac{hc}{\varepsilon} = 398 \text{ nm}$$

**29.2**  For green light at the absorption maximum of rhodopsin ($\lambda = 510$ nm), the absolute energy threshold of the dark-adapted human eye is $E = 4.0 \times 10^{-17}$ J; that is, the flash of duration $\Delta t = 100$ ms should have at least enough energy to generate a sense of vision in the eye. The number of rod cells in the illuminated area of the eye is 500. What is the power of the light source at the threshold of seeing? How few incident photons of green light does the eye respond to? How few rod cells should be excited to cause a visual sensation if the following losses are to be taken into account: 4% for reflection at the cornea, 50% for absorption or deflection in the vitreous humor, and 20% for the yield of photoelectric conversion in rhodopsin?

**Solution**  The optical power of the light source is the energy emitted over the duration of the flash: $P = E/\Delta t$. Numerically, $P = 4 \times 10^{-16}$ W. The minimal number of incident photons required for a person to experience a visual event is

$$N = \frac{E}{h(c/\lambda)}$$

where $h = 6.6 \times 10^{-34}$ J · s is Planck's constant and $c = 3 \times 10^8$ m/s is the speed of light in a vacuum. Numerically, we find that $N = 103$. Of the 103 incident photons, about 4 will be reflected from the surface of the cornea. Of the remaining 99 photons, about 50 will be absorbed in the eye while traveling from the cornea to the retina. Of the remaining 49 photons, 20% can evoke an electric potential in the rhodopsins (i.e., about 10 photons). Because of the very low number of photons relative to the number of illuminated rod cells in this experiment, each rod is activated by only one photon; that is, the 10 excited rhodopsin pigments belong to 10 different rod cells.

**29.3**  The first observable effect of the harmful effects of the sun's UV radiation is erythema (skin reddening). The incident threshold energy inducing erythema is wavelength dependent. For example, it has a value of $E_1 = 3.7$ mJ at $\lambda_1 = 254$ nm and of $E_2 = 13$ mJ at $\lambda_2 = 300$ nm for a skin surface area $A = 1$ cm$^2$.
  (a) What is the minimal light intensity in watts per square meter that induces erythema during radiation for $t = 1$ h at the wavelengths given above?
  (b) What is the energy of photons in electron-volts at the above given wavelengths?
  (c) What is the ratio of the photon numbers that induce erythema at the two wavelengths?

**Solution**

**(a)** The light intensity is given as

$$I = \frac{E}{At}$$

For the given data, we obtain $I_1 = 1 \times 10^{-2}$ W/m$^2$ and $I_2 = 3.6 \times 10^{-2}$ W/m$^2$.

**(b)** Photon energies can be calculated as $\varepsilon = hc/\lambda$. We find that $\varepsilon_1 = 7.8 \times 10^{-19}$ J = 4.9 eV and $\varepsilon_2 = 6.6 \times 10^{-19}$ J = 4.1 eV.

**(c)** The number of incident photons per square centimeter is $n = E/\varepsilon$; thus $n_1 = 4.7 \times 10^{15}$, $n_2 = 1.97 \times 10^{16}$ and their ratio $n_2/n_1 = 4.2$.

**29.4**  The beam of a surgical laser of power $P = 20$ W is focused onto a circular surface of diameter $d = 0.4$ mm. What is the light intensity at the focus?

**Solution**  The cross-sectional area of the laser beam at the focus is

$$A = \tfrac{1}{4}\pi d^2 = 1.26 \times 10^{-7} \text{ m}^2$$

Hence, the light intensity is

$$I = \frac{P}{A} = 1.6 \times 10^8 \text{ W/m}^2$$

**29.5**  In the coincidence photon counter model of the eye, visual sensation is experienced if at least four photons are absorbed by the rod cells within the flash duration. In an experiment, 1000 flashes of light, each containing 100 photons on average, are presented to a person. How many are expected to be not visible if 4% of the photons directed at the pupil of the eye are absorbed by the rod cells?

**Solution**  The number of rods activated per flash follows the Poisson distribution. If the average number of photons at the retina per flash is $\langle n \rangle$, the probability that $n$ photons will be absorbed by the rod cells per flash is

$$P(n) = e^{-\langle n \rangle} \frac{\langle n \rangle^n}{n!}$$

where $n = 0, 1, 2, \ldots$ and $P(n)$ is the probability that exactly $n$ photons will be absorbed if the average is $\langle n \rangle$. The sum of the probabilities taken for all integers should be 1. In the given case, the average number of photons absorbed by the rod cells per flash is $\langle n \rangle = 4$. The numerical values of the probabilities that only $n = 0, 1, 2, 3$ rod cells are activated per flash are $P(0) = 0.018$, $P(1) = 0.073$, $P(2) = 0.146$, and $P(3) = 0.195$, respectively. The probability that fewer than four rod cells will be activated in one of the flashes is the sum of these probabilities: 0.432. This means that if the average number of absorbed photons per flash is 4 and the coincidence of four activated rod cells within the duration of the flash is required to evoke visual sensation, then the person will give no response after 432 of the 1000 flashes.

# EXERCISES

**29.1**  Estimate the number of atoms comprising a bacterium such as *Escherichia coli*. Assume that the average mass of an atom in the bacterium is 10 times the mass of a hydrogen atom.

**29.2** The cell membrane is measured to be 100 nm thick. Assume that the average atom comprising it is twice the size of the hydrogen atom. How many atoms thick is this membrane?

**29.3** A single electron in a TV tube typically carries the energy of $4.0 \times 10^{-15}$ J. To break one DNA strand requires an energy on the order of $10^{-19}$ J. How many DNA molecules can therefore be damaged by such an electron? Does this explain the need for shielding?

**29.4** A $CO_2$ laser is used in surgery by emitting a 1-ms pulse of infrared light. Its wavelength is 1060 nm. As a result, 1 cm$^3$ of tissue has been evaporated by heating it to 100°C. Calculate the number of photons employed by this effect, assuming that the heat of vaporization of the tissue is the same as that of water.

# CHAPTER 30

# NUCLEAR PHYSICS AND RADIOACTIVITY

## PHYSICAL BACKGROUND

The nucleus of an atom consists of protons and neutrons. By convention, the atomic number of an element consists of $Z$ protons and $N$ neutrons and its mass number $A = Z + N$. Nuclei that contain the same number of protons but a different number of neutrons ($N \neq Z$) are called isotopes. The approximate radius $r$ (in meters) of a nucleus is given by

$$r \approx (1.2 \times 10^{-15} \text{ m})A^{1/3} \tag{30.1}$$

The strong nuclear force is the force of attraction between nucleons and is one of the four fundamental forces of nature. This force balances the electrostatic force of repulsion between protons and holds the nucleus together. The strong nuclear force has a very short range of action and is almost independent of electric charge.

The binding energy of a nucleus is the energy required to separate the nucleus into its constituent protons and neutrons. The binding energy is equal to $(\Delta m)c^2$, where c is the speed of light in a vacuum and $\Delta m$ is the mass defect of the nucleus. The mass defect is the amount by which the sum of the individual masses of the constituent protons and neutrons exceeds the mass of the parent nucleus.

When specifying nuclear masses, it is customary to use the atomic mass unit (a.m.u.), which is one-twelfth of the mass of a $^{12}_{6}C$ atom. One atomic mass unit has a mass of $1.6605 \times 10^{-27}$ kg and is equivalent to an energy of 931.5 MeV (mega-electron-volt).

Unstable radionuclide nuclei spontaneously decay by breaking apart or rearranging their internal structures in a process called radioactivity. Radioactive decay is a nuclear phenomenon exhibited by radioisotopes of some elements across the periodic table whose atomic number $Z$ is greater than that of lead ($Z = 82$). In these elements, which contain generally more $Z$ than $A$, the repulsive electric forces in the nucleus become greater than the attractive nuclear forces, making the nuclei unstable. The radioactive element, in nature, strives toward a stabilized state of existence and, in the process, spontaneously emits particles (photons and charged and uncharged particles) in the transformation to a different nucleus and hence a different element. This process of radioactive decay is dependent on the amount and identity of the radioactive element.

Radionuclides undergo radioactive decay typically according to three common types:

1. Alpha decay, which is caused by the repulsive electric forces between the protons, involves the emission of an alpha particle (or a helium nucleus that consists of two protons and two neutrons) by nuclei with many protons. In alpha decay, the radioactive nucleus decreases in $A$ by 4 and decreases in $Z$ by 2.

2. Beta decay, which occurs in nuclei that have too many neutrons, can occur by emission of a $\beta$ particle. A $\beta$ particle exists in two forms. A $\beta^-$ particle is an electron, and a $\beta^+$ particle is a positively charged electron, known as a positron. In $\beta^-$ de-

cay, the radioactive nucleus remains unchanged in $A$ and increases in $Z$ by 1. In $\beta^+$ decay, the radioactive nucleus remains unchanged in $A$ and decreases in $Z$ by 1.

**3.** Gamma decay occurs by the emission of highly energetic photons. In gamma decay, the radioactive nucleus remains unchanged in both $A$ and $Z$.

If a radioactive parent nucleus disintegrates into a daughter nucleus that has a different atomic number, one element has been converted into another element, the conversion being referred to as a transmutation. An induced nuclear transmutation is the process whereby an incident particle or photon strikes a nucleus and causes the production of a new element.

Nuclear fission occurs when a massive nucleus splits into two fragments. Fission can be induced by the absorption of a thermal (low-energy) neutron. When a massive nucleus fissions, energy is released, because the binding energy per nucleon is greater for the fragments than for the original nucleus. Neutrons are also released during nuclear fission. These neutrons can, in turn, induce other nuclei to fission and lead to a process known as a chain reaction. A nuclear reactor is a device that generates energy by a controlled chain reaction.

In a fusion process, two nuclei with smaller masses combine to form a single nucleus with a larger mass. Energy is released by fusion when the binding energy per nucleon is greater for the larger nucleus than for the smaller nuclei.

Given a radioactive element originally with $N_0$ number of atoms, the number of atoms present at any time $t$ is (see Figure 30.1) $N(t)$, given by

$$N(t) = N_0 e^{-\lambda t} \tag{30.2}$$

where $\lambda$ is a decay constant defined by

$$\lambda = \frac{0.693}{T_{1/2}} \tag{30.3}$$

Here $T_{1/2}$ is the half-life of the radioactive element and represents the time required for one-half of the radioactive atoms to remain unchanged. Half-lives for radioactive elements range from fractions of seconds (e.g., for polonium, $T_{1/2} = 3 \times 10^{-7}$ s) to millions of years (e.g., $T_{1/2} = 704$ million years for uranium-235). (See Table 30.1 for more examples.) The SI unit for activity is the becquerel (Bq), one becquerel being one disintegration per second (dps). Radioactive decay obeys the following relation:

$$\frac{\Delta N}{\Delta t} = -\lambda N \tag{30.4}$$

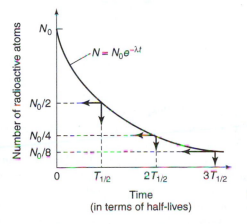

Time
(in terms of half-lives)

**FIGURE 30.1**   Schematic diagram of an exponential decay process.

**TABLE 30.1   Some Half-Lives $\tau$ for Radioactive Decay**

| Isotope | | Half-life $\tau$ | Decay mode |
|---|---|---|---|
| Polonium | $^{214}_{84}$Po | $1.64 \times 10^{-4}$ s | $\alpha, \gamma$ |
| Krypton | $^{89}_{36}$Kr | 3.16 min | $\beta^-, \gamma$ |
| Radon | $^{222}_{86}$Rn | 3.83 days | $\alpha, \gamma$ |
| Strontium | $^{90}_{38}$Sr | 28.5 yr. | $\beta^-$ |
| Radium | $^{226}_{88}$Ra | $1.6 \times 10^3$ yr. | $\alpha, \gamma$ |
| Carbon | $^{14}_{6}$C | $5.73 \times 10^3$ yr. | $\beta^-$ |
| Uranium | $^{238}_{92}$U | $4.47 \times 10^9$ yr. | $\alpha, \gamma$ |
| Indium | $^{115}_{49}$In | $4.41 \times 10^{14}$ yr. | $\beta^-$ |

where $\lambda$ is the decay constant of equation (30.2) and $\Delta N$ is the change in the number of radioactive nuclei in time $\Delta t$. An alternative method of expressing radioactivity is by the activity or disintegration rate. The activity $A$ is the number of disintegrations per second that occur. In other words, the activity is the magnitude of $\Delta N/\Delta t$. Hence,

$$A = \lambda N \qquad (30.5)$$

which is expressed in units of curie (Ci), where

$$1 \text{ Ci} = 3.7 \times 10^{10} \text{ Bq} \qquad (30.6)$$

Natural radioactivity in the human body is 10 nCi, nuclear medicine scanning techniques use 10 $\mu$Ci to 10 mCi, and $\gamma$-ray sources for radiotherapy are 1000 Ci (see Chapter 31 for more information). The sequential decay of one nucleus after another is called a radioactive decay series.

# TOPIC 30.1   Isotopes and the Human Body

Ionizing radiation is made up of either moving particles or photons which have sufficient energy to remove an electron from an atom or molecule to form an ion or a charged molecule. The moving particles can either be $\alpha$ or $\beta$ particles emitted during decay. The $\alpha$ particles consist of positively charged particles each one being the $^4_2$He nucleus of helium. They are the least penetrating, being blocked by a thin sheet of lead which is about $10^{-2}$ mm thick. The $\beta$ rays are deflected by a magnetic field in a direction opposite to that of positively charged $\alpha$ rays, and experiment shows that $\beta^-$ particles are electrons. Beta rays penetrate lead to a greater distance, of order 0.1 mm. The $\gamma$ rays are the most penetrating and can pass through 100 mm of lead. Gamma rays are photons which usually lie in the ultraviolet, X-ray, or $\gamma$-ray region of the electromagnetic spectrum. To ionize an atom or molecule, an energy of roughly 1 to 35 eV is needed. The particles and $\gamma$ rays emitted during nuclear disintegration often have energies of several million electron-volts. Hence a single $\alpha$ particle, $\beta$ particle or $\gamma$ ray can ionize thousands of molecules.

Radiation, including $\alpha$, $\beta$, $\gamma$, and X-rays as well as protons, neutrons and other particles like pions, when it passes through matter, can do considerable damage. Metals and other structural materials become brittle and their strength is weakened if the radiation is very intense (e.g., as in nuclear power stations and for space vehicles which may pass through areas of intense cosmic radiation). In biological organisms the radiation damage produced is due primarily to ionization in cells. Ions or radicals are produced that are highly reactive and, when they take part in chemical reactions, can interfere with the normal operation of the cell. Ionization of atoms can be brought about by all forms of radiation when electrons are excited out of their atoms. If these electrons are of bonding type, a molecule's structure may be altered so that it does not perform its normal function—or it performs a harmful function. The loss of one molecule, in the case of proteins, may not be serious if there are copies of that particular one in the cell and additional ones can be made from its gene. It could happen, however, that large doses of radiation may damage so many molecules that new copies cannot be made quickly enough, and the cell dies. If DNA is damaged, this could be more serious since a cell may have only one copy. In fact, each alteration in the DNA affects a gene so that needed proteins or other materials may not be made at all. Once again, the cell may die. Since the body can replace a dead single cell, this is usually not a problem. However, the loss of neurons is very serious since they are not replaceable. The organism may not be able to recover if many cells die. If a cell does survive, it may be defective, in which case it may go on dividing and producing more defective cells, to the detriment of the whole organism. Thus radiation can cause cancer, that is, the rapid production of defective cells.

There are two types of radiation damage to biological organisms, namely somatic and genetic damage. Somatic damage refers to any part of the body except the reproductive cells. Somatic damage can cause cancer at high doses of radiation, radiation sickness (characterized by nausea, fatigue, loss of body hair, and other symptoms), or even death. Genetic damage refers to damage to reproductive cells and so affects an individual's offspring. Mutations result from damage to genes, the majority of which are harmful. If they occur in reproductive cells, they are transmitted to future generations. The diagnostic use of X-rays is widespread, and the possible damage done by the radiation must be balanced against the medical benefits.

The body removes substances by excretion in an amount of time that depends on the chemical compound. This process is often like the decline of radiation due to nuclear decay; that is, half the substance is excreted in a certain amount of time, half of what remains in a similar amount of time, and so on. In such instances the substance is said to have a biological half-life. The biological half-life is the time the radioisotope resides in the organ before being excreted. The actual half-life to be expected in a radioisotope experiment with a patient—or the effective half-life—must be computed as a combination of the nuclear (physical) and biological half-lives via the formula

$$\frac{1}{T(\text{effective})} = \frac{1}{T(\text{nuclear})} + \frac{1}{T(\text{biological})} \tag{30.7}$$

Table 30.2 gives nuclear, biological, and effective half-lives in days for selected isotopes. Note that the effective half-life is always shorter than either the nuclear or biological half-lives, as equation (30.7) adds reciprocals.

One area of major concern from the point of view of the effects of radiation on the human body is nuclear contamination. This can be due to nuclear power plant malfunctions, nuclear warfare, or transport of hazardous materials. Grass on which fallout has settled is contaminated with the isotope $^{90}$Sr, which can be eaten by animals. Strontium is also taken up by plants from soil into which it has been washed. It is ingested by humans principally from milk and milk products, cereals, and vegetables and laid down in bones along with calcium. It persists there once in the bones and, in large concentrations, can produce bone cancer and aplastic anemia due to the destruction of bone marrow. Once it has been ingested, there is no way of dealing with it. The isotope $^{137}$Cs is acquired mainly from meat and milk but is distributed throughout the body and so does not produce the

**TABLE 30.2  Nuclear, Biological, and Effective Half-Lives of Selected Isotopes (days)**

| Isotope | $T$(nuclear) | $T$(biological) | $T$(effective) |
| --- | --- | --- | --- |
| $^3$H | $4.5 \times 10^3$ | 12 | 12 |
| $^{14}$C | $2.1 \times 10^6$ | 40 | 40 |
| $^{22}$Na | 850 | 11 | 11 |
| $^{32}$P | 14.3 | 1155 | 14.1 |
| $^{35}$S | 87.4 | 90 | 44.3 |
| $^{36}$Cl | $1.1 \times 10^8$ | 29 | 29 |
| $^{45}$Ca | 165 | $1.8 \times 10^4$ | 164 |
| $^{59}$Fe | 45 | 600 | 42 |
| $^{60}$Co | $1.93 \times 10^3$ | 10 | 10 |
| $^{65}$Zn | 244 | 933 | 193 |
| $^{86}$Rb | 18.8 | 45 | 13 |
| $^{90}$Sr | $1.1 \times 10^4$ | $1.8 \times 10^4$ | $6.8 \times 10^3$ |
| $^{99m}$Tc | 0.25 | 1 | 0.20 |
| $^{123}$I | 0.54 | 138 | 0.54 |
| $^{131}$I | 8.0 | 138 | 7.6 |
| $^{137}$Cs | $1.1 \times 10^4$ | 70 | 70 |
| $^{140}$Ba | 12.8 | 65 | 10.7 |
| $^{198}$Au | 2.7 | 280 | 2.7 |
| $^{210}$Po | 138 | 60 | 42 |
| $^{226}$Ra | $5.8 \times 10^5$ | $1.6 \times 10^4$ | $1.5 \times 10^4$ |
| $^{235}$U | $2.6 \times 10^{11}$ | 15 | 15 |
| $^{239}$Pu | $8.8 \times 10^6$ | $7.3 \times 10^4$ | $7.2 \times 10^4$ |

concentrated effects of strontium. It is also eliminated more quickly. Its half-life is longer than $^{90}$Sr. It persists in soil but is unlikely to pose a great threat except in areas of massive contamination or to people with unusual diets. After the Chernobyl nuclear power plant disaster in 1986, a high concentration of cesium was found in reindeer meat and, and as this represents a major portion of the diet of Laplanders and Inuits, it made them highly vulnerable.

## TOPIC 30.2  Measurement of Radiation: Dosimetry

Ionizing radiation which passes through the body can cause considerable damage, but it can also be used to treat certain diseases, particularly cancer. Very narrow beams are used and directed at the cancerous tumor to destroy it. The subject of dosimetry is the quantitative determination of the dose of radiation. The specification of the strength of a source at a given time is given by stating the source activity, or how many disintegrations occur per second. Suppliers of radioactive nucleotides—or radionuclides—specify activity at a given time. Since the activity decreases over time, particularly for short-lived isotopes, it is important to take this into account. The source activity $A = \Delta N/\Delta t$ is related to the half-life by [see also equation (30.5)].

$$A = \frac{\Delta N}{\Delta t} = \lambda N = \frac{0.693}{T_{1/2}}N \tag{30.8}$$

# TOPIC 30.3 Radioactive Radon Gas in Houses

When $^{226}_{88}$Ra undergoes $\alpha$ decay, the naturally occurring radioactive gas $^{224}_{88}$Ra is produced. Radioactive radon is considered to be a health hazard because radon in the soil is gaseous, and it is possible for it to enter the basement of homes through cracks in the foundation. Depending on the type of house construction and the concentration of radon in the surrounding soil, the concentration of radon, once inside a home, can markedly rise. Daughter nuclei, which are formed when radon gas decays, are also radioactive. Unfortunately, the radioactive nuclei can attach themselves to dust and smoke particles, which can be inhaled. Under these circumstances they remain in the lungs and release tissue-damaging radiation. This may lead to lung cancer if there is prolonged exposure to high levels of radon. The concentration of radon gas can easily be measured with inexpensive equipment, so all homes should be tested for radon.

# SOLVED PROBLEMS

**30.1** The standard radiopharmaceuticals for thyroid scans contain $^{131}$I, which has a nuclear half-life of 8.0 days and an effective half-life of 7.6 days. Calculate the dose received in the first half hour of exposure using the following information. The patient is given approximately 0.50 MeV per decay. Half the energy emitted is absorbed in a total mass of 0.15 kg.

**Solution** Since the effective half-life is 7.6 days, the activity will not decrease significantly in the first half hour. Therefore,

$$(0.050 \times 10^{-3} \text{ Ci})\left(\frac{3.7 \times 10^{10} \text{ decays/s}}{\text{Ci}}\right) = 1.85 \times 10^6 \text{ decays/s}$$

**30.2** A patient is given an injection containing $1.0 \times 10^{-12}$ g of $^{99m}$Tc.
**(a)** Calculate the effective half-life.
**(b)** Calculate the activity in curies.
**(c)** What activity remains in the patient 3.0 days after injection?

**Solution**

**(a)** Effective half-life can be calculated using the equation

$$T_{\text{eff}} = \frac{T_{1/2}T_{\text{bio}}}{T_{1/2} + T_{\text{bio}}}$$

Using values given in Table 30.2 yields

$$T_{\text{eff}} = \frac{(6.0 \text{ h})(24 \text{ h})}{30.0 \text{ h}} = 4.8 \text{ h}$$

consistent with the value of $T_{\text{eff}}$ in Table 30.2.
**(b)** Activity is given by $A = (0.693/T_{1/2})N$ [see equation (30.5)], so that

$$A = \frac{0.693}{6.02 \text{ h}}\left(\frac{6.022 \times 10^{23} \text{ nuclei/mol}}{99 \text{ g/mol}}\right)(1.0 \times 10^{-12} \text{ g})(1 \text{ hr}/3600 \text{ s})$$

$$= (1.94 \times 10^5 \text{ decays/s})\left(\frac{1 \text{ Ci}}{3.7 \times 10^{10} \text{ decays/s}}\right)$$

$$= 5.2 \text{ } \mu\text{Ci}$$

(c) A period of 3.0 days is 72 hours, or 15 times the effective half-life. The activity remaining in the patient after 3.0 days is therefore obtained by dividing 5.2 $\mu$Ci in half 15 times. After 3.0 days, $A = 1.6 \times 10^{-10}$ Ci.

**30.3** In an experiment $5 \times 10^{-6}$ m$^3$ of water, with $^3$H replacing the normal hydrogen isotope, was injected into a vein of an individual. The activity of $10^{-6}$ m$^3$ of the injected sample was 30,200 counts per second. After $\frac{1}{2}$, 1, 2, and 3 h blood samples were drawn from the same vein and the plasma was separated out. Samples of volume $10^{-6}$ m$^3$ were taken, the activities being 30.2, 3.46, 3.28, and 3.40 counts per second, respectively. Calculate the total volume of labeled water in the person's body.

**Solution** It is clear that equilibrium has been achieved before 1 h, and it was assumed that complete mixing has taken place in that time. The average of the last three activities is 3.38 counts per second. The activity has decreased by a factor of 3.38/30200. The total volume finally occupied by the labeled water is therefore

$$V = 5 \times 10^{-6} \text{ m}^3 \times \frac{30,200}{3.38} = 4.47 \times 10^{-2} \text{ m}^3$$

This is the total volume of the person's body occupied by water.

**30.4** In a certain experiment, 0.016 $\mu$Ci of $^{32}_{15}$P is injected into a medium containing a culture of bacteria. After 1 h, the cells are washed and a detector that is 70% efficient (it counts 70% of emitted $\beta$ rays) records 720 counts per minute from all the cells. What percentage of the original $^{32}_{15}$P is taken up by the cells?

**Solution** The total number of disintegrations per second originally was $(0.016 \times 10^{-6})$ $(3.7 \times 10^{10}) = 592$. The counter could be expected to count 70% of this, or 414 s$^{-1}$. Since it counted $720/60 = 12$ s$^{-1}$, $12/414 = 0.029$, or 2.9%, was incorporated into the cells.

**30.5** After the nuclear reactor accident in Chernobyl, a large amount of the radioactive isotope $^{131}$I was emitted, which was deposited on the soil and plants. Thus, the milk of cows that grazed on the contaminated land contained a higher level of $^{131}$I, and the milk could therefore be used only for cheese production instead of direct consumption. What storage time is needed to decrease the $^{131}$I content of the cheese to 5% of the original level? The half-life of $^{131}$I is $T = 8$ days.

**Solution** The number of radioactive atoms decreases exponentially:

$$N = N_0 e^{-\lambda t}$$

where the coefficient $\lambda = \ln 2/T$ and $t$ is the time for the number of radioactive atoms to decrease from $N_0$ to $N = 0.05N_0$. From these values, we have

$$0.05 \, N_0 = N_0 e^{-(\ln 2/T)t} \quad \text{or} \quad \log 20 = \lambda t \log e$$

The necessary storage time $t = 34.5$ days.

**30.6** The half-life of the radioactive isotope $^{32}$P is $T = 14.3$ days. If, after injection of this isotope, the observed count rate of a whole-body detector is $A_1 = 8 \times 10^4$ counts per second, how long will it take to fall to $A_2 = 1 \times 10^4$ counts per second? The half-life of the metabolic passage of phosphorus is 1155 days.

**Solution**   Since the biological process of metabolism of phosphorus is much slower than the radioactive decay of the isotope, the half-life of the exponential decay of activity $A$ can be taken as identical to that of the radioactive disintegration process, $T = 14.3$ days:

$$A = A_0 e^{-(\ln 2/T)t}$$

where $A_0$ is the activity at $t = 0$. The time required for the activity to fall from $A_1$ to $A_2$ can be calculated from the exponential law above. After division of the exponential expressions of the radioactive activity at times $t_1$ and $t_2$ and rearrangement, we obtain

$$\Delta t = t_2 - t_1 = T \ln \frac{A_1}{A_2} \frac{1}{\ln 2}$$

Numerically, this yields $\Delta t = 42.9$ days.

## EXERCISES

**30.1**   The strontium isotope $^{90}_{38}$Sr has a half-life of 28.5 years and enters the body through the food chain, collecting in the bones and posing a particularly serious health hazard. How long will it take for 99.99% of the $^{90}_{38}$Sr released in a nuclear reactor accident to disappear?

**30.2**   What is the activity of an adult of mass 70 kg in consequence of the natural radioactivity of $^{40}$K? The total potassium content is 0.35% of the body mass. The incidence rate and the half-life of the radioactive isotope $^{40}$K are 0.012% and $T = 1.25 \times 10^9$ years, respectively.

**30.3**   Radioactive isotope $^{131}$I, with a physical half-life of 8.1 days, is ingested by patients with healthy or with hyperactive (diseased) thyroids. On two consecutive days, a counter positioned over the gland reads 10000 and 9136 counts per second (cps) for the healthy person and 9500 and 7340 cps for the sick person. What are the biological half-lives of the isotope in the thyroids of the patients?

**30.4**   Calculate the increase in body temperature due to the energy deposited by a large dose of X-rays received accidentally by a technician. Assume the energy of the X-rays is 200 keV, $3.13 \times 10^{14}$ of them are absorbed per kilogram of tissue, and the specific heat of the tissue is 0.83 cal/g °C.

# CHAPTER 31

# DOSE OF IONIZING RADIATION, NUCLEAR DIAGNOSTICS, AND RADIATION THERAPY

## PHYSICAL BACKGROUND

Exposure is a measure of the ionization produced in air by ionizing radiation such as X-rays or $\gamma$ rays (see also Chapter 30). In what follows we show how exposure may be defined. Suppose a beam of X-rays or $\gamma$ rays is sent through a mass $m$ of dry air at standard temperature and pressure (STP: 0°C, 1 atm pressure). This beam will produce positive ions whose total charge is $q$. Exposure is defined as the total charge per unit mass of air:

$$\text{Exposure} = \frac{q \ (\text{C})}{m \ (\text{kg})} \tag{31.1}$$

Having defined exposure in this way, however, the first radiation unit to be defined was the roentgen (R), which is still used today. If $q$ is expressed in coulombs and $m$ is in kilograms, the exposure in roentgens is given by

$$\text{Exposure (R)} = \left(\frac{1}{2.58 \times 10^{-4}}\right) \frac{q}{m} \tag{31.2}$$

Hence, when X-rays or $\gamma$ rays produce an exposure of 1 R, $q = 2.58 \times 10^{-4}$ C of positive charge is produced in $m = 1$ kg of dry air:

$$1 \text{ R} = 2.58 \times 10^{-4} \text{ C/kg} \quad \text{(dry air at STP)} \tag{31.3}$$

## TOPIC 31.1 Absorbed Dose and Relative Biological Effectiveness

The way exposure is defined in terms of ionizing abilities in air does not inform us of the effect of radiation on living tissue. The absorbed dose is a more suitable quantity for biological purposes because it is the energy absorbed from the radiation per unit mass of absorbing material:

$$\text{Absorbed dose} = \frac{\text{energy absorbed}}{\text{mass of absorbing material}} \tag{31.4}$$

The SI unit of absorbed dose is the gray (Gy), where

$$1 \text{ Gy} = 1 \text{ J/kg} \tag{31.5}$$

Another unit which is often used for absorbed dose is the rad, related to the gray by

$$1 \text{ rad} = 0.01 \text{ Gy} \tag{31.6}$$

**TABLE 31.1  Relative Biological Effectiveness for Various Types of Radiation**

| Type of radiation | RBE |
| --- | --- |
| 200-keV X-rays | 1 |
| $\gamma$ Rays | 1 |
| $\beta^-$ Particles (electrons) | 1 |
| Protons | 10 |
| $\alpha$ Particles | 10–20 |
| Neutrons | |
|    Slow (low energy) | 2 |
|    Fast (high energy) | 10 |

Ionizing radiation produces different amounts of biological damage for different kinds of radiation. For example, a 1-rad-dose of neutrons is much more likely to produce eye cataracts than a 1-rad-dose of X-rays. To compare the damage caused by different types of radiation, we define relative biological effectiveness (RBE). The RBE of a particular type of radiation compares the dose of that radiation needed to produce a certain biological effect to the dose of 200-keV X-rays needed to produce the same biological effect. Thus

$$RBE = \frac{\text{dose of 200-keV X-rays that produces a certain biological effect}}{\text{dose of radiation that produces the same biological effect}} \qquad (31.7)$$

Obviously the RBE depends not only on the nature of the ionizing radiation and its energy but also on the type of tissue being irradiated. In Table 31.1 we list some typical RBE values for different kinds of radiation, assuming that an "average" biological tissue is being irradiated. For example, a value of RBE = 1 for $\gamma$ and $\beta^-$ rays indicates that they produce the same biological damage as do 200-keV X-rays. The larger values of RBE for protons, $\alpha$ particles, and fast neutrons indicate that they cause substantially more damage. The absorbed dose and the RBE are often used in conjunction to reflect the damage-producing character of the radiation. The product of the absorbed dose in rads (not in grays) and the RBE is called the biological equivalent dose (BED):

$$BED \text{ (rem)} = \text{absorbed dose (rad)} \times RBE \qquad (31.8)$$

where the unit for the biological equivalent dose is the rem, which is an acronym for roentgen equivalent man. The definition is rem = (RBE)(rad).

## TOPIC 31.2  Biological Effects of Ionizing Radiation

Ionizing radiation has two key effects on living cells. It can interfere with cell reproduction by either inhibition or mutation or it can lead to cell death. Both effects occur for any exposure to ionizing radiation, but there will be more of the first than the second. These two effects of ionizing radiation at the cellular level can be used to explain all the macroscopic effects of radiation, such as cancer, radiation sickness, and cataract inducement. The biological effects of ionizing radiation are directly proportional to the amount

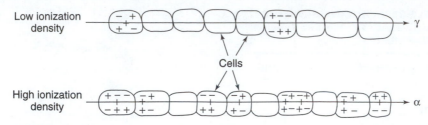

**FIGURE 31.1** Ionization densities created in tissue by a gamma ray and an alpha ray.

of ionization produced in living tissue. The amount of ionization produced is in turn proportional to the energy deposited.

Since most biological matter is composed of water, it is worth starting our discussion with the effects of radiation on water. The irradiation of $H_2O$ molecules can produce the hydrogen atom (H), the neutral OH group, or free hydrated electrons. This process is known as the radiolysis of water. These free radicals can start chemical reactions (e.g., two HO groups can form $H_2O_2$), but more importantly, free radicals produced by radiolysis can react with DNA. They can cause breaks in the sugar phosphate backbone of DNA or alterations in the base-pair sequence. The break, if isolated, may be corrected by DNA repair enzymes, but a double rupture of the two sugar phosphate strands is harder to correct. Experience shows that a dosage of 1 Gy corresponds to about 1000 changes in the base-pair sequence, 1000 simple fractures of the sugar phosphate background, and 100 double fractures. Radiation levels of 1 Gy or more stop cell production and a dose of 5 Gy is lethal for the cells.

The effects of radiation on tissue can be quantified by measuring the ionization density along the path of radiation. While the number of cells adversely affected by gamma and alpha rays are the same, alpha rays produce more concentrated damage which is more difficult to repair; see Figure 31.1.

Immediate biological effects of ionizing radiation are observable only for moderate to large doses; see Table 31.2. The most easily produced observable effects are changes in blood count. Because radiation interferes with cell reproduction, the systems with the greatest cell division, such as bone marrow, are most sensitive to radiation. The long-term effects are inducement of cancer and genetic defects.

**TABLE 31.2  Immediate Effects of Radiation on Adults**

| Dose (rem) | Whole body, single exposure |
|---|---|
| 0–100 | No observable effect |
| 10–100 | Slight decreases in white blood cell counts |
| 35–50 | Temporary sterility; 35 for women, 50 for men |
| 100–200 | Significant reduction in blood cell counts, brief nausea, and vomiting; rarely, if ever, fatal |
| 200–500 | Nausea, vomiting, hair loss, severe blood damage, hemorrhage |
| 450 | Lethal to 50% ($LD_{50}$), within 30 days if untreated |
| 500–2000 | Worst effects due to malfunctions of small intestine and blood systems; survival possible, if treated |
| >2000 | Fatal within hours from collapse of central nervous system and gastrointestinal system |

## TOPIC 31.3   Medical Diagnostics Based on Nuclear Effects

### Emission Tomography

**(i) Single-Photon Emission Tomography (SPET)**   It is possible to image radioactive emissions in a single plane or slice through the body using computer tomography (CT) techniques. A basic gamma camera is moved around the patient to measure the radioactive intensity from a radioactive tracer source within the volume of the body. This is repeated at many points and angles and the data are processed in much the same way as for X-ray CT scans.

**(ii) Positron Emission Tomography (PET)**   This technique makes use of positron emitters such as $^{11}_{6}C$, $^{13}_{7}N$, $^{15}_{8}O$, and $^{18}_{9}F$. When inhaled or injected, molecules incorporating these isotopes accumulate in the organ or region of the body to be studied. When a nuclide of this sort $\beta^+$ decays, the emitted positron travels at most a few millimeters before it collides with a normal negative electron. These two particles annihilate one another and produce two $\gamma$ rays; that is, $e^+ + e^- = 2\gamma$, each ray having an energy of 510 keV. The $\gamma$ rays travel in opposite directions ($180° \pm 0.25°$) because they must have almost equal and opposite momenta (the original momenta of the $e^+$ and the $e^-$ are approximately zero compared to the momenta of the $\gamma$ rays). These $\gamma$-ray photons travel along the same line but in opposite directions so their detection in coincidence by rings of detectors around the patient (see Figure 31.2) establishes the line along which the emission took place. If the difference in the time of arrival of the two photons could be accurately determined, the actual position of the emitting nuclide, along that line, could be calculated. Positron emission tomography has one big advantage in that no collimators are needed. Fewer photons are wasted and lower doses can be administered with PET.

Both SPET and PET systems can give images related to biochemistry, metabolism, and function. This is compared to X-ray CT scans, whose images reflect shape and structure or the anatomy of the imaged system.

### CAT Scans

CAT scanning utilizes X-rays to provide images from specific locations within the body. The acronym CAT means computerized axial tomography or computer-assisted tomography. A series of images obtained from X-ray beams form a "fanned-out" array of radiation which passes simultaneously through the patient. Each beam is detected by a separate detector which records the beam intensity, the various beam intensities being different

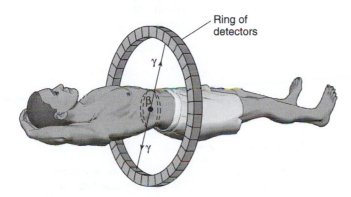

**FIGURE 31.2**  Positron emission tomography system showing a ring of detectors to detect the two $\gamma$ rays, produced by the annihilation of $e^+$ and $e^-$, emitted at 180° to each other ($e^+ + e^- \rightarrow 2\gamma$).

depending on the nature of the body material through which the beams have passed. Dramatic improvements over conventional techniques can be obtained with a CAT scan because the X-ray source can be rotated into different orientations so the fan array of each beam is recorded as a function of orientation, and how the intensity of a beam changes from one orientation to another is fed into a computer. The computer then constructs a highly resolved image of the cross-sectional slice of the body. Details as small as 1 mm are observable in CT scans. There is the added benefit that only the tissue being imaged is exposed to radiation since a very narrow beam of X-rays is used.

## Nuclear Magnetic Resonance

Since its discovery in 1946, NMR has become a powerful research tool in physics, chemistry, and biology. More recently, it has developed into an important and widely used imaging technique. In this section we discuss the physical principles behind NMR and in the next we look at its applications.

A loop of wire placed in a magnetic field which carries an electric current experiences a torque. The loop is said to have a magnetic moment. The torque tends to align the loop so that the direction of its magnetic moment, which is perpendicular to the plane of the coil, is parallel to the magnetic induction field, **B**. This arises because when the coil is so aligned, it has its lowest potential energy. Atomic electrons, in their orbital motion, act as tiny current loops. Furthermore, they have an intrinsic angular momentum called spin, which in some ways acts like the magnetic moment of a current loop. When atoms are in a magnetic field, the atomic energy levels split into several closely spaced levels.

Nuclei too can have magnetic moments. Many nuclides have this property, but we will examine the simplest, the hydrogen (H) nucleus, since it is the one most often used in medical imaging. The ${}^1_1$H nucleus consists of a single proton and its spin angular momentum, like that of the electron, can take on only two values in a magnetic field; see Figure 31.3. We call these "spin-up" (parallel to the field) and "spin-down" (antiparallel to the field). When a magnetic field is present, the energy of the ${}^1_1$H nucleus splits into two levels, as shown in Figure 31.4, with the spin up having the lowest energy (this is very like the Zeeman effect for an atomic electron in a magnetic field). The energy difference, $\Delta E$ between the two levels is proportional to the total magnetic field, $B_T$, at the nucleus. Thus

$$\Delta E = kB_T \tag{31.9}$$

where $k$ is a proportionality constant which is different for different nuclides.

The sample to be examined is placed in a static magnetic field in a standard NMR arrangement. A radiofrequency (RF) pulse of electromagnetic radiation is applied to the sample, and if its frequency is $f$ and such that $hf$ precisely corresponds to the energy difference, $\Delta E$, so that

$$hf = \Delta E = kB_T \tag{31.10}$$

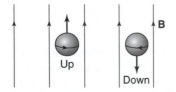

Up

Down

B

**FIGURE 31.3** Schematic picture of a proton represented in a magnetic field **B** (pointing upward) with its two possible states of spin, up and down.

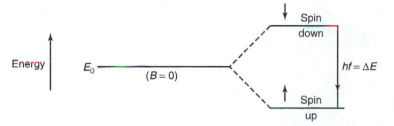

**FIGURE 31.4** Energy of a proton in the absence of a magnetic field ($E_0$), that is, $B = 0$, and the two energies into which the $E_0$ level splits in the presence of a magnetic field, $B \neq 0$.

then the photons of the RF beam will be absorbed, exciting many nuclei from the lower to the upper state. There is a significant absorption only if $f$ is very near $f = kB_T/h$ and is a resonance phenomenon (hence the name nuclear magnetic resonance). For free ${}_1^1$H nuclei, the frequency is 42.58 MHz for a field $B_T = 1.0$ T. On the other hand, if the H atoms are bound in a molecule, the total magnetic field, $B_T$, at the H nuclei will be the sum of the external applied field plus the magnetic field due to electrons and nuclei of neighboring atoms. Since $f$ is proportional to $B_T$, the value of $f$ will be slightly different for the bound H nuclei than for free H nuclei in the same external field. This change in frequency can be measured and is called the chemical shift. Much has been learned about the structure of molecules and bonds using such NMR measurements. The NMR spectroscopy provides information about biological activity and ongoing chemical reactions in a noninvasive way. The method works so well because it is possible to measure NMR spectra with sufficiently high precision to access chemical shifts.

## Magnetic Resonance Imaging

Useful NMR images obtained through the principles outlined above and adapted for medical purposes can be generated by the commonly used magnetic resonance imaging (MRI) technique. The element most used is again hydrogen, since it is the most common element in the human body and gives the strongest NMR signals. The apparatus is shown in Figure 31.5. The large coils set up the static magnetic field and the RF coils produce the RF pulse that causes the nuclei to make a transition from the lower state to the upper state in Figure 31.4. These same coils—or another coil—can detect the absorption of energy or emitted radiation—also at frequency $f = \Delta E/h$—when the nuclei become deexcited and fall back down into the lowest energy state. The formation of a two- or three-dimensional image can be done using similar techniques to those in CAT. What is measured is the intensity of absorbed and/or reemitted radiation from many different points of the body, and this would be a measure of the density of H atoms at that point. To do this, we need to be able to determine from which part of the body a given photon comes. One technique is to add a gradient to the static magnetic field (e.g., using shaped pole pieces). Thus, instead of using a uniform magnetic field $B_T$, it is made to vary with position across the width of the sample or patient. The frequency absorbed by the H atoms is proportional to $B_T$, and only one plane within the body will have the correct value of $B_T$ to absorb photons at a frequency $f$. Alternatively, if the field gradient is applied after the RF pulse, the frequency of emitted photons will be a measure of where they are emitting; see Figure 31.6. When a magnetic field gradient is applied in one direction during excitation (absorption of photons) and photons of a single frequency are transmitted, only the H atoms in one thin slice will be excited. Applying a gradient in another direction, perpendicular to the first, during reemission, the frequency $f$ of the reemitting radiation will represent depth

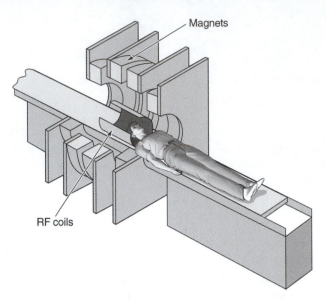

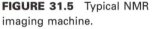

**FIGURE 31.5** Typical NMR imaging machine.

in that slice. To correlate NMR frequency with position, other ways of varying the magnetic field throughout the volume of the body can be used.

A reconstructed image based on the density of H atoms (or intensity of absorbed or emitted radiation) is not particularly interesting. However, more useful images based on the rate at which the nuclei decay back to the ground state (or lowest energy state) can be obtained and can produce a resolution of 1 mm or better. This NMR method is sometimes called spin echo and can produce images which have considerable diagnostic value both in the study of metabolic processes and in anatomy. Such an NMR image is given in Figure 31.7. Table 31.3 summarizes the medical imaging techniques and their resolution capabilities.

## Radiopharmaceuticals

Radiopharmaceuticals have found widespread use in medical diagnostics. A radiopharmaceutical is any drug that contains a radioactive isotope. Radiopharmaceuticals can be designed to be very organ or system specific. Table 31.4 summarizes diagnostic uses of radiopharmaceuticals.

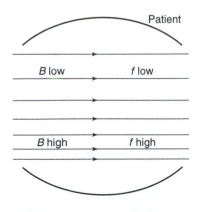

**FIGURE 31.6** Static field that is stronger at the bottom than at the top. The frequency of absorbed or emitted radiation is proportional to $B$ in NMR.

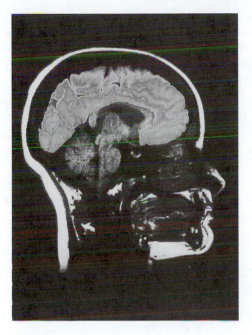

**FIGURE 31.7** An NMR image of a vertical section through the head showing structures in the normal brain. (Photo credit: Scott Camazine/Photo Researchers.)

The radionuclide commonly used for medical diagnosis is $^{99m}_{43}$Tc, a long-lived excited state of technetium-99 (m is the superscript symbol which stands for "metastable" state). It is formed when $^{99}_{42}$Mo decays. It has a convenient half life of 6 h (short but not too short) and can combine with a large variety of compounds. A compound is chosen because it concentrates in the organ or region of the anatomy to be studied and is then labeled with the radionuclide. Detectors outside the body also record the distribution of the radioactivity-labeled compound. Detection can also be achieved by a single detector which is moved across the body and measures the intensity of radioactivity at a large number of points. The image recorded then represents the relative intensity of radioactivity at each point and is used in diagnosis. High or low radioactivity may represent overactivity or underactivity of an organ or part of an organ, or, in another case, it may represent a lesion or tumor. Gamma detector cameras, which are much more complex, make use of many detectors which simultaneously record radioactivity at many points and display the result on a CRT (a TV monitor or an oscilloscope screen) which allows dynamic studies to be performed (studies which utilize images which change with time).

**TABLE 31.3  Medical Imaging Techniques**

| Technique | Resolution |
|---|---|
| Conventional X-ray | $\frac{1}{2}$ mm |
| CT scan, X-ray | $\frac{1}{2}$ mm |
| Nuclear medicine (tracers) | 1 cm |
| SPET | 1 cm |
| PET | 3–5 mm |
| NMR | $\frac{1}{2}$–1 mm |
| Ultrasound | 2 mm |

**TABLE 31.4  Diagnostic Uses of Radiopharmaceuticals**

| Procedure and agent | Typical activity (mCi) | Radiation Dose (rem) |
|---|---|---|
| Brain scan | | |
| $^{99m}$Tc-pertechnetate | 7.5 | 1.5 (colon) |
| $^{113m}$In-DTPA | 7.5 | 4 (bladder) |
| Lung Scan | | |
| $^{99m}$Tc-MAA | 2 | 0.7 (lung) |
| $^{133}$Xe | 7.5 | 0.4 (lung) |
| Cardiovascular blood pool | | |
| $^{131}$I-HSA | 0.2 | 3 (blood) |
| $^{99m}$Tc-HSA | 2 | 0.08 (blood) |
| Placental localization | | |
| $^{99m}$Tc-pertechnetate | 0.7 | 0.1 (colon) |
| $^{113m}$In-transferrin | 1 | 0.1 (blood) |
| Thyroid scan | | |
| $^{131}$I | 0.05 | 75 (thyroid) |
| $^{123}$I | 0.07 | 1.5 (thyroid) |
| Liver scan | | |
| $^{198}$Au-colloid | 0.1 | 5 (liver) |
| $^{99m}$Tc-sulfur colloid | 2 | 0.6 (liver) |
| Bone scan | | |
| $^{85}$Sr | 0.1 | 4 (bone) |
| $^{99m}$Tc-STPP | 10 | 0.5 (bone) |
| Kidney scan | | |
| $^{197}$Hg-chlormerodrin | 0.1 | 1.5 (kidney) |
| $^{99m}$Tc-iron ascorbate | 1.5 | 0.8 (kidney) |

DTPA = diethylenetniaminepentaacetic acid; HSA = human serum albumin; MAA = macro albumin aggregated; STPP = sodium tripoly phosphate.

In agricultural research the tracer technique is also useful, for example, in determining the best method of administering fertilizer to a plant. In the fertilizer, a certain material such as nitrogen can be tagged with one of its radioactive isotopes. The fertilizer is then sprayed on one group of plants, sprinkled on the ground for a second group, and raked into the soil for a third. A Geiger counter can then be used to track the nitrogen through the three groups of plants.

## Single-Emission Computed Tomography

Single-emission computed tomography (SPECT) is a particularly useful technique of applying radiopharmaceuticals. This technique utilizes radioactive agents, and in a SPECT procedure, a specific type of radioisotope chemically attached to a drug known to specifically target a particular organ is first administered into the bloodstream of the patient. The radioisotope is a unique type of radioactive source that is a single-photon emitter or emits photons as a primary emission upon decay. The source distributes itself and localizes within the body and is continually emitting photons, which have sufficient energy to penetrate surrounding tissues and escape from the body. Photons which escape the body are registered and collected by a detector assembly. This latter consists of a collimator which acts to restrict unwanted scattered photons from reaching the detector and increases image quality and sharpness. After passing the collimator, photons impinge on a gamma ray detector which is coupled to photomultiplier tubes. A snapshot image of the object or organ is then computed from the two-dimensional position and intensity of the detected gamma rays. These images are found at angular increments with a circular or elliptical orbit about the organ of interest. The snapshots are then mathematically reconstructed to provide the physician with a three-dimensional distribution of activity within the organ.

## Tracers in Medicine and Biology

Radioactive isotopes are commonly used in biological and medical research as tracers. Artificially synthesized compounds are made using a radioactive isotope such as $^{14}_{6}C$ or $^{3}_{1}H$. Molecules which are tagged in this way can then be traced as they move through the organism or as they undergo a chemical reaction. As they, or parts of them, undergo chemical change, they can be detected by a Geiger or scintillation counter. How food molecules are digested and to what parts of the body they are diverted can be traced in this way. How essential components and amino acids are synthesized by organisms have been investigated using radioactive tracers. Furthermore, the permeability of cell walls to various ions can be determined using radioactive isotopes. In this case the tagged molecule or ion is injected into the extracellular fluid, and the radioactivity present inside and outside the cells is measured as a function of time.

The position of radioactive isotopes is detected on film in a technique known as autoradiography. For example, the distribution of carbohydrates produced in leaves of plants from absorbed $CO_2$ can be observed by keeping the plant in an atmosphere where the carbon atom in the $CO_2$ is $^{14}_{6}C$. The leaf is placed firmly on a photographic plate, and after a time, the emitted radiation darkens the film most strongly where the isotope is most strongly concentrated.

## Nonradioactive Tracers

Tracer techniques can be used without using radioactive atoms. The only requirement is that the tracer atoms can be identified. If an unusual isotope is used, its presence can be detected using a mass spectrograph or more usually by centrifugation. For example, one can use a nonradioactive isotope of nitrogen, $^{15}N$, which is present in about 0.36% quantities in normal samples. Abundances greater than this are indication of the added tracer. One can feed the amino acid glycine to an individual, and the glycine is synthesized using this isotope. As nitrogen is incorporated into hemoglobin, the presence or otherwise of $^{15}N$ in samples of blood can be investigated as a function of time. It is found in such studies that the average lifetime of red cells is about 4 months; see Figure 31.8. This technique was also employed in the classic experiment of Matthew Meselon and Franklin Stahl which confirmed the Watson–Crick method of DNA replication. A medium in which almost all the N atoms were of the isotope $^{15}N$ was used to grow bacteria for generations. After this the DNA was extracted, purified, and placed in a cesium chloride solution in an ultracentrifuge spun at 14000$g$. The DNA was stabilized at a different level from that of DNA obtained from bacteria grown in a normal $^{14}N$ medium because of the different mass characteristics.

Those bacteria grown in the $^{15}N$ environment were then allowed to reproduce in a normal $^{14}N$ medium. After one reproduction, DNA from the bacteria was spun in a centrifuge stabilized in a single layer at a level intermediate between those for heavy and normal DNA. Two stabilized layers appeared after a second reproduction, one corresponding to the intermediate level found after one reproduction and the other corresponding to

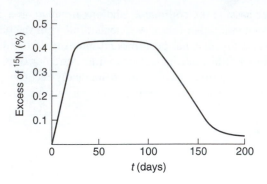

**FIGURE 31.8** Excess of $^{15}N$ in heme of human erythrocytes after feeding $^{15}N$-labeled glycine for three days.

normal DNA. Thereafter, after each successive further reproduction, the latter increased in density in comparison with the former. See Figure 31.9.

In replication these results indicate that the two strands of the DNA come apart, each one duplicating from the surrounding medium the strand it has lost. After one reproduction there can only be hybrid DNA present containing one strand of the original $^{15}N$-DNA and one strand of $^{14}N$-DNA formed from the medium. After further division, there can be both hybrid DNA and DNA with both strands containing $^{15}N$. The amount of hybrid DNA can never increase for the number of strands of $^{15}N$-DNA is fixed. The amount of normal DNA goes on increasing with each replication, and in each successive sample spun in the centrifuge the fraction of normal DNA becomes greater. Tracer techniques using labeled nucleotides as components of DNA have revealed much about the details of DNA replication.

## Chromosome Division

The study of replication has also been greatly enhanced by the use of radioactive tracers. Consider, for example, the chromosome division in broad bean root tips. These are often used since the chromosomes can be made clearly visible in a microscope. When thymidine labeled with $^3H$ is used in growing specimens, the $^3H$ is only incorporated into the chromosomes. Thus autoradiographs of cells will therefore clearly show chromosomes or portions of them.

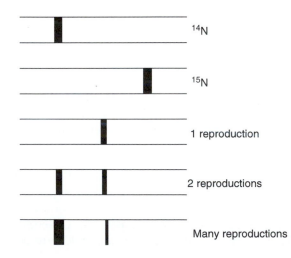

**FIGURE 31.9** Ultracentrifuge data for the DNA from bacteria grown in a $^{15}N$ medium and those from a $^{14}N$ medium.

Colchicine may inhibit cell division but does not affect the division of chromosomes. When cells containing labeled chromosomes are treated with colchicine, chromosome division proceeds to different stages in different cells. After completion of the process, some cells will contain two sets of chromosomes, some four sets, and so on. If an autoradiography film is placed over the preparation, pictures of these cells can be obtained and the distribution of $^3$H after one, two, three, and so on, divisions can be followed. Thus information may be gained about the manner in which the actual material of the chromosome divides and how it replicates from the medium.

## Metabolic Uptake

The absorption of vitamin $B_{12}$ is low in patients suffering from pernicious anemia. Using normal methods, it is difficult to determine exactly how much of the vitamin is being absorbed by the body. This vitamin contains a cobalt atom, and it is possible to synthesize the vitamin with the radioactive isotope $^{58}$Co and then feed a known dose to a patient. The dose given together with subsequent analysis of the feces allows the calculation of the dose which has been absorbed by the gut.

Nearly all iodine which is ingested and absorbed by the gut is taken up by the thyroid. The hormones thyroxine and di-iodo-tyrosine are formed from the stored iodine and are then circulated through the body. A measured dose of the radioactive $^{131}$I can be given by the mouth to a normal individual and the activity can be measured at the neck near the thyroid gland. After 48 h a blood sample will also show activity because of the labeled iodine in the hormones. A patient with an overactive thyroid (i.e., suffering from hyperthyroidism) absorbs too much iodine too quickly and therefore produces an excessive amount of hormone. The complaint can be quickly diagnosed by giving the patient with hyperthyroidism radioactive iodine orally and comparing the activity of the thyroid with data from normal individuals.

## Isotopic Dilution

Tracer techniques can be utilized to determine the total volume occupied by a fluid. Suppose the activity of a known quantity of suitably labeled fluid is measured and this fluid is then injected into the volume. When the radioactive tracer has dispersed throughout the whole region, the activity of an extracted sample indicates by how much the tracer has been diluted. The total volume can then be calculated.

## Location of Hemorrhages

It is quite often difficult to ascertain whether an internal hemorrhage is taking place or, if it is, where it is located. The chromium isotope $^{51}$Cr has been used in various studies of blood because it is taken up by red blood cells. The procedure here is to inject $^{51}$Cr-labeled blood into the patient. In a normal blood circulation the radioactivity will be distributed throughout the circulatory system. However, if a hemorrhage is occurring, the radioactivity will markedly increase at some region of the body, and the rate at which activity increases is a measure of the volume of blood being lost.

## Radiocardiography

One method of investigating heart conditions is the insertion of a catheter into the bloodstream, from where it is passed into the heart. This is not without attendant risks and takes skill and time. Radiography is a much simpler procedure which provides the same information routinely, and furthermore there are no risks, relatively speaking. Information concerning pulmonary conditions may also be obtained in this way.

Barium-137 has a half-life of 127 s and is used as a tracer element. When 10 mm$^3$ of a solution of $^{137}$Ba is injected rapidly into the subclavian vein, it enters the right ventricle almost immediately. The tracer is detected via a counter directed from above at the heart. It is found that there is a dip in the recording as the tracer is pumped out to the lungs and a rise when it returns to the heart. A counter collimated to pick up radiation from the aorta, at the back, follows the flow from the heart. If the heart and lungs are functioning normally, a recording obtained from the counters will have some typical form. However, a blockage or malfunction will lead to a nonstandard pattern on the recording. The features of this latter recording can be interpreted in terms of various known conditions. Thus various congenital and pathological lung conditions can be rapidly and easily diagnosed.

Another application of the tracer technique is for determining the condition of the human circulatory system. Here, salt solution containing radioactive sodium is injected into a vein in the leg. The time at which the radioisotope arrives at another part of the body is detected with an appropriate counter. The elapsed time is a good indication of the presence or absence of constrictions in the circulatory system.

The technique of gamma knife radiosurgery is a medical procedure for treating problems of the brain which might include cancerous tumors as well as blood vessel malformation. The procedure uses a powerful, highly focused beam of $\gamma$ rays aimed at the tumor or malformation, the $\gamma$ rays being generated by a radioactive $^{60}$Co source administered to the patient who wears a protective metal helmet that is perforated with many small holes. The holes focus the $\gamma$ rays to a single target within the brain. A very intense dose of radiation is received by the target which is destroyed, while surrounding healthy tissue is undamaged. The method is noninvasive, painless, and requires no blood to be taken out of the patient.

## TOPIC 31.4   Radiation Therapy

Although radiation can cause cancer, it can also be used to treat it. Cancer cells which are rapidly growing are especially susceptible to destruction by radiation. Nevertheless, large doses are needed to kill cancerous cells, and inevitably some of the surrounding normal cells are killed as well. Thus patients who undergo radiation therapy often suffer side effects characteristic of radiation sickness. To try and minimize the destruction of normal cells, a narrow beam of gamma or X-rays is often used when the cancerous tumor is well localized. The beam is directed at the tumor and the source—or the body—is rotated so that the beam passes through various parts of the body to keep the dose at any one place as low as possible—except at the tumor and its immediate surroundings, where the beam passes at all times (see Figure 31.10). The radiation can come from a $^{60}_{27}$Co radioactive source or from an X-ray machine that produces photons in the range 200 keV to 5 MeV. Particles such as protons, neutrons, electrons, and pions, which are produced in particle accelerators, are also used in cancer therapy.

In some instances a tiny radioactive source may be inserted directly inside a tumor, which will eventually kill the majority of cells. In the treatment of cancer of the thyroid a similar technique is used using the radioactive isotope $^{131}_{53}$I. The thyroid gland tends to concentrate any iodine present in the bloodstream so when $^{131}_{53}$I is injected in the blood, it

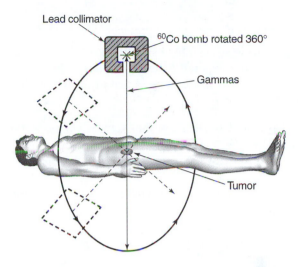

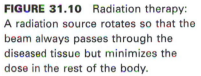

**FIGURE 31.10** Radiation therapy: A radiation source rotates so that the beam always passes through the diseased tissue but minimizes the dose in the rest of the body.

becomes concentrated in the thyroid, particularly in any area where abnormal growth is taking place. The intense radioactivity emitted can then destroy the defective cells.

Radiotherapy is sometimes used in combination with the other two major types of cancer treatment, surgery and chemotherapy. Chemotherapy uses chemicals that, like radiation, inhibit cell division. As a result many of the side effects of chemotherapy are similar to those produced by radiation. It is usually possible to localize radiation better than chemicals, so the side effects of radiotherapy tend to be more localized than those of chemotherapy.

The central problem in radiotherapy is to concentrate radiation in abnormal tissue, giving as little dose as possible to normal tissue. The ratio of abnormal cells killed to normal cells killed is called the therapeutic ratio, and all radiotherapy techniques are designed to enhance this ratio. Table 31.5 lists typical doses given to cancerous tissue depending on the organ targeted.

**TABLE 31.5  Typical Doses Given to Cancerous Tissue in Cancer Radiotherapy**

|  | Typical dose* (rem) |
| --- | --- |
| Lung | 1000–2000 |
| Hodgkin's disease | 4000–4500 |
| Skin | 4500 |
| Ovarian | 5000–7500 |
| Breast | 5000–8000 |
| Bladder | 7000–7500 |
| Head (brain) | 8000+ |
| Neck | 8000+ |
| Bone | 8000+ |
| Soft tissue | 8000+ |

*Usually given at 200 rem/treatment from three to five times per week.

## SOLVED PROBLEMS

**31.1** A biological tissue is irradiated with $\gamma$ rays that have an RBE of 0.70. The absorbed dose of $\gamma$ rays is 850 rads. The tissue is then exposed to neutrons whose RBE is 3.5. The biologically equivalent dose of the neutrons is the same as that of the $\gamma$ rays. What is the absorbed dose of neutrons?

**Solution** The biologically equivalent dose (BED) of the $\gamma$ rays is the product of the absorbed dose (in rads) and the RBE:

$$\text{BED of } \gamma \text{ rays} = (850 \text{ rads})(0.70) = 595 \text{ rem}$$

For the neutrons (RBE = 3.5), the BED is the same. Therefore, 595 rem = (absorbed dose of neutrons)(3.5) and

$$\text{Absorbed dose of neutrons} = \frac{595 \text{ rem}}{3.5} = 170 \text{ rads}$$

**31.2** Suppose a person is accidentally exposed to gamma radiation from a $^{60}$Co source. Calculate the radiation dose received in rads, given the following information. The person's mass is 75 kg, and his entire body is exposed to $5.0 \times 10^9$ gamma rays per second for 1 min. The gamma rays have an average energy of 1.25 MeV and are 50% absorbed.

**Solution** The energy absorbed in joules is found as

$$\text{Energy absorbed} = \frac{\text{energy}}{\gamma} \times \frac{\gamma}{\text{s}} \times \text{time (s)} \times \% \text{ absorbed}$$

$$= (1.25 \times 10^6 \text{ eV}/\gamma)(1.6 \times 10^{-19} \text{ J/eV})(5 \times 10^9 \text{ }\gamma/\text{s})(60 \text{ s})(0.50) = 0.030 \text{ J}$$

which is then divided by the mass of tissue exposed to yield the dose as

$$\text{Dose} = \frac{0.030 \text{ J}}{75 \text{ kg}} = 0.04 \text{ rad}$$

**31.3** Calculate the radiation dose in rem due to exposures of
(a) 0.10 rad of alpha radiation (RBE = 20) and
(b) 2.0 rads of gamma radiation (RBE = 1).

**Solution** We use the definition of rem and multiply the RBE by the exposure in rads to obtain the dose:
(a) Dose = 20 × 0.1 rad = 2.0 rem.
(b) Similarly, dose = 1 × 2.0 rads = 2.0 rem.

**31.4** What whole-body dose is received by a 70-kg laboratory worker exposed to a 40-mCi $^{60}_{27}$Co source, assuming that the person's body has cross-sectional area 1.5 m$^2$ and is, on average, 4.0 m from the source for 4.0 h/day? The isotope $^{60}_{27}$Co emits $\gamma$ rays of energy 1.33 MeV and 1.17 MeV in quick succession. Approximately 50% of the $\gamma$ rays interact in the body and deposit all their energy.

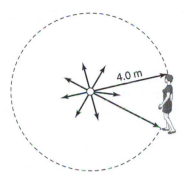

**Solution**  The total γ-ray energy per decay is $1.33 + 1.17$ MeV $= 2.50$ MeV, so the total energy emitted by the source is

$$(0.0400 \text{ Ci})(3.7 \times 10^{10} \text{ decays/Ci s})(2.50 \text{ MeV}) = 3.7 \times 10^9 \text{ MeV/s}$$

The proportion of this absorbed by the body is its 1.5-$m^2$ area divided by the area of a sphere of radius 4.0 m:

$$\frac{1.5 \text{ m}^2}{4\pi r^2} = \frac{1.5 \text{ m}^2}{4\pi(4.0 \text{ m})^2} = 0.0075$$

So the rate energy is deposited in the body (remembering that only half of the γ rays interact with the body) is

$$E = \tfrac{1}{2}(0.0075)(3.7 \times 10^9 \text{ MeV/s})(1.6 \times 10^{-13} \text{ J/MeV}) = 2.2 \ \mu\text{J/s}$$

Since 1 Gy = 1 J/kg, the whole-body dose rate for this 70-kg person is $(2.2 \times 10^{-6} \text{ J/s})/$ 70 kg $= 3.1 \times 10^{-8}$ Gy/s. Over the period of 4.0 h, this amounts to a dose of (4.0 h) (3600 s/h)(3.1 × 10⁻⁸ Gy/s) = 4.5 × 10⁻⁴ Gy. Since relative biological effectiveness (RBE) is approximately 1 for gamma rays, the effective dose of 450 $\mu$Sv or (100 rads/Gy) $(4.5 \times 10^{-4} \text{ Gy})(1) = 45$ mrem.

# EXERCISES

**31.1**  A particular radioactive source produces 150 mrads of 2.5-MeV gamma rays per hour at a distance of 1.0 m. (a) How long could a person stand at this distance before accumulating an intolerable dose of 1 rem? (b) Assuming the gamma radiation is emitted isotropically, at what distance would a person receive a dose of 15 mrads/h from this source?

**31.2**  A 2.5-kg tumor is being irradiated by a radioactive source. The tumor receives an absorbed dose of 15 Gy in a time of 1000 s. Each disintegration of the radioactive source produces a particle that enters the tumor and delivers an energy of 0.45 MeV. What is the activity $A = \Delta N/\Delta t$ of the radioactive source?

**31.3**  The isotope $^{60}$Co is used for radiotherapy. A radioactive source of this isotope has an activity of $4 \times 10^{13}$ decay events per second (becquerels). What is the mass of the source? Suppose that in the previous problem each decay event produces a gamma ray of energy 2.5 MeV. Five kilograms of living tissue close by the cobalt source absorbs about half the radiation. How long does it take for the tissue to absorb a lethal dose of 5 Gy?

**31.4**  (a) How many $^{14}$C nuclei decay per second in a 70-kg person, assuming that the person's body contains 12% of carbon and that there is one $^{14}$C nucleus for every $10^{12}$ $^{12}$C nuclei?

(b) How much time must pass after this person dies before the $^{14}$C activity decreases to one decay per second?

**31.5**   Calculate the effective half-life of $^3$H, $^{90}$Sr, and $^{131}$I.

**31.6**   (a) A dose of 0.06 mCi of $^{131}$I is given to a patient for a thyroid scan. Calculate the activity of the iodine remaining in the patient after 20 days. (b) A dose of 0.12 mCi of $^{198}$Au is given to a patient for a liver scan. Calculate the activity of the gold remaining in the patient after 30 days.

**31.7**   Naturally occurring $^{40}$K gives the average person a radiation dose of 16 mrem/yr. Calculate the mass of $^{40}$K that must be inside a person's body to give this dose, assuming that the mass is 70 kg. Each decay of $^{40}$K produces a 1.31-MeV beta ray. Assume that 40% of the total energy is absorbed by the person.

# BIBLIOGRAPHY

Alloca, J. A., and Levenson, H. E., *Electrical and Electronic Safety* (Reston, Reston, VA, 1982).

Almgren, F. J., and Taylor, J. E., "Geometry of Soap Films and Soap Bubbles," *Scientific American* (July 1976), pp. 82–93.

Avery, M. E., Wang, N.-S., and Taeusch, H. W., "The Lung of the Newborn Infant," *Scientific American* (April 1973), p. 74.

Beebe, G. W., "Ionizing Radiation and Health," *American Scientist* (January–February 1982), p. 35.

Benade, A. H., *Fundamentals of Musical Acoustics* (Oxford University Press, New York, 1976).

Benedek, G. B., and Villars, F. M. H., *Physics with Illustrative Examples from Medicine and Biology* (Springer Verlag, New York, 2000).

Berg, H. C., *Random Walks in Biology* (Princeton University Press, Princeton, 1993).

Birnholz, J. C., and Farrell, E. E., "Ultrasound Images of Human Fetal Development," *American Scientist* 72 (1984), p. 608.

Blake, T. M., *Introduction to Electroencephalography*, 2nd ed. (Prentice-Hall, Englewood Cliffs, NJ, 1972).

Blakemore, R. P., and Frankel, R. B., "Magnetic Navigation in Bacteria," *Scientific American* (December 1981), p. 58.

Blodgett, D., *Manual of Respiratory Care Procedures* (Lippincott, Philadelphia, 1980).

Bruinsma, R., *Physics 6B* (Holt, Rinehart and Winston, Los Angeles, 1999).

Butler, H. H., *How to Read an ECG* (Delmar, Albany, NY, 1973).

Cameron, J. R., and Skofronick, J. R., *Medical Physics* (Wiley-Interscience, New York, 1978).

Childress, S., *Mechanics of Swimming and Flying* (Cambridge University Press, Cambridge, 1981).

Clements, J. A., "Surface Tension in the Lungs," *Scientific American* (December 1962), p. 121.

Cohen, D., "Magnetic Fields of the Human Body," *Physics Today* (August 1975), p. 34.

Cutnell, J. D., and Johnson, K. W., *Physics,* 5th ed. (Wiley, New York, 2000).

Dalziel, C. F., "Electric Shock Hazard," *IEEE Spectrum* (February 1972), p. 41.

Davidovits, P., *Physics in Biology and Medicine,* 2nd ed. (Harcourt/Academic, New York, 2000).

Denes, P. B., and Pinson, E. N., *The Speech Chain* (Bell Telephone Laboratories, Murray Hill, N.J. 1963).

Denton, E., "The Buoyancy of Marine Animals," *Scientific American* (July 1960), p. 118.

Devey, G. B., and Wells, P. N. T., "Ultrasound in Medical Diagnostics," *Scientific American* (May 1978), p. 98.

Dudrick, S. J., and Rhoads, J. E., "Total Intravenous Feeding," *Scientific American* (May 1972), pp. 73–80.

Freidman, H. H., *Diagnostic Electroencephalography and Vectorcardiography* (McGraw-Hill, New York, 1971).

Friedlander, G. D., "Electricity in Hospitals, Elimination of Lethal Hazards," *IEEE Spectrum* (September 1971), p. 40.

Giancoli, D. C., *Physics*, 2nd ed. (Prentice-Hall, Englewood Cliffs, NJ, 1985).

Goldsmith, T. H., "Hummingbirds See Near Ultraviolet Light," *Science* 207 (1980), p. 786.

Guyton, A. C., *Physiology of the Human Body*, 5th ed. (Saunders, Philadelphia, 1979).

Hall, E. J., *Radiobiology for the Radiologist*, 2nd ed. (Harper & Row, New York, 1978).

Hayward, A. T., "Negative Pressure in Liquids: Can It Be Harnessed to Serve Man?" *American Scientist* 59 (1971).

Hobbie, R. K., *Intermediate Physics for Medicine and Biology* (Wiley, New York, 1978).

Hoenig, S. A., and Scott, D. H., *Medical Instrumentation and Electrical Safety, the View from the Nursing Station* (Wiley, New York, 1977).

Hudspeth, A. J., "The Hair Cells of the Inner Ear," *Scientific American* (January 1983), p. 54.

Jackson, C. Lee, "The Allocation of the Radio Spectrum," *Scientific American* (February 1980), p. 34.

Jaffe, C. C., "Medical Imaging," *American Scientist* 70 (1982), p. 576.

Johansen, K., "Aneurysms," *Scientific American* (July 1982), p. 110.

Jones, R. C., "How Images Are Detected," *Scientific American* (September 1968), p. 111.

Kane, J. W., and Sternheim, M. M., *Physics*, 2nd ed. (Wiley, New York, 1983).

Knudsen, E. I., "The Hearing of the Barn Owl," *Scientific American* (December 1981), p. 112.

Kooi, K. A., Tucker, R. P., and Marshall, R. E., *Fundamentals of Electroencephalography*, 2nd ed. (Harper & Row, New York, 1978).

Land, E. H., "The Retinex Theory of Color Vision," *Scientific American* (December 1977), p. 108.

Levine, J. S., and MacNichol, E. F., "Color Vision in Fishes," *Scientific American* (February 1982), p. 140.

Lewis, H. W., "The Safety of Fission Reactors," *Scientific American* (March 1980), p. 53.

Llinas, R. R., "Calcium in Synaptic Transmission," *Scientific American* (October 1982), p. 56.

Lock, J. A., "The Physics of Air Resistance," *Physics Teacher* 20 (1982).

Loeb, G. E., "The Functional Replacement of the Ear," *Scientific American* (February 1985), p. 104.

Lounasmaa, O. V., "New Methods for Approaching Absolute Zero," *Scientific American* (December 1969), p. 26.

MacDonald, S. G. G. and Burns, D. M., *Physics for the Life and Health Sciences* (Addison-Wesley, New York, 1975).

Mandoli, D. F., and Briggs, W. R., "Fiber Optics in Plants," *Scientific American* (August 1984), p. 90.

Maugh, T. H., "A New Microscopic Tool for Biology," *Science* 206 (1979), p. 918.

Metcalf, H., *Topics in Classical Biophysics* (Prentice-Hall, Englewood Cliffs, NJ, 1980).

Morell, P., and Norton, W. T., "Myelin," *Scientific American* (May 1980), p. 88.

Morrison, A. R., "A Window on the Sleeping Brain," *Scientific American* (April 1983), p. 94.

Nassau, K., "The Causes of Color," *Scientific American* (October 1980), p. 124.

Neuweiler, G., "How Bats Detect Flying Insects," *Physics Today* (August 1980), p. 34.

Newman, E. A., and Hartline, P. H., "The Infrared 'Vision' of Snakes," *Scientific American* (March 1982), p. 106.

Overheim, R. D., and Wagner, D. L., *Light and Color* (Wiley, New York, 1982).

Pennington, J. A. T., and Church, H. Nichols, *Bowes and Church's Food Values of Portions Commonly Used*, 13th ed. (Lippincott, Philadelphia, 1980).

Poggio, T., "Vision by Man and Machine," *Scientific American* (April 1984), p. 106.

Popovic, V., and Popovic, P., *Hypothermia in Biology and Medicine* (Grune and Stratton, New York, 1974).

Purcell, E. M., "Life at Low Reynolds Numbers," *American Journal of Physics* 45 (1977), pp. 3–11.

Pykett, I. L., "NMR Imaging in Medicine," *Scientific American* (May 1982), p. 78.

Raskin, M. M., and Viamonte, M., Eds., *Clinical Thermography* (American College of Radiology, Chicago, 1977).

Roller, D., Ed., *Early Development of the Concepts of Heat and Temperature* (Harvard University Press, Cambridge, MA, 1950).

Romer, R. H., "Energy," *Energy—An Introduction to Physics* (W. H. Freeman, New York, 1976).

Rubin, P., Ed., *Clinical Oncology for Medical Students and Physicians*, 5th ed. (American Cancer Society, Washington, DC, 1978).

Ruch, T. C., and Patton, H. D., Eds., *Physiology and Biophysics*, Vol. 3 (Saunders, Philadelphia, 1973).

Ruis, M. J., "The Physics of Visual Acuity," *The Physics Teacher* 18 (1980), p. 457.

Sabbagha, R. E., Ed., *Diagnostic Ultrasound Applied to Obstetrics and Gynecology* (Harper & Row, New York, 1980).

Safir, A., Ed., *Refraction and Clinical Optics* (Harper & Row, New York, 1980).

Sassin, W., "Energy," *Scientific American* (September 1980), p. 118.

Schmidt-Nelson, K., "Locomotion: Energy Cost of Swimming, Flying, and Running," *Science* 177 (1972), p. 222.

Schulam, R. G., "NMR Spectroscopy of Living Cells," *Scientific American* (January 1983), p. 86.

Schwartz, J. H., "The Transport of Substances in Nerve Cells," *Scientific American* (April 1980), p. 152.

Sears, F. W., *Optics* (Addison-Wesley, Reading, MA, 1949).

Shapiro, J., *Radiation Protection*, 2nd ed. (Harvard University Press, Cambridge, MA, 1981).

Shepherd, J. T., and Vanhoutte, P. M., *The Human Cardiovascular System, Facts and Concepts* (Raven, New York, 1979).

Smith, F. D., "How Images Are Formed," *Scientific American* (September 1968), p. 97.

Spooner, R. B., *Hospital Electrical Safety Simplified* (Instrument Society of America, Research Triangle Park, NC, 1980).

Tortora, G. J., and Anagnostakos, N. P., *Principles of Anatomy and Physiology*, 4th ed. (Harper & Row, New York, 1985).

Tubis, M., and Wolf, W., Eds., *Radiopharmacy* (Wiley-Interscience, New York, 1976).

Upton, A. C., "The Biological Effects of Low-Level Ionizing Radiation," *Scientific American* (February 1982), p. 41.

Urone, P. P., *College Physics* (Brooks/Cole, Pacific Grove, CA, 2001).

Von Bekesy, G., *Experiments in Hearing* (McGraw-Hill, New York, 1960).

Walker, J. S., *Physics* (Prentice-Hall, Upper Saddle River, NJ, 2002).

Whipp, B. J., and Wiberg, D. M., Eds., *Modeling and Control of Breathing* (Elsevier Biomedical, New York, 1982).

Williamson, S. J., and Cummins, H. Z., *Light and Color in Nature and Art* (Wiley, New York, 1983).

Wolfe, J. M., "Hidden Visual Processes," *Scientific American* (February 1983), p. 94.

Wu, C. H., "Electric Fish and the Discovery of Animal Electricity," *American Scientist* (November/December 1984), p. 598.

Wurtman, R. J., "The Effects of Light on the Human Body," *Scientific American* (July 1975), p. 69.

# INDEX